Metamaterial for Microwave Applications

Mohammad Tariqul Islam
Professor, Department of Electrical, Electronic and Systems Engineering
Universiti Kebangsaan Malaysia, Malaysia

CRC Press is an imprint of the
Taylor & Francis Group, an **informa** business

A SCIENCE PUBLISHERS BOOK

First edition published 2023
by CRC Press
6000 Broken Sound Parkway NW, Suite 300, Boca Raton, FL 33487-2742

and by CRC Press
4 Park Square, Milton Park, Abingdon, Oxon, OX14 4RN

© 2023 Mohammad Tariqul Islam

CRC Press is an imprint of Taylor & Francis Group, LLC

Reasonable efforts have been made to publish reliable data and information, but the author and publisher cannot assume responsibility for the validity of all materials or the consequences of their use. The authors and publishers have attempted to trace the copyright holders of all material reproduced in this publication and apologize to copyright holders if permission to publish in this form has not been obtained. If any copyright material has not been acknowledged please write and let us know so we may rectify in any future reprint.

Except as permitted under U.S. Copyright Law, no part of this book may be reprinted, reproduced, transmitted, or utilized in any form by any electronic, mechanical, or other means, now known or hereafter invented, including photocopying, microfilming, and recording, or in any information storage or retrieval system, without written permission from the publishers.

For permission to photocopy or use material electronically from this work, access www.copyright.com or contact the Copyright Clearance Center, Inc. (CCC), 222 Rosewood Drive, Danvers, MA 01923, 978-750-8400. For works that are not available on CCC please contact mpkbookspermissions@tandf.co.uk

Trademark notice: Product or corporate names may be trademarks or registered trademarks and are used only for identification and explanation without intent to infringe.

Library of Congress Cataloging-in-Publication Data (applied for)

ISBN: 978-1-032-41452-2 (hbk)
ISBN: 978-1-032-41454-6 (pbk)
ISBN: 978-1-003-35815-2 (ebk)

DOI: 10.1201/9781003358152

Typeset in Times New Roman
by Radiant Productions

Preface

Material property based field exploration research is not a new concept but 'Metamaterial' has developed a step ahead of exploration by enriching the knowledge of property variations of conventional materials in microwave applications. This book is focused on specific metamaterial structures and designs developed with benchmark solutions both in terms of theoretically and experimentally validated results intended for application in different microwave range frequencies. Conventionally material identification is based on atomic structure or physical characteristics.

The metamaterial structural unit is rationally designed to achieve negative dielectric properties and it can be customized in terms of shape, size and substrate materials. Therefore, a wide range of research scope is emerging with metamaterial absorber based remote sensing and upcoming technologies are going to extract the best performance from it.

Metamaterial sensors with high sensitivity and quality factors are designed for microwave sensing because metamaterials are promising for high-sensitivity sensors. Metamaterials can be used to build passive wireless sensors like microwave devices, whereas present technology needs measuring and transmitting data. Metamaterial sensor effects vary with resonator shape, thickness, substrate size, thickness, and material. This finding allows people to utilize a sensor to quantify the permeability and effective permittivity of specified sensing areas.

Nowadays, millimetre-wave (mm Wave) technology is thought to be the most critical factor in effectively accelerating the development of 5G and beyond communication systems due to the sufficient saved transmitting power and spectrum resources in high-frequency bands, whereas metamaterial (MTM) loaded antennas show a significant effect on the performance such as high isolations and gain particularly in the design of mm Wave or Terahertz antennas due to their unique properties which are not readily available. For instance,, Multiple-Input Multiple-Output (MIMO) technology based on MTM is also becoming increasingly important in 5G wireless communications. Compact MIMO systems frequently use MTM-based decoupling structures to isolate the array elements in a dense scattering environment. For optimum reception, Near-Zero-Index (NZI) metamaterial system categories are appropriate for indoor localization systems, 3D systems-in-packages, off-device deployment, and 5G smart devices.

An increasing demand of metamaterial for applications in wireless communication systems requires a compact structural design with better performance enhancement capabilities of miniaturized microwave devices. Moreover, frequency tuning metamaterial is being explored to provide the flexibility to easily adjust its properties. Various methods are employed to achieve these goals though each has its own advantages and disadvantages. Specific frequency targeted designs with easy frequency tuning properties are also being explored by researchers targeting various microwave applications and as well terahertz frequencies.

Metamaterial loaded antennas are one of the strongest contenders for UWB applications in terms of achieving good impedance bandwidth, higher radiation efficiencies, and stable Omni-directional radiation patterns due to their simple structure and ease of fabrication. Compared to the conventional antennas, the planar metamaterial antennas can be printed onto a piece of PCB which makes them suitable for embedding into portable devices. That's why, industry and academia have put enormous efforts on research to study, design and develop planar metamaterial antennas for UWB applications.

Development of flexible communication systems is one of the most significant parts for their future and also it is the biggest challenge to meet the future flexible electronic market demands. An unavoidable part of such a system is flexible antennas and metamaterials, which are required to ensure signal transfers depending on the application. So, it is necessary to deal with the preparation of nanoparticle-based flexible microwave substrate materials for developing flexible metamaterials and antennas. The nanoparticle-based flexible materials can also reduce the human body's influence on wearable devices in case of wireless body area network (WBAN) applications. Thus, flexible materials can be potential candidates instead of conventional materials to meet the demands for flexible electronics fields.

A portable, IoT-enabled, 3-D antenna head imaging system is one that operates in the microwave range and can detect, identify, and monitor brain strokes. The primary cause of mortality and disability worldwide is brain stroke, one of the prevalent cardiovascular disorders. It has been recognized as a severe issue with healthcare resources throughout the last few decades. The currently available technologies, such as magnetic resonance imaging, computed tomography, positron emission tomography, ultrasound, and X-ray imaging, are very expensive, large, and difficult for patients in remote hospitals to acquire and afford. Therefore, a quick, non-ionizing, lightweight, affordable, and high image resolution electromagnetic (EM) head imaging system has become a dependable method that can play a crucial role in diagnosing stroke inside the human brain.

The metamaterial inspired stacked antenna is realistically designed to achieve high performance in terms of high bandwidth, gain, and directional radiation properties. It can be utilized in microwave brain imaging systems to reconstruct microwave brain images. So, the research opportunity is promising for detecting and classifying brain abnormalities from the reconstructed microwave brain images, which may open a new era in medical applications.

Communication between a nanosatellite and the Earth significantly depends upon antenna systems. Lower UHF nanosatellite antenna design is a crucial challenge due to the physical size constraint, electromagnetic coupling with structure and the need for solar panel integration. To overcome this problem, epsilon-and-mu-near-zero (EMNZ) metamaterial has been developed to reduce the EM coupling with the nanosatellite structure of the conventional meander line antenna, while maintaining good impedance matching and efficiency for the UHF communication system. In this chapter, design and performance analysis of compact EMNZ inspired lower UHF antennas that are strategically integrated with the satellite body and do not require mechanical deployment are conducted for the nanosatellite communication systems. Moreover, this chapter also emphasizes on the nanosatellite communication test using the antenna that proves the validity of the antenna's real-time performance.

Acknowledgements

First of all, an undoubted acknowledgement goes to the family. The unparalleled support from my family members made it possible to complete the book. A significant effort and mental support from them made it possible to make progress towards its completion. I also thank my parents as well for their motivation.

Secondly, an acknowledgement goes to all the co-authors of this book: Dr. Ahasanul Hoque, Dr. Touhidul Alam, Dr. Samsuzzaman Sobuz, Dr. Samir Al Bawri, Dr. Moniruzzaman Mr. Rashedul Islam, Mr. Atiqur Rahman, Mr. Amran Hossain and Mr. Mohammad Shahidul Islam, who have contributed to the diverse technical metamaterial content for microwave applications. Furthermore, all co-authors worked wholeheartedly with their knowledge, drafting suggestions for completing the book.

Next, I would like to acknowledge the support from Spacecraft Environment Interaction Engineering (LaSEINE), Kyushu Institute of Technology (Kyutech), Kitakyushu-shi, Fukuoka, Japan and Microwave Research Laboratory, Universitiy Kebangsaan Malaysia (UKM), Bangi, Selangor, Malaysia in providing their laboratory facilities.

Finally, I want to thank CRC Press for accepting my proposal for the book and helping me with the editing and publishing processes.

Contents

Preface iii
Acknowledgements vi

1. Metamaterial and Sensing 1

   1.1 Introduction 1
   1.2 Concept of metamaterial & microwave sensing 4
   1.3 Metamaterial based microwave sensor design Principle: resonator structure 5
   1.4 Geometry design, fabrication and performances 7
   1.5 Measurement method for metamaterial properties assessment 11
   1.6 Metamaterial based microwave sensor design Principle: sandwich structure 12
   1.7 Unit cell design with circuit modelling and performance 12
   1.8 Sensitivity performance with absorption and polarization affects 14
   1.9 Stacked DNG metamaterial for biosensing 16
   1.10 Stacked DNG MTM scattering parameters and dielectric characteristics 18
   1.11 Blood glutamate concentration sensing performance of the MTM unit cell 21
   1.12 Summary 24

2. Liquid and Solid Material Sensing Using Metamaterial 26

   2.1 Introduction 26
   2.2 Metamaterials and microwave sensing 28
   2.3 Importance of metamaterial sensor 29
   2.4 Tri Circle SRR (TCSRR) shape metamaterial sensor design and analysis 30
   2.5 Electric field and surface current distribution investigation 31
   2.6 Double negative (DNG) metamaterial characteristics investigation 33
   2.7 Parametric study 34
   2.8 Mathematical modelling of the TCSRR based metamaterial sensor 36

viii *Metamaterial for Microwave Applications*

2.9	Sensing performance evaluation	39
	2.9.1 Dielectric constant investigation of oil samples	39
	2.9.2 Loss tangent analysis of oil samples	41
	2.9.3 Detection and evaluation of different LUTS using TCSRR structure	42
2.10	Design and analysis of complementary square SRR-based metamaterial	45
2.11	Effective parameters extraction methods	47
	2.11.1 Metamaterial characteristics of the designed unit cell	48
	2.11.2 Surface current, E-field, and H-field analysis	49
2.12	Equivalent circuit analysis	51
2.13	Parametric analysis	52
	2.13.1 Effect of change of split gap	52
	2.13.2 Effect of change of resonator width	53
	2.13.3 Effect of change of substrate material	54
	2.13.4 Metamaterial structure fabrication and experimental results analysis	55
	2.13.5 Effective medium ratio (EMR) analysis	57
2.14	Materials and thickness sensing using the proposed sensor	57
	2.14.1 Quality factor and sensitivity analysis	61
2.15	Summary	61

3. Metamaterial for Future Generation Wireless Communications — **65**

3.1	Introduction	65
3.2	Metamaterial's background	67
3.3	Metamaterial based mm-wave	69
3.4	Metamaterial particle design	70
	3.4.1 Principle of metamaterial working	72
	3.4.2 Metamaterial properties extraction	73
3.5	Metamaterial based Sub 6 GHz	74
	3.5.1 Metamaterial example	75
3.6	Analysis, design and simulation of 5G metamaterial	79
3.7	Emerging metamaterial applications	83
	3.7.1 Antennas - gain improving	83
	3.7.2 Miniaturizing of the size of antennas	84
	3.7.3 Antenna's bandwidth enhancement	85
	3.7.4 Metamaterials for multiband generation	86
	3.7.5 Metasurface	86
3.8	Summary of existing research related to DNG metamaterial	87
3.9	Summary	90

4. Metamaterial Structure Exploration for Wireless Communications — **93**

4.1	Introduction	93
4.2	Metamaterial with asymmetrical resonator	95

	4.3	Metamaterial with single axis symmetric resonator with applications	97
		4.3.1 Design of MTM unit cell	97
		4.3.2 Result analysis of the metamaterial	98
		4.3.3 Application of metamaterial in antenna for performance enhancement	98
		4.3.4 Application of the metamaterial as an absorber	101
	4.4	Two axes symmetric metamaterial and it's application	102
		4.4.1 MTM design and result analysis	102
		4.4.2 Antenna gain enhancement using MTM superstrate	104
	4.5	Mirror symmetric resonator based tuned metamaterial	105
		4.5.1 Design of the metamaterial	105
		4.5.2 Frequency tuning of the metamaterial	107
	4.6	Rotating symmetric resonator based metamaterial absorber	108
		4.6.1 Metamaterial absorber (MMA) design	108
		4.6.2 Frequency tuning of the MMA	110
		4.6.3 Metamaterial property analysis of the MMA	112
		4.6.4 Power and current distribution analysis of MMA	113
		4.6.5 Angular stability and polarization insensitivity study of MMA	116
		4.6.6 Experimental result of the MMA	117
	4.7	Summary	119
5.	**Metamaterial Antennas for Ultra-wideband Applications**		**123**
	5.1	Introduction	123
	5.2	Metamaterial and UWB antennas	125
		5.2.1 Metamaterials	125
		5.2.2 Ultra wideband (UWB) technology	127
	5.3	Metamaterial based antenna design	130
		5.3.1 Unit cell configuration	130
		5.3.2 Antenna geometry	133
		5.3.3 Experimental validation	135
		5.3.4 Surface current distribution	135
		5.3.5 Time domain performance	136
	5.4	Summary	139
6.	**Flexible Metamaterials for Microwave Application**		**142**
	6.1	Introduction	142
	6.2	Importance of flexible substrate materials for microwave application	144
	6.3	Development of flexible substrate material	146
		6.3.1 Synthesis of $Mg_xZn_{(1-x)}Fe_2O_4$ nanoparticles	147
	6.4	Characterization of $Mg_xZn_{(1-x)}Fe_2O_4$ nanoparticles	148
		6.4.1 Structural analysis	148
		6.4.2 Morphological analysis	151
		6.4.3 Dielectric properties analysis	153

	6.4.4	Optical and photoluminescence analysis	155
	6.4.5	Magnetic properties analysis	156
6.5	Flexible metamaterial design technique	158
	6.5.1	Metamaterials on $Mg_xZn_{(1-x)}Fe_2O_4$ nanoparticles-based flexible substrate	158
	6.5.2	Metamaterial measurement method	161
	6.5.3	Performance of flexible metamaterial with $Mg_xZn_{(1-x)}Fe_2O_4$ nanoparticles with Mg_{40}	163
	6.5.4	Electromagnetic properties analysis of the flexible metamaterial with Mg_{40}	165
	6.5.5	Electromagnetic field interaction of the metamaterial properties with Mg_{40}	167
	6.5.6	Performance analysis of flexible metamaterial properties with Mg_{60}	168
	6.5.7	Comparison of $Mg_xZn_{(1-x)}Fe_2O_4$ nanoparticles-based proposed flexible metamaterials with Mg_{40} and Mg_{60}	170
	6.5.8	Comparison of $Mg_xZn_{(1-x)}Fe_2O_4$ nanoparticles-based proposed flexible metamaterials with existing metamaterials	170
6.6	Summary	171

7. Microwave Head Imaging and 3D Metamaterial-inspired Antenna   175

7.1	Introduction	175
7.2	CCSRR based metamaterial structure design	178
	7.2.1 Design and analysis of CCSRR unit cell structure	178
	7.2.2 Effective medium parameters of CCSRR unit cell	179
	7.2.3 Equivalent circuit model of CCSRR unit cell structure	181
	7.2.4 Parametric study of CCSRR unit cell structure	183
	7.2.5 Design and analysis of CCSRR loaded 3D antenna	184
	7.2.6 Mathematical modeling of the CCSRR loaded 3D antenna	185
	7.2.7 Parametric study of CCSRR loaded 3D antenna	188
	7.2.8 Antenna fabrication and measurement	189
	7.2.9 Head phantom fabrication and measurements	190
	7.2.10 Preparation and fabrication of tissue mimicking head phantom	190
	7.2.11 Electrical properties measurement technique	193
7.3	EM head imaging system	194
	7.3.1 Imaging setup with nine antennas	194
	7.3.2 Antenna phase center optimization	195
7.4	Image reconstruction technique	196
	7.4.1 IC-CF-DMAS image reconstruction algorithm	196
	7.4.2 Matching medium consideration	198
	7.4.3 Internet of things framework for em head imaging system	199
7.5	CCSRR loaded 3D antenna with head model	200
	7.5.1 Specific absorption rate (SAR) analysis	204

		7.5.2	SAR analysis of CCSRR loaded 3D antenna	204
		7.5.3	Measurements of CCSRR loaded 3D antenna fabricated prototype	205
	7.6	Electrical properties of tissue mimicking head phantom		208
	7.7	Em imaging results		209
		7.7.1	Imaging results with CCSRR loaded 3D antenna setup	212
	7.8	Sensitivity analysis		214
		7.8.1	Internet of things based image transfer	215
	7.9	Summary		216

8. Metamaterial Inspired Stacked Antenna Based Microwave Brain Imaging — 221

	8.1	Introduction	221
	8.2	Importance of deep learning in current brain imaging technologies	224
	8.3	Metamaterial loaded stacked antenna	225
		8.3.1 Stacked antenna structure design and analysis	225
		8.3.2 Stacked antenna geometry and design evolution analysis	226
		8.3.3 Parametric analysis of MTM loaded stacked antenna	230
	8.4	Stacked antenna prototype fabrication and performance analysis	231
	8.5	Radiation characteristic analysis of the stacked antenna	233
	8.6	Performance analysis of the stacked antenna with head model	236
	8.7	Specific absorption rate (SAR) analysis of stacked antenna	239
	8.8	Microwave brain imaging (MBI) system implementation method	240
	8.9	Image data collection pre-processing and augmentation techniques	242
		8.9.1 Image data collection	245
		8.9.2 Image pre-processing and input size requirement	245
		8.9.3 Image augmentation technique	245
	8.10	Deep learning based tumor segmentation and classification models	246
	8.11	Microwave segmentation network (MSegNet)-brain tumor segmentation model	247
		8.11.1 Architecture of MSegNet segmentation model	247
		8.11.2 Training experiment of MSegNet model	250
		8.11.3 Evaluation matrix for the MSegNet segmentation model	251
		8.11.4 Brain tumor segmentation performances	252
	8.12	BrainImageNet (BINet)-brain tumor classification model	253
		8.12.1 Mathematical analysis of the classification model	253
	8.13	Architecture of BINet classification model	254
	8.14	Training experiment of BINet classification model	255
	8.15	Evaluation matrix for the BINet classification model	256
	8.16	Brain images classification performances of the BINet model	257
	8.17	Receiver operating characteristic of BINet model	258
	8.18	Miss classification performance analysis of BINet model	259
	8.19	Summary	260

9. Lower UHF Metamaterial Antenna for Nanosatellite Communication System	263
9.1 Introduction	263
9.2 Antennas for nanosatellite	265
9.3 EMNZ metamaterial design and characterization	267
9.4 EMNZ inspired UHF antenna	270
9.5 Summary	276
Index	**279**

Chapter 1

Metamaterial and Sensing

Mohammad Tariqul Islam[1,][*] *and Ahasanul Hoque*[2]

1.1 Introduction

Electromagnetic devices have opened a new horizon in modern technology. Standard materials are used in current devices that have high losses, narrow bandwidth, low detecting responses and they do not have negative effective properties. Therefore, standard materials based on electromagnetic devices are complex in design and larger in electrical length. The metamaterial is a new affordable technology to overcome restrictions that are the main resistance to upgrade electromagnetic devices. Sensors significantly impact the detection process and quality control of any analyte used in daily life. Sensors need a sharp response and their operating frequency undoubtedly is low enough to reduce the absorption in the sensor device. The operating frequency can be controlled by the metamaterial sensor whereas, the standard material-based sensors' operating frequencies cannot be reduced because of their limited area. The artificial material is used to efficiently increase the sensor's effective medium ratio by replacing the narrow shape resonator embedded in the metamaterial structure [1, 2].

Besides, the metamaterial can create strong transmission shift behaviour, high **Q-factor** and sharp, readable sensitivity that is not required by the conventional sensing devices. Therefore, depending on these criteria, the metamaterial Split Ring Resonator (SRR) structure is ideal for proving the concept based on its subwavelength dimensions and its high sensitivity quality factor and resolution at the resonance frequency. The following Figure 1.1(a, b) has been approximated for GHz range operation based on the principle described in [3, 4]. Authors have reported the THz operation design, but a similar design was studied for lower frequencies and gave an interesting response for the quality factor and sensitivity. The particular geometry

---

[1] Department of Electrical, Electronic and Systems Engineering, Universiti Kebangsaan Malaysia, Bangi, Selangor, Malaysia.
[2] Institute of Climate Change (IPI), Universiti Kebangsaan Malaysia, Bangi, Selangor, Malaysia.
Email: ahasanul@ukm.edu.my
* Corresponding author: tariqul@ukm.edu.my

2 *Metamaterial for Microwave Applications*

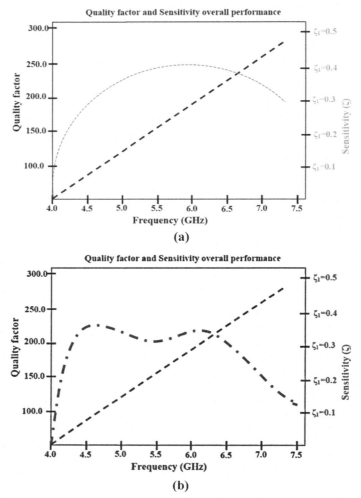

Figure 1.1. Derived response of a typical metamaterial structure for sensing (a) For horizontal geometrical orientation (b) For vertical geometrical orientation, however, some limitations can still be removed by narrowing the **resonator** structure. The SRR resonators are bigger in size and the designed structure can show resonance at higher frequencies, as a result, the electromagnetic wave absorption rate is high, and the produced loss is also high in **sensing** applications. The narrow resonator shape can gradually change the straight resonator arm without changing the split gap in the metamaterial structure. These modified resonators decrease **resonance** frequency and increase the structure's effective medium ratio and exhibit higher sensitivity and large transmission dip that can significantly enhance the sensing device's performance.

significantly affects the sensing performance due to the change of resonator structure and horizontal and vertical movements. It is worth noting that the horizontal resonator geometry (Figure 1.1a) is more like an arc shape over the spectrum, whereas the vertical orientation (Figure 1.1b) is much more stable between 4.5 GHz to 6 GHz for similar a quality factor and achieved sensitivity. Hence, the subwavelength

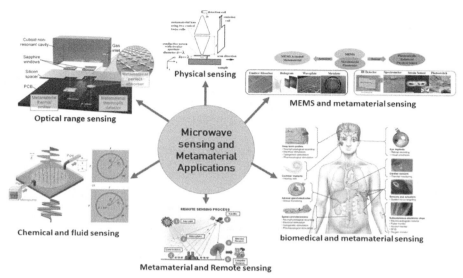

Figure 1.2. Diverse applications of metamaterial-based microwave sensing.

dimensions with specific metamaterial properties have potential and support the anticipation of metamaterial sensing applications.

The design flexibility offered by metamaterials can create numerous technological possibilities as shown in Figure 1.2. Metamaterials promise a great advancement in electronics and motivate research to propose a wide range of metamaterial-based devices. The metamaterial inspired sensors are commonly used in devices for their low cost, high sensitivity and compactness advantages. The conventional sensor's general limitations are low sensitivity, high absorption, low Q-factor, more effort to reduce the operating frequency and higher cost In the design of the sensor, sensitivity depends on the resonator coupling. The electric and magnetic field distribution in the overall sensor can strongly enhance or reduce the sensitivity. As a result, the sensor's performance is affected by the size and shape of its structure. The metamaterial characteristics can offer a new degree of freedom for sensor design.

The use of artificial materials to improve the sensor characteristics presents a novel manner of overcoming the restrictions and exhibits a technique to reduce the sensor size. The research focuses not only on the **metamaterial** structure design, which produces unusual characteristics, but also the desired high sensitivity, compact size, good transmission dip, proper field enhancement and higher effective medium ratio. The effective medium ratio is the ratio of the wavelength of the applied field and the geometrical size of the metamaterial unit cell structure according to the metamaterial transmission line theory that should be below or equal to $\lambda/4$. Based on that condition, the designed Oval, mirror-reflected C and Split L-I metamaterial geometries are larger than the $\lambda/4$ dimension. The finite integration technique and finite-difference time-domain method based electromagnetic simulator, Computer Simulation Technology Microwave Studio (CST MWS) are adopted for the numerical simulations. The Nicolson-Rose-Weir (NRW) and Direct Refractive Index (DRI) methods are used to characterize the metamaterial structures.

In this chapter we represent the basic metamaterial concept and design approach with a short background study, analytical derivation for characterization and finally sensing the potential using the proposed metamaterial in the X and Ku bands. The design concept has been **mathematically** explained and experimentally verified for justifying the concept. The sensing performance of the proposed metamaterial unit cell was demonstrated both in numerical environments as well as experimental setups.

1.2 Concept of metamaterial & microwave sensing

The concept of metamaterials started with the exploration of artificial dielectrics for microwave applications during the second world war. This was started by exploring artificial material development by modifying the electromagnetic properties at the end of the nineteenth century. Metamaterials are now recognized as an enabling technology and one of the most buoyant and exciting research disciplines in microwave sensing, antenna technology, photonics, nanoscience and so on. The remarkable success and impact of the international metamaterial research programs has resulted in the emergence of a mighty and highly skilled research community, and substantial funding across the globe. Evolution in the design of sensing applications has also played a significant role [5–7].

Almost all **isotropic materials** existing in nature have positive values of permittivity and permeability greater than unity. They are thus determined as DPS (double-positive) materials. Materials with a negative ε or μ only are termed as SNG (single negative) materials and divided into two classes depending on the **negative effective parameter**: ENG (epsilon-negative) and MNG (mu-negative). It should be noted that if one constant takes on a negative value, the refractive index of the incident beam becomes imaginary and only damped (evanescent) electromagnetic waves can propagate in such a material. The material is opaque to radiation if its thickness is greater than the characteristic attenuation length of the electromagnetic wave.

The following Figure 1.3 describes these properties in terms of the dielectric characteristics of these materials. A medium showing both positive dielectric properties ($\varepsilon > 0$, $\mu > 0$) is called **double-positive (DPS)**, i.e., crystals, semiconductors, air, water, and dielectrics (mostly dielectrics are designated as DPS).

Such coordination or materials are imposed intentionally and engineered through prior mathematical calculations. Nowadays, this technique is updated with an additional state-of-the-art method to expedite the device or prototype key parameter. However, ENG medium ($\varepsilon < 0$, $\mu > 0$) is a certain frequency region where many plasmas illustrate such features. Besides, Noble metals, such as silver and gold, act as ENG media in the infrared (IR) and visible spectra. Gyrotropic materials, i.e., magneto-optic materials at some frequencies, demonstrate the MNG feature ($\varepsilon > 0$, $\mu < 0$). Finally, a medium of both negative ($\varepsilon < 0$, $\mu < 0$) is termed, Double Negative (DNG). Until now, no naturally available material has shown DNG features which have been demonstrated only by artificial constructs.

Metamaterial inspired resonators provide localized electromagnetic fields that are sensitive to variations in dielectric properties and their geometries. The sensitivity

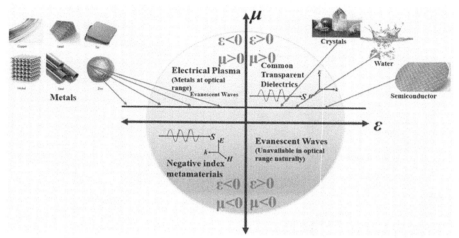

Figure 1.3. Category of materials and metamaterials based on effective medium parameters.

to the geometry can be used to design rotation and displacement detectors. These two parameters can be sensed using the SRR **symmetric** and asymmetric structures. When the symmetry axis of the SRR is aligned with the symmetry axis of the coplanar waveguide (CPW), the total magnetic flux passing through the SRR is zero, and the resonator will not get excited. If the symmetry is broken by displacing the resonator, then the CPW gets excited by a spectral notch from the transmission coefficient. The same principle also works for rotation sensing with a circular resonator structure rather than a rectangular shape inspite of the high sensitivity compared to very small rotation and displacement between the SRR and the CPW. Still a few limitations exist. For example, a small dynamic range resonance shift for undesired displacement or rotation was proposed, which influences the prototype's accuracy.

1.3 Metamaterial based microwave sensor design Principle: resonator structure

The design and development technique and workflow are adopted to realize the transmission line principle-based metamaterial absorber for sensing. Based on the sensing principle (Figure 1.4), a metamaterial absorber based sensing unit can detect precise parameter variation, at least in a lab environment. Additionally, a sensing prototype operating in a particular frequency spectrum with a significant performance indicator is a challenge. Compared to a conventional sensing technique, metamaterial absorber based sensing requires a more precise method while developing the prototype. Typically sensing devices are mostly electronic or microelectronic circuits where size, accuracy, power output, efficiency, and measurement criteria are quite plugged and played. At first, the microstrip transmission line model was analyzed to design the unit cell sensor because the inductance and capacitance formed for patching were specifically dependent on lumped components for **modeling** the prototype. Lumped elements are especially suitable for broadband hybrid microcircuits where

6 *Metamaterial for Microwave Applications*

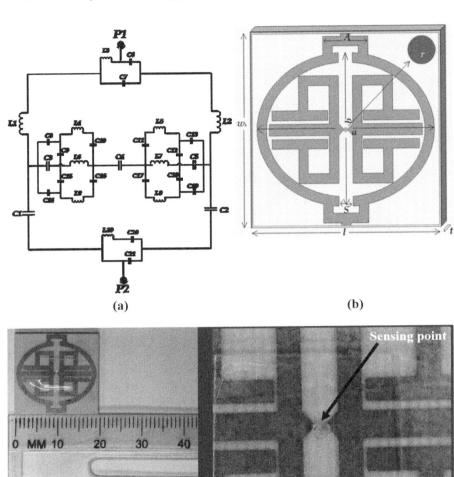

Figure 1.4. Metamaterial sensor design: (a) **Equivalent circuit** of unit cell design (b) Unit cell with major dimensions (c) Fabricated cell with a sensing point.

the Q factor, smaller size and lower cost are of prime importance. Proposed designs have considered such issues according to the study along with a comprehensive mathematical model developed in the presence of ground planes, proximity effects, fringing fields and parasitics among others. Optimization and necessary formulation of the microstrip patches regarding lumped elements of the design was carried out on the relationship between the wavelength and operating frequency. Furthermore, these components show notable value and dependence characteristics with a low Q factor and resonance frequency characteristic of S-parameters. If we see Figure 1.3(a), the transmission line model and our design are intended to achieve an oval or semi-circular shape that can have higher Q factors and provide a higher inductance. Similarly, for a microstrip circuit capacitance formed by a conductor patch on a dielectric substrate, a low value is preferable due to practical considerations like per unit area **capacitance**. Hence, let's numerically determine the value of the total equivalent inductance and capacitance suggested in reference [8]

$$L(nH) = 1.257 \times 10^{-3} a \left[ln\, ln\left(\frac{a+b}{w+t}\right) + 0.078 \right] K_g \tag{1.1}$$

$$C(pf) = \frac{10^{-3} \varepsilon_{rd} w l}{36\pi d} \tag{1.2}$$

where, a = major axis length, b = minor axis length, l = length of micro strip line, w = width of micro strip line, t = thickness of microstrip line, ε_{rd} = dielectric constant of dielectric film, d = Average distance between mutual striplines, correction factor $K_g = 0.57 - 0.145\, ln\frac{w_s}{h'}$, w_s = substrate width, h' = substrate thickness. Using equations (1.1) and (1.2), the lumped inductance and capacitance are 3.439×10^{-5} nH and 1.933×10^{-5} pF respectively.

1.4 Geometry design, fabrication and performances

The proposed metamaterial biosensor consists of an oval shaped patching with a **sandwich** T strip inside. Also, a nozzle patch connects the round sensing area of 0.25 mm radius. The sensing area is simply a cavity in the dielectric substrate and the purpose is to hold the sample as well as forming a capacitance dependent of the amount of glucose. The unit cell consists of a Rogers RO4350B substrate of 1.575 mm thickness with a 0.035 mm copper patch layer. Geometrical dimensions' details are shown in Table 1.1. RO4350B materials are proprietary woven glass reinforced hydrocarbon/ceramics with an electrical performance close to that of PTFE/woven glass and the manufacturability of epoxy/glass provides a tight control on the dielectric constant and allows for low losses while utilizing the same processing method as standard epoxy/glass. The design and simulation was performed through the commercially available CST Microwave studio 2017 software. Figure 1.4(b) shows the proposed design with the major dimensions of the fabricated unit cell biosensor in Figure 1.4(c).

EM field distribution in Figure 1.4 at the resonance frequency of the transmission coefficient, have to be explained from a physics stand point and a mathematical point of view. As we know, according to 'Helmholtz equation' the E-field and H-field both depend on the **propagation constant** γ of the medium. As is known, γ has frequency dependent characteristics along with the permittivity, permeability, and conductivity of the material under consideration (Rogers RO4350B). Moreover, the loss tangent is

Table 1.1. Oval shape metamaterial sensor.

Parameter	l	w_s	A	b	r	S	A
Size (mm)	20	20	19	16	0.25	1.5	4.80

8 *Metamaterial for Microwave Applications*

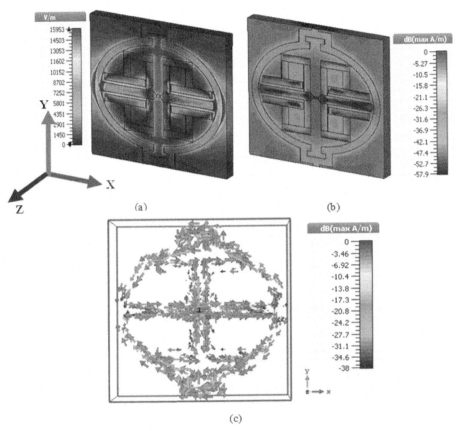

Figure 1.5. At resonance Frequency (3.914 GHz) (a) E-field distribution (b) H-field distribution (c) Surface current distribution.

0.0037 and the dielectric constant is, $\varepsilon r = 3.48$. Now, the transmission line equivalent circuit Figure 1.5(a) clearly shows consecutive capacitance formations along two edges of the oval shape. So, at the resonance frequency (3.914 GHz) these capacitors create a strong electric field Figure 1.5(a) according to linear homogeneous differential equations derived from the 'Helmholtz equation' because the above parameters have a non-**linear** relationship with the propagation constant γ. A similar solution applies for the H-field distribution Figure 1.5(b) at the resonance frequency because the magnetic field components of the mutual coupling distributes its energy according to the homogeneous solution of the field equation. Moreover, the vertical split gap forms an additional LC resonance to accelerate the H-field formation.

Surface current distributions shown in Figure 1.5(c) are significant with a high observed density at the resonance frequency. At 3.914 GHz, the outer and inner resonators have a significant amount of surface current roaming around the edges. An important point to remember is that the microstrip pattern uses Cu (thickness of 0.035 mm) on top of the RO4350B substrate. Therefore, a **traveling wave** may face a situation where at a low frequency the wave has a good conductor medium but with increasing frequency, the dielectric property of the same medium decreases. Additionally, the vertical and horizontal portion of the unit cell have more area compared to other portions. As a result, conduction current density and displacement current density both contribute to getting a **higher density** current. However, the loss tangent becomes higher as charges travel through the oval shaped wall. An interesting point to note is that at a nozzle point the sensing layer that has been placed shows a significant number of charge elements traveling back and forth. Due to the narrower split gap, an overlapping charge distribution may hamper the uniform charge flow and this results in a dense surface current. Though the "hot" area shown in Figure 1.5(c) does not have any significant current element the sandwich T strip has a certain amount of surface current elements which are more than 24 dB. Hence, during sample injection in the sensing area this can make a change in capacitance (C4) and help to identify the parametric changes for glucose sensing in the proposed design.

Herein the metamaterial biosensor that we design and is analyzed according to the **finite-integration technique** (FIT) based simulation method in the CST microwave studio regarding the basic transmission and reflection coefficients. After design completion, boundary conditions are applied on the waveguide port 1 and 2 without imposing any polarization angle. For optimum results from the simulation, material properties like permittivity, permeability, thermal and electrical properties are maintained in a normal condition. The unit cell boundaries were given an electric field along the X-axis, magnetic field along the Y-axis and the Z-axis was kept as an open space for field propagation. For the time being, the split gap remains 1.5 mm and gradually this gap change effect will be discussed in this section. Now, the reflection (S11) and the transmission (S21) coefficients are scattering parameters to identify the characteristics of this design sensor concerning EM field interaction within the targeted frequency range. Figure 1.6, depicts that the transmission coefficient is –0.603 whereas the reflection coefficient is +0.296 at a resonance frequency of 3.914 GHz. In addition Figure 1.5 shows the Smith chart for a transmission coefficient with a normalized impedance of 376.7 and 50 ohm. Starting from 2 GHz while the inductive property of the line swept with increasing frequencies at a frequency of 3.914 GHz a quarter wave LC resonance occurred. For a broadband response optimizing the match at a single frequency can be suboptimal across a band. After that, the transmission line gives a capacitive reactance of 376.7 ohms at 5 GHz and a remaining inductive reactance of 50-ohm for the normalized impedance line. This change indicates whether we have a sensing ability or not through the sensing layer permittivity change. In glucose measurements, permittivity depends on its dielectric constant. So, we intend to change the layer permittivity ε_r to observe a shift in the resonance frequency. As we know the s-parameters from the simulation allow us to numerically analyze the designed sensor using the refractive index parameter. To do so, we can choose the NRW, TR or DRI method. However, as we consider the simple unit cell it is better to go for the DRI method [9] since, the previous two methods are convenient for that metamaterial which is composed of metallic arrays with complex

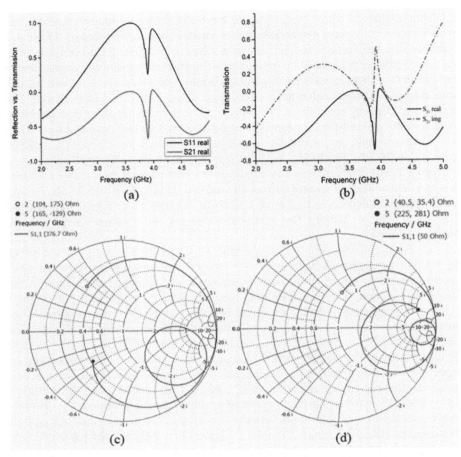

Figure 1.6. Reflection and Transmission properties in simulated design (a) Reflection vs. Transmission, (b) Transmission characteristics (c) & (d)Transmission characteristics using Smith chart of unit cell along propagation (Normalized impedance 376.7 and 50 ohms).

shapes. Also, the TR method requires material **impedance** determination and the correct branch index with a trigonometric function. So,

$$\eta \approx \frac{c}{j\pi f d}\left\{\frac{(S_{21}-1)^2 - S_{11}^2}{(S_{21}+1)^2 - S_{11}^2}\right\}^{\frac{1}{2}} \quad (1.3)$$

Figure 1.7 follows, indicating the metamaterial property of the sensor as well as its absorbing capacity though it is less significant from (a)–(c), wherein the **simulated** unit cell sensor's basic material property is extracted using S11 and S21 parameters. For the clarification to be of interest to the metamaterial based sensor, it is noteworthy to mention that, traditional **micro-ring resonators** have a lower sensitivity compared to DNG metamaterial based resonators since these material types amplify the evanescent wave [10]. Moreover, the sensitivity resolution also

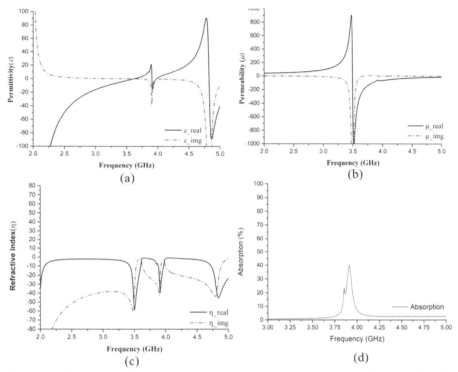

Figure 1.7. Simulation performance of proposed sensor (a) permittivity (ε) (b) permeability (μ) (c) Refractive index (η) (d) Absorption with the focused frequency spectrum.

enhances the use of such metamaterials. Hence, the DNG property and the surface current amount in the sensing layer region at the resonance frequency indicate the sensing potential of this proposed design since the variation of permittivity on the dielectric cavity shows a shift in this frequency. However, the absorption loss of the proposed structure can be revealed using numerical computations as described in the reported article [11].

1.5 Measurement method for metamaterial properties assessment

The **waveguide** measurement method or WMM is a technique for measuring the complex permittivities of materials. The exact electrodynamical model of the sensor similar to a dielectric tube in a rectangular waveguide is used for the solution of an inverse **electrodynamical** task that allows for reducing the error of measurement results. Coaxial, circular or rectangular waveguides are implemented in transmission/ reflection measurement techniques which are directly shown in Figure 1.8. The resonator cavity is filled with the sample under testing. Microwave Systems and Applications are in contact with the sample. Although various measurement techniques are available for use, when choosing the appropriate one, some other factors such as accuracy, cost, sample shape and operating frequency require consideration in technique selection.

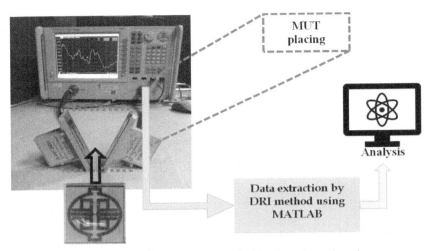

Figure 1.8. Waveguide measurement method for absorption and sensing.

1.6 Metamaterial based microwave sensor design Principle: sandwich structure

Effective microwave sensing requires field interactions between the sample and the sensing element, like resonator structure, microstrip lines or conducting element. Obviously, a **non-invasive** or contactless measurement system needs to be highly precise and the EM field must pass through the sample. Therefore, a combination of the resonator structure during analysis or measurement must ensure this fact. The earlier approach was based on the unit cell and sample interactions mainly occurring through the resonator structure. Compared to the sample size, a zero back plane of the unit cell and narrowband absorption in a different approach such as a sandwich structure was explored during this research. This approach will describe the metamaterial absorber unit cell for microwave sensing using the sandwich combination sample sensing.

1.7 Unit cell design with circuit modelling and performance

A metamaterial absorber for the pressure sensing application in the X and **Ku-bands is proposed**. The metamaterial absorber has been fabricated on FR-4 (lossy) substrate (thickness t = 1.575 mm) which is a glass-reinforced NEMA grade epoxy laminated material. These bands are used for satellite communications especially tracking data relay satellites used for both space shuttles and the International Space Station (ISS). Splitting the band into multiple segments by varying geographical regions can be used for services such as radio astronomy, space research and mobile satellites. Relative permittivity and permeability are −0.4213 and −1.251, respectively, at the highest absorption peak which is 99.6% (at a frequency of 13.78 GHz). The design was intended to create a left-handed metamaterial (LHM) in the X and Ku-bands with maximum absorption and, hence, the reflection coefficient is also negative. The same Transmission Line Theory and **Epsilon-near-zero** (ENZ) property of the material

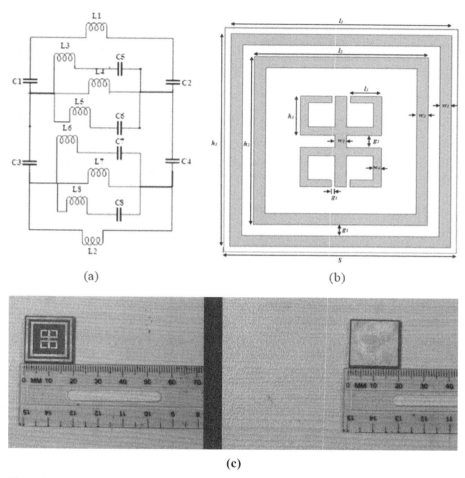

Figure 1.9. Metamaterial absorber design (a) Equivalent prototype circuit (b) Unit cell absorber with dimensions, and (c) Fabricated **prototype.**

[8, 12] gave a simplified analytical model in the GHz range and performed as an absorber. The advantage of this method is very straightforward and the conventional circuit theory was used for the analysis. The absorber consists of two ring shape resonators and the dielectric materials inside the two-mirror reflexed C shaped resonators inside are embossed with a solid ground plane behind the substrate. The structural representation of the front view of the proposed absorber with measurement details is depicted in Figure 1.9(a–c). Metallic elements of the absorber are made of copper with thickness 0.035 mm. Optimized geometrical dimensions of the proposed unit cell MM absorber are represented in Table 1.2.

14 *Metamaterial for Microwave Applications*

Table 1.2. Dimensions of Ring C-shaped quasi-TEM absorber.

Parameter	l_1	l_2	l_3	h_1	h_2	h_3	w_1	w_2	w_3	w_4	g_1	g_2	g_3
Size (mm)	19	15	2.70	19	15	3.50	1	1	1	0.70	0.30	1.10	1
	\multicolumn{13}{c}{Substrate length & width, S}
	\multicolumn{13}{c}{20}

1.8 Sensitivity performance with absorption and polarization affects

The unit cell design, analysis, and simulation were performed in the CST microwave studio using the FIT method with the **boundary condition** applied periodically on the waveguide port. All other conditions like substrate thickness, patching pattern, outer and inner resonator spacing have been analyzed based on the design methodology. In Figure 1.10(a), simulated absorption as well as a **reflection** of EM energy of concerned frequencies are shown. As from equation (1.4),

$$A(\omega) = 1 - T(\omega) - R(\omega) = 1 - R(\omega) = 1 - \left| \frac{Y_0 - Y_{in}}{Y_0 + Y_{in}} \right|^2 \qquad (1.4)$$

it is evident that absorption depends on the admittance form of a patch from an equivalent circuit considered for the prototype. Peak absorptions (Figure 1.10a) were rated at 99.6% and 99.14% with respective frequencies of 13.78 GHz and 15.3 GHz in the presence of the LHM property. Theoretically, a perfect absorber should have zero reflection (R = 0) to get an ideal absorption of 100% at the resonance frequency. In the experimental measurement, the simulated and measured absorption have a proximity as depicted in Figure 1.10(b). It is evident that measured data has roughly 5~10% variation compared to the simulated one at resonance frequency. For microstrips or patches, the TEM properties have some slight propagation differences. As we know, the TEM has transverse characteristics of an EM wave, but as for the general microstrip, the EM wave propagates through the air above the top pattern and the dielectric substrate. As a result, two different media have different resistivity waves propagating at different speeds in both the regions. This is referred to as the quasi-TEM mode.

Extensive study has been done on metamaterial absorbers in the last few years. For example, single nanoparticle & nanorod device for biosensing applications, **Radar Cross Section** (RCS) control for antenna designs based on metamaterials, optical nano circuits inspired by metamaterials, surface wave cloaking from gradient index materials to use nanocomposites in EM wave controlling, ultrathin multicolour meta surface for high resolution spatial light modulators [13–15].

Despite such contributions, this unit cell has a new design with inner and outer resonators with **polarization** independent characteristics. This feature was studied applying quasi-TEM mode wave propagation which makes this design more versatile in X and Ku band applications. Also, dual band resonance with a high absorption rating and sensing characteristics in a simulation environment is a notable point for this design. Moreover, the second resonator inside the lattice reduces the form factor

Metamaterial and Sensing 15

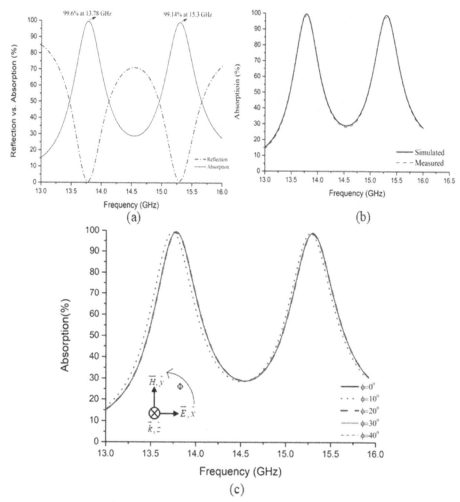

Figure 1.10. (a) Reflection vs. absorption performance, (b) simulated and measured absorption of proposed MM absorber (c) Absorption of **quasi-TEM** polarization independence of the proposed absorber.

whereas the design absorber has a significant value. On the other hand, dimensions of the unit cell are optimized for X and Ku band pressure sensing applications. Figure 1.11(a) shows a simulation arrangement where two unit cells sandwich with the sensor layer. Each unit cell has a dielectric substrate FR-4, metamaterial absorber (MMA) Ring with mirror reflexed C-shape and copper ground plane. Rogers RT 5880 ($\varepsilon = 2.2$, $\mu = 1.0$), RT 5870 ($\varepsilon = 2.33$, $\mu = 1.0$) and RT 6202 ($\varepsilon = 2.94$, $\mu = 1.0$) are three high frequency circuit materials that have been studied in the simulation as the sensor layer. RT 5880/5870 substrates are filled with **polytetrafluroethene** (glass or ceramic) composites laminated for high reliability in aerospace and defence systems specifically for millimeter wave applications, space satellite transceivers and more. Waveguide ports 1 and 2 are used for field excitation. To evaluate the

16 *Metamaterial for Microwave Applications*

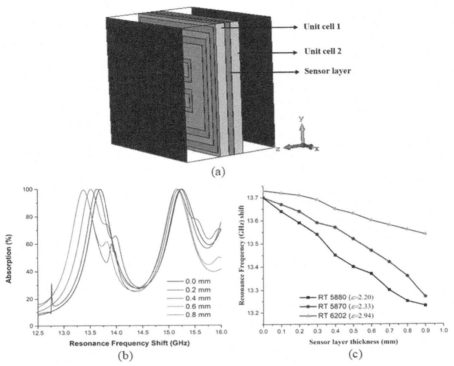

Figure 1.11. Unit cell absorber for pressure sensitivity in simulation (a) Simulation setup for pressure sensor using metamaterial absorber (b) Absorption (%) amount and Resonance frequency shift with respect to sensor layer thickness (c) Sensor layer relationship with Resonance frequency.

unit cell for pressure sensitivity through simulation, sensor layer thicknesses from 0.0 mm are considered and gradually increased by 0.1 mm/observation up to 0.9 mm. Figure 1.11(b) depicts that the sensor layer thickness change is inversely proportional to the resonance frequency shift. It is notable to mention that, for the simplicity of the shifting sequence representation data plotting was performed with 0.2 mm/observation. **Data extraction** from the simulation was based on 0.1 mm/observation. Hence, this unit cell shows an average resonance frequency shift of approximately 52 MHz/observation for the Rogers RT 5880 sensor layer material. A similar step has been followed for RT 5870 and RT 6202 sensing layer materials and plotted in Figure 1.11(c).

1.9 Stacked DNG metamaterial for biosensing

The split-Ω MTM resonator consists of two mirror reflexed resonating patches embossed on the dielectric material and a solid ground plane behind the substrate. The structural representation in isometric and side views with major dimensional details of the proposed sensing element is depicted in Figure 1.12(a, b). The fabricated unit cell with a **stacking** arrangement (Figure 1.12c) and an experimental setup for MTM

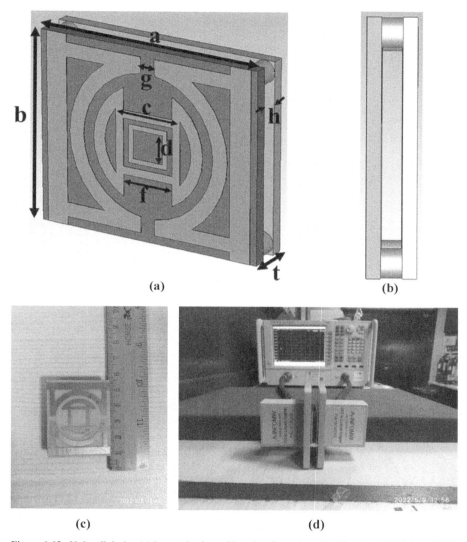

Figure 1.12. Unit cell design (a) Isometric view with major dimensions (b) side view (c) Fabricated Unit cell with stacking (d) Measurement setup.

performance is shown in Figure 1.12d. Metallic elements of the resonator are made of copper 0.035 mm thick. As a result, cost can be minimized from a fabrication point of view and the overall unit cell preparation. Moreover, results have been demonstrated on a simulation and measurement basis to enhance the judgment, versatility and scope of work of the proposed design. **Optimized** geometrical dimensions of the proposed unit cell MM absorber are represented in Table 1.3.

For the microstrip MTM sensing element design, a transmission line model was first analyzed for the prototype which is shown in Figure 1.13. Following a typical

Table 1.3. Mirror reflexed split-Ω MTM unit cell major dimension.

Parameter	a	b	c	d	g	f	t	h
Size (mm)	45	36	13.55	7.20	2.90	10.20	7.53	4.47

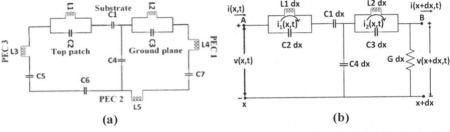

Figure 1.13. Analytical representation of the proposed MTM (a) Equivalent circuit of the MTM (b) Simplified equivalent circuit of the MTM for mathematical modelling.

approach for the microstrip circuit, inductor and capacitor specifically depends on the lumped components referred to for modelling the design. Additionally, **lumped components** have significant parasitic characteristics related to their values as they are closely related to the Q factor and resonance frequency. Moreover, as the prototype contains a ground plane, the effect of the inductance value, i.e., a close ground plane decreases the inductance of the MTM unit cell which basically has three different layers, the ground plane of the sensing element consisting of copper (Cu) 0.035 mm thick which gives a zero transmission of the incident wave during propagation. However, in this design a finite-integration technique (FIT) based simulation method was used for the structural analysis of the MTM unit cell sensing. Commercially available CST Microwave Studio 2019 was used for numerical verification.

The proposed MTM mathematical modelling starts developing the equivalent circuit based on the **transmission line** principle [8]. It is worth noting that all the components have been approximated by following standard lumped element parameter equations mentioned in [16, 17]. Figure 1.13(a) shows the equivalent circuit model where L1, C2, L2, and C3 represent the top and ground layers. The substrate and the air gap of the MTM structure is equivalent to the C1 and C4 capacitors, respectively. The copper plate parasitic elements along with the horizontal and vertical sections are characterized by three consecutive LC series circuits from L3 to L5 and C5 to C7, respectively.

1.10 Stacked DNG MTM scattering parameters and dielectric characteristics

As mentioned in the previous section, the numerical setup for excitation of the MTM sensing element was further analyzed using MATLAB for **characterization**. The basic MTM unit cell S-parameter response is as shown in Figure 1.15. The transmission coefficient depicts three significant resonance frequencies at f_1 = 2.55 GHz (−40.98 dB), f_2 = 4.034 GHz (−68.64 dB) and f_{3s} = 4.56 GHz

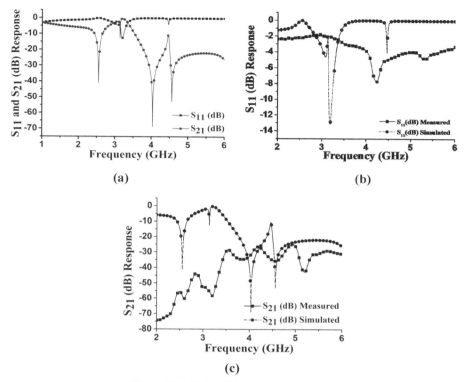

Figure 1.14. **Unit cell S-parameter** response.

(−52.74 dB). Hence, further analysis based on these resonance point reveals that DNG properties do exist during the EM wave propagation. Figure 1.15a illustrates that both S11 and S21 (real part) have a magnitude greater than or close to zero. However, the imaginary part is mostly negative for transmission, i.e., the magnetic resonance, particularly during the inward propogation of resonance frequencies. So, the theoretical prediction is approximately same for the transmission and reflection behaviour. On the other hand, the measured results of S11 and S21 (Figure 1.15b,c) compare the simulated and **experimental** performance. The difference between the simulated and measured S parameters especially in the 2–4 GHz range is due to the calibration limitation of the VNA and air interference during measurement.

We know a relationship from [18] that was being used to study the **Nicolson-Ross-Weir (NRW)** method

$$S_{21} = js_0 S_{11} M \qquad (1.5)$$

where, $s_0 = \pm 1$ and $M = \sqrt{\dfrac{1-|S_{11}|^2}{|S_{11}|^2}}$

this is so as there is a polarity change of s_0 with the real value of S_{11} remaining unchanged producing four distinct symmetries with polarization angle φ

20 *Metamaterial for Microwave Applications*

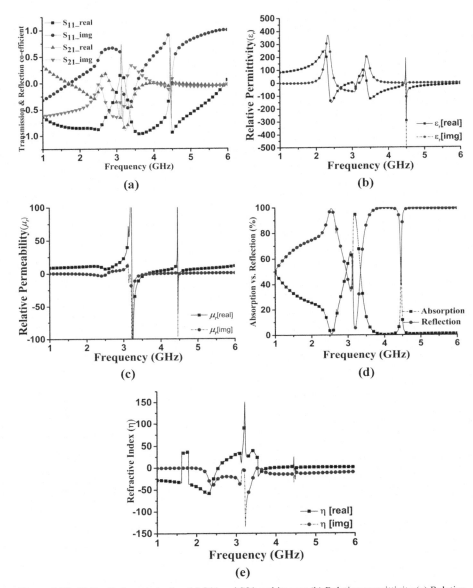

Figure 1.15. Unit cell characterization (a) S11 and S21 real images (b) Relative permittivity (c) Relative permeability (d) Absorption vs. Reflection performance (e) Refractive index.

varying from $-\pi$ to $+\pi$ having no identical figures. As a result, equation (1.5) and Figure 1.14(a) illustrate that at resonance frequency especially 2.55 GHz real parts of both transmission (S_{21}) and reflection (S_{11}) possess negative values of ε (–1.48) and μ (–0.36) (in Figure 1.15c, d) signifying the DNG metamaterial properties. Similarly, at 4.034 GHz and 4.56 GHz, positive values of transmission and reflection give negative permittivity and permeability, but at this point permittivity becomes very close to positive. On the other hand, based on the s-parameter **retrieval** from the simulated environment of the proposed MM absorber, the refractive index (η) performance is shown in Figure 1.15e. The traditional Direct Refractive Index (DRI) method [19] is used for calculating the parameter which is,

$$\eta \approx \frac{c}{j\pi fh'}\sqrt{\frac{((S_{21}-1)^2 - S_{11}^2)}{((S_{21}+1)^2 - S_{11}^2)}} \qquad (1.6)$$

where, f represents frequency and h' is thickness of the substrate. At resonance frequencies of 4.034 GHz and 4.56 GHz, we get negative refractive index values of −0.887 and −0.460 respectively.

1.11 Blood glutamate concentration sensing performance of the MTM unit cell

The introduction briefly describes the importance of blood glutamate for normal brain operation. For example, **converging** evidence indicates that dysfunctions in glutamatergic neurotransmission and in the glutamate-glutamine cycle play a role in the pathophysiology of schizophrenia. Here, we investigated glutamate and glutamine levels in the blood of patients with recent onset of schizophrenia or chronic schizophrenia compared to healthy controls. Compared with healthy controls, patients with recent onset of schizophrenia showed an increased glutamine/glutamate ratio, while patients with chronic schizophrenia showed a decreased glutamine/glutamate ratio. Results indicate that circulating glutamate and glutamine levels exhibit a dual behaviour in schizophrenia, with an increase in the glutamine/glutamate ratio at the onset of schizophrenia followed by a decrease with a progression of the disorder. Further studies are warranted to elucidate the mechanisms and consequences of changes in circulating glutamate and glutamine in schizophrenia.

Another important part is **Traumatic brain** injury-induced acute lung injury (TBI-ALI) a serious complication after brain injury for which predictive factors are lacking. In this study, we found significantly elevated blood glutamate concentrations in patients with TBI or multiple peripheral trauma (MPT), and patients with more severe injuries showed higher blood glutamate concentrations and longer durations of elevated levels. Although the increase in amplitude was similar between the two groups, the duration was longer in the patients with TBI. There were no significant differences in blood glutamate concentrations in the patients with MPT with regard to ALI status, but the blood glutamate levels were significantly higher in the patients with TBI-ALI than in those without ALI. Moreover, compared to patients without ALI, patients with TBI showed a clearly enhanced inflammatory response that was closely correlated with the blood glutamate levels. The blood glutamate concentration was also found to be a risk factor (adjusted odds ratio, 2.229; 95% CI, 1.082–2.634) and was a better predictor of TBI-ALI than the Glasgow Coma Scale (GCS) score. These results indicated that dramatically increased blood glutamate concentrations were closely related to the occurrence of TBI-ALI and could be used as a predictive marker for "at-risk" patients.

22 *Metamaterial for Microwave Applications*

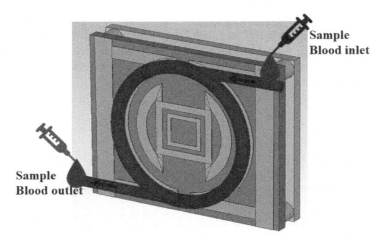

Figure 1.16. Simulation setup for blood glutamate analysis.

So, our numerical analysis basically focuses on the collection of blood or plasma glutamate samples from patients and flow through the inlet and outlet of the proposed MTM sensing element. As shown in Figure 1.16, the blood or plasma sample interacts with the most sensitive part of the resonating element since the EM field signifies the maximum wave absorption and surface current on those parts. Two different conditions have been taken as grounds for a numerical analysis as mentioned in [20]: Normal concentration between 5–100 uM/L and Abnormal concentration between 150–300 uM/L. If a patient's brain **glutamate concentration** is normal, then observe the resonance frequency and compare it with the abnormal brain glutamate concentration value.

In Figure 1.17, all the **transmission coefficient** responses depict a significant observed change for normal and abnormal brain blood glutamate concentration changes. The third resonance point of the proposed MTM sensing unit cell (4.56 GHz), gradually shifts from –46 dB to –52.8 dB and inversely changes with respect to the normal blood glutamate concentration. This inverse relation coincides approximately at –50 dB. On the other hand, under abnormal conditions, the concentration changes from 150 uM/L to 300 uM/L with a step size of 50 uM/L. Similar trends have been seen in this condition too but the coinciding point drops down to –58 dB approximately. So, on average, –2 dB/(50 uM/L) shift of the resonance point for the abnormal blood glutamate range changes. Hence, the MTM sensing elements have a potential to estimate the blood glutamate variation when a patient suffers from an increased level of the blood brain barrier region.

The comparison table shown in Table 1.4, represents relevant articles that were reported for analysis regarding biosensing parameters. Microwave sensors in glutamate concentration sensing have not been extensively explored based on metamaterial, therefore an exact performance comparison would be irrational. Similar research such as blood glucose monitoring [21–24] or Tissue characteristics identification shows a performance of ≤–1 dB whereas the proposed MTM sensing element shows a performance of –2 dB/(50 uM/L). Moreover, the resonance shows

Metamaterial and Sensing 23

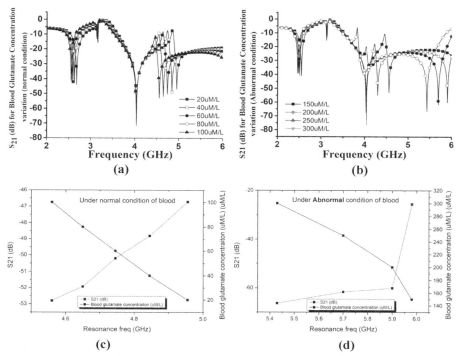

Figure 1.17. Resonance frequency operation and blood glutamate concentration variation observations (a) S21 (dB) response for normal operation (b) S21 (dB) response for **abnormal** operation (c) Resonance frequency variation for normal operation (d) Resonance frequency variation for abnormal operation.

Table 1.4. Comparative features of relevant biosensing element.

Research Article	Size (mm)	Substrate	Bandwidth (GHz)	Application/ Op. band	Performance (per uM/L)	Remarks
[21]	60 × 20	GML 1000	1–2	S band	−1 dB	Blood Glucose Monitoring
[23]	20 × 20	RO3003	NR	HF range	NR	Tissue Characteristics
[25]	0.4 × 0.4	GST	NR	IR range	0.1 dB	No fabrication of Sensing element
[22]	Chemical based	NR	NR	NR	0.01 dB	Glutamate concentration minimal detection
[24]	30 × 20	FR-4	0.63	L band	−10 dB	Blood Glucose Monitoring
Proposed sensing unit	45 × 36	RT5880	3.85–4.25	S and C band	−2 dB	Glutamate concentration in blood

the **ISM band** operation numerically, which represents the potentiality of the proposed MTM.

1.12 Summary

Resonator and sandwich structure metamaterial absorber based sensing with two different designs have been described in this chapter. The **Geometrical** dimensions of the prototypes clearly indicate that a gradual low profile unit cell design has been developed. Both designs have a significant EM field response that creates concentrated mutual field interactions between the sensing region and the unit cell. Besides, negative indexed features (SNG and DNG) illustrate a specific operating frequency and represent the potential EM wave modification capability of each prototype. Additionally, the DNG property with slight absorbing characteristics make the proposed design more appealing for biosensing applications. The field concentration (E-field and H-field) around the split-Ω arm is quite significant and enhances the sensing performance of the unit cell. Besides, at 3.19 GHz the unit cell MTM absorbs more than 90% of the EM wave and shows the unit cell potentiality for microwave absorption. In terms of sensing, the prototype with a sandwich structure and contactless microstrip flow channel verified the identity of the microwave sensor application scope using the metamaterial structure. Despite the simulation, fabrication procedure was conducted for more accurate result verification. **Biosensing** like blood glutamate concentration will be further investigated using different blood samples following the standard protocol. The proposed metamaterial structures' analytical and experimental performances indicate significant potential to be used as an absorber and sensing element in the S- and X-, Ku-band applications.

References

[1] Buriak, I. A., V. Zhurba, G. Vorobjov, V. Kulizhko, O. Kononov and O. Rybalko. 2016. Metamaterials: Theory, classification and application strategies. *Журнал нано-та електронної фізики*, pp. 04088-1–04088-11.

[2] Hoque, A., M. Tariqul Islam, A. F. Almutairi, T. Alam, M. Jit Singh and N. Amin. 2018. A polarization independent quasi-TEM metamaterial absorber for X and Ku band sensing applications. *Sensors,* 18: 4209.

[3] Cen, C., Z. Chen, D. Xu, L. Jiang, X. Chen, Z. Yi et al. 2020. High quality factor, high sensitivity metamaterial graphene—perfect absorber based on critical coupling theory and impedance matching. *Nanomaterials,* 10: 95.

[4] Li, D., Y. Tang, D. Ao, X. Xiang, S. Wang and X. Zu. 2019. Ultra-highly sensitive and selective H2S gas sensor based on CuO with sub-ppb detection limit. *International Journal of Hydrogen Energy,* 44: 3985–3992.

[5] Ferrer González, P. J. 2015. Multifunctional metamaterial designs for antenna applications.

[6] Liberal, I. and N. Engheta. 2017. Near-zero refractive index photonics. *Nature Photonics,* 11: 149–158.

[7] Vakil, A. and N. Engheta. 2011. Transformation optics using graphene. *Science,* 332: 1291–1294.

[8] Bahl, I. J. 2003. *Lumped Elements for RF and Microwave Circuits*. Artech House.

[9] Islam, S. S., M. R. I. Faruque and M. T. Islam. 2015. A new direct retrieval method of refractive index for the metamaterial. *Current Science,* pp. 337–342.

[10] Yang, J. J., M. Huang and J. Sun. 2011. Double negative metamaterial sensor based on microring resonator. *IEEE Sensors Journal,* 11: 2254–2259.

[11] Sethi, K. K., G. Palai and P. Sarkar. 2018. Realization of accurate blood glucose sensor using photonics based metamaterial. *Optik,* 168: 296–301.
[12] La Spada, L. and L. Vegni. 2016. Metamaterial-based wideband electromagnetic wave absorber. *Optics Express,* 24: 5763–5772.
[13] Iovine, R., L. La Spada and L. Vegni. 2013. Nanoparticle device for biomedical and optoelectronics applications. *COMPEL: The International Journal for Computation and Mathematics in Electrical and Electronic Engineering.*
[14] Engheta, N. 2007. Circuits with light at nanoscales: optical nanocircuits inspired by metamaterials. *Science,* 317: 1698–1702.
[15] Liu, Y., Y. Hao, K. Li and S. Gong. 2015. Radar cross section reduction of a microstrip antenna based on polarization conversion metamaterial. *IEEE Antennas and Wireless Propagation Letters,* 15: 80–83.
[16] Bahl, I. J. J. M. 1977. A designer's guide to microstrip line. pp. 1–380.
[17] Garg, R., P. Bhartia, I. J. Bahl and A. Ittipiboon. 2001. *Microstrip Antenna Design Handbook.* Artech House.
[18] Rothwell, E. J., J. L. Frasch, S. M. Ellison, P. Chahal and R. O. Ouedraogo. 2016. Analysis of the Nicolson-Ross-Weir method for characterizing the electromagnetic properties of engineered materials. *Progress in Electromagnetics Research,* 157: 31–47.
[19] Islam, S. S., M. R. I. Faruque and M. T. Islam. 2015. A new direct retrieval method of refractive index for the metamaterial. *Current Science (00113891),* vol. 109.
[20] Leibowitz, A., M. Boyko, Y. Shapira and A. Zlotnik. 2012. Blood glutamate scavenging: insight into neuroprotection. *International Journal of Molecular Sciences,* 13: 10041–10066.
[21] Vrba, J. and D. Vrba. 2015. A microwave metamaterial inspired sensor for non-invasive blood glucose monitoring. *Radioengineering,* vol. 24.
[22] Shadlaghani, A., M. Farzaneh, D. Kinser and R. C. Reid. 2019. Direct electrochemical detection of glutamate, acetylcholine, choline, and adenosine using non-enzymatic electrodes. *Sensors,* 19: 447.
[23] Pokharel, R. K., A. Barakat, S. Alshhawy, K. Yoshitomi and C. Sarris. 2021. Wireless power transfer system rigid to tissue characteristics using metamaterial inspired geometry for biomedical implant applications. *Scientific Reports,* 11: 1–10.
[24] Ali, D., T. A. Elwi and S. Özbay. 2021. metamaterial based printed circuit antenna for blood glucose level sensing applications.
[25] Patel, S. K., J. Parmar, V. Sorathiya, T. K. Nguyen and V. Dhasarathan. 2021. Tunable infrared metamaterial-based biosensor for detection of hemoglobin and urine using phase change material. *Scientific Reports,* 11: 1–11.

Chapter 2

Liquid and Solid Material Sensing Using Metamaterial

Mohammad Tariqul Islam[1,*] and *Md. Rashedul Islam*[2]

2.1 Introduction

A metamaterial is an artificial composition having unique features not found in nature [1, 2]. An example of such a composition includes a structure with **negative permittivity and permeability** within the RF and microwave frequencies [3]. A structure with negative permeability or permittivity is called a single negative (SNG) structure, while one with negative permittivity and permeability is usually called a double negative (DNG) or left-handed (LH) structure. LH materials were theoretically introduced in 1967 by the Russian physicist V. Veselago [4]. Three decades later, a group of researchers was able to realize a 3D LH structure consisting of conductive square split ring resonators (SRR) and conductive wire strips deposited on a fiberglass circuit board [5, 6]. After that, two-dimensional metamaterials (widely known as planar metamaterials or meta surfaces) were realized using a meta-resonator loaded on conventional microwave transmission lines [7, 8]. The introduction of such types of planar metamaterials was a remarkable advancement in the metamaterials field as it allows the realization of a metamaterial-based system using PCB technology and transmission lines. A transmission line-based metamaterial is designed using a planar transmission line such as a microstrip transmission line (MTL) and coplanar waveguide (CPW) loaded with a meta-resonator in the form of either a defected ground structure (DGS) or a surface substrate inclusion [9].

In-depth studies and state-of-the-art employment of the metamaterial idea for microwave sensing are extremely promising [10–12]. Because of Maxwell's

---

[1] Department of Electrical, Electronic and Systems Engineering, Faculty of Engineering and Built Environment, Universiti Kebangsaan Malaysia, Bangi 43600, Selangor, Malaysia.

[2] Department of Electrical, Electronic and Communication Engineering, Faculty of Engineering and Technology, Pabna University of Science and Technology, Pabna 6600, Bangladesh.
Email: rashed.85@pust.ac.bd

* Corresponding author: tariqul@ukm.edu.my

equation for nanoparticle structure analysis for sensing, the dielectric property at zero or negative conductivity is very important. In the visionary range, a novel material development path is indicated by non-uniform conductivity. Metamaterial-based remote sensing is opening a wide variety of research possibilities. Because metamaterial structure-based **microwave sensing** has remarkable material characteristics, it can overcome traditional remote sensing performance constraints in a hostile environment. Therefore, research field applications like biomedical engineering, optics, photonic devices, chemical, and microfluid technology are gradually developing significant breakthroughs in sensing technology. A simple example can demonstrate the significance, such as the dielectric constant of a conventional material is inversely proportional to the **resonance frequency**. A triple ring resonator metamaterial structure can easily detect fuel adulteration using frequency shifts [13]. Another metamaterial sensor is utilized to detect chemical liquids as well as the state of transformer oil with a quality factor [14]. Dielectric property-based metamaterial sensing has been implemented both numerically and experimentally [15, 16]. The split ring resonator (SRR) geometric shapes such as circular, square, or cylindrical are applied for microfluid, rotation, and strain sensing. Liquid permittivity sensing was also reported using a rectangular complimentary ring resonator (RCRR) though the Q factor of the resonator was relatively low [17]. The substrate integrated waveguide (SIW) reported by [18] shows a single resonance, which was controlled by the amount of dielectric material loaded in the resonator. Metamaterial qualities may be utilized in electromagnetics (EM) from microwave frequencies (MHz-GHz) to Optics (THz), leading to discoveries and applications, such as subsequent utilisation in acoustics, as illustrated in Figure 2.1.

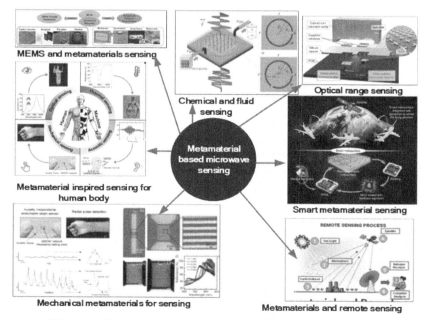

Figure 2.1. Versatile applications of metamaterial-based microwave sensing.

Metamaterials allow the creation of perfect magnetic conductors (PMCs), which are of interest in constructing smaller and more compact antenna systems made of one or more antennas, among other innovative and specific EM applications, such as the negative refractive index (NRI) application. Because metamaterials may overcome and minimize the $\frac{\lambda}{4}$ distance requirement of a linear antenna positioned above (or opposite) a perfect electric conductor (PEC), that is, a reflecting metallic surface, the antenna system can be downsized. The Image Current Theory predicts PMCs in theory, but the scientific community has paid little attention to how to properly realize such structures at microwave frequencies (MHz-GHz). One detriment of metamaterials is that the resultant PMC condition can only be attained within a small frequency spectrum of operation [19].

2.2 Metamaterials and microwave sensing

Metamaterials are artificially engineered materials with exotic **electromagnetic properties** that do not commonly exist in natural materials. Metamaterials have been a rapidly emerging interdisciplinary field including physics, electrical engineering, materials science, optics, and nanoscience since the early 2000s. Metamaterial qualities can be modified by changing their internal **physical structure**. This distinguishes them from natural materials, whose qualities are primarily dictated by chemical ingredients and bonds. The distinctive impact of light traveling through metamaterials is the major reason for the intense interest in them. When the structures are exposed to regulated interactions with electromagnetic radiation, however, they develop completely new features. Some naturally occurring materials, such as opal and vanadium oxide, have been named "natural metamaterials" because of their distinctive properties when they interact with electromagnetic radiation. Metamaterials are frequently referred to as artificially occurring materials.

The concept of metamaterials began with the discovery of artificial dielectrics for microwave applications during the second world war. At the end of the nineteenth century, researchers started investigating artificial material development by altering the electromagnetic characteristics. The fundamental theory of metamaterials was utilized to re-engineer and create various types of MTM structures based on the concept of defining metamaterial properties through their structural arrangement. Metamaterials are being increasingly recognized as an enabling technology and one of the most important and exciting study subjects in microwave sensing, **antenna technology**, photonics, nanoscience, and other domains. With the worldwide success and impact of metamaterial research, the emergence of a robust and highly qualified research community and significant funding from around the world has occurred. Advancement plays a key role in the design of sensing applications. The demand for microwave sensing has been steadily expanding as industrial atomization has progressed. A lot of measurement problems have been solved by using different kinds of microwave sensors [20, 21].

Microwaves are electromagnetic waves with a wavelength in the range of 1 mm to 1 m, which corresponds to the region between the infrared and the short-wave radio wavelengths [22, 23]. Microwave sensing was originally introduced in the

1950s when permittivity measurement and the relationship between permittivity and material physical attributes were investigated. Microwave sensing was initially only used in limited circumstances due to its scarcity, expensive nature, big component size, and low signal processing power. Microwave sensing became simple to develop and inexpensive as **solid-state components** were introduced in the 1970s and microprocessors as measuring devices were available in the 1980s [20], since then, the application of microwave sensing has risen quickly. The benefits and drawbacks of using microwave sensors are very dependent on the application for which they are being utilized. However, there are certain general benefits of using microwave sensors.

1. Microwave sensors are stable due to resonance frequency which is dependent on the physical dimensions.
2. The results of microwave sensors do not vary with the environmental conditions like the presence of pressure, vapor, dust, and high temperature.
3. Sensors are safe from radiation when measuring at the power level with a microwave.
4. Microwave sensors provide fast results.
5. Microwave sensors do not affect the material under test.

However, microwave sensors have some disadvantages also which can be listed as,

1. Microwave sensors are calibrated separately for different materials.
2. The sensors are often designed for specific applications resulting in limited universal applicability.
3. The spatial resolution is limited because of the relatively **long wavelength.**

Despite their drawbacks, microwave sensors are successfully used in a variety of applications such as moisture, humidity, pressure, temperature and medical imaging sensing. Microwave sensors are low profile, lightweight, and **measurement system** development costs are low.

2.3 Importance of metamaterial sensor

Metamaterials' design flexibility can open a slew of new technological possibilities. Metamaterials promise significant advancements in electronics and encourage research in a variety of metamaterial-based technologies. Because of their low cost, outstanding sensitivity, and compactness, metamaterial-inspired sensors are widely employed in devices. The conventional sensor's general limitations are low sensitivity, high absorption, low Q-factor, larger operating frequency, and higher cost. In the design of the sensor, sensitivity depends on the resonator coupling [24, 25]. The electric and magnetic field distribution in the overall sensor can strongly enhance or reduce the sensitivity. As a result, the sensor's performance is affected by the size and shape of the structure. The metamaterial properties may provide a new level of sensor design flexibility. The use of artificial materials to improve sensor properties is a novel way of getting around the limitations [26] and demonstrates an approach to reducing sensor size [27].

The study focuses on not only the metamaterial structure design, which creates exceptional properties, but also the desire for high sensitivity, compact size, good transmission notch, correct field enhancement, and a better effective medium ratio. According to the metamaterial transmission line theory, the effective medium ratio is the ratio of the wavelength of the applied field to the geometrical size of the metamaterial unit cell structure, which should be less than or equal to ($\lambda/4$) [28, 29]. For numerical simulations, the electromagnetic (EM) simulator **Computer Simulation Technology (CST)** Microwave Studio uses the finite integration technique and the finite-difference time-domain method. The metamaterial structures are characterized using the Nicolson-Rose-Weir (NRW) and Direct Refractive Index (DRI) methods.

2.4 Tri Circle SRR (TCSRR) shape metamaterial sensor design and analysis

Figure 2.2 shows a metamaterial structure in the shape of a tri circle SRR. The EM high-frequency solver computer simulation technology (CST-2019) microwave studio has been used for the MTM's structural analysis. Three suitable layers are used in this MTM sensor design; these are the substrate, resonator, and sensor layer of the acrylic sheet. As a substrate, **flame retardant-4 (FR-4)** was used because of its minimal loss, low cost, and superior mechanical intensity. The relative permittivity and loss tangent of the FR-4 substrate is 4.3, 0.025, and its thickness is 1.5 mm. The resonator is made of copper metal which is printed on both sides of the substrate. The thickness and conductivity of the copper are 0.035 mm, and 5.95 × 10^7 S/m, respectively. The sample holder is attached to the backside of the substrate, where the thickness of the sample holder is 7 mm. The sample holder has been made with several layers of acrylic sheets. The thickness of the acrylic sheet for the side wall is 1.5 mm, and for the front and backside layers it is 1 mm; the tangential loss and permittivity of the sensor layers are 0.004 and 2.4, respectively. The sample holder was used to keep the samples under examination in place as they were being tested. There are three key layers to the MTM sensor. Figure 2.2 signifies the unit cell dimension; it is well-suited for the waveguide (X-band), which is consistent with the investigational assessment. The whole area of the MTM sensor is 22.86 × 10.16 mm²; these dimensions are appropriate for the

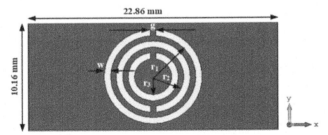

Figure 2.2. Proposed TCSRR MTM sensor structure.

Table 2.1. Design parameters of the MTM structure for microwave sensing.

Parameters	Value (mm)	Parameters	Value (mm)
r_1	3.9	w	0.6
r_2	2.8	g	0.5
r_3	1.7	-	-

mentioned waveguide. The sensor was created to work in the X-band frequency range for the ease of access and fabrication of a sensitive arrangement that allows for a consistent sample holder. The resonance frequency shift has occurred due to the variation of the dielectric constant. Parametric research and genetic algorithms have been used to optimize the required resonator dimensions in the X-band zone; several models have been investigated. Table 2.1 depicts the contained features of the resonator utilized in the proposed research.

The goal of achieving waveguide measurements in the simulation procedure for the designed structure with an effective size is to apply different **boundary conditions**. Because of the side-wall waveguide's metallic composition, it is appropriate to take the boundary requirements into consideration, containing free space, periodic distributions, **perfect electric conductor (PEC)**/perfect magnetic conductor (PMC), and PEC. Intrinsically, the WR-90 waveguide together with a corresponding sample holder is used to calculate transmission response (S_{21}). As shown in Figure 2.3, the boundary conditions of the PEC were assigned to the x- and y-axes, while the z-axis (added space) along the propagation path is assumed to be free. The simulation is carried out in the 8–12 GHz frequency range.

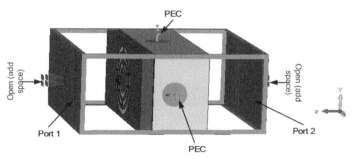

Figure 2.3. Design of MTM structure: PEC-PEC **open add spaces** boundary conditions.

2.5 Electric field and surface current distribution investigation

The electric field distribution has been investigated to understand the designed MTM structure's mechanism. The **electric field distribution** provides information about the energy contained in the device.

The electric field is created across the gap between the CSRR capacitive plate and the circled resonator during the resonance, making the region adjacent to and inside the CSRR sensitive to dielectric changes. As a result, this region inside the CSRR can be used to test the dielectric characteristics of materials. In the empty sensor layer,

32 *Metamaterial for Microwave Applications*

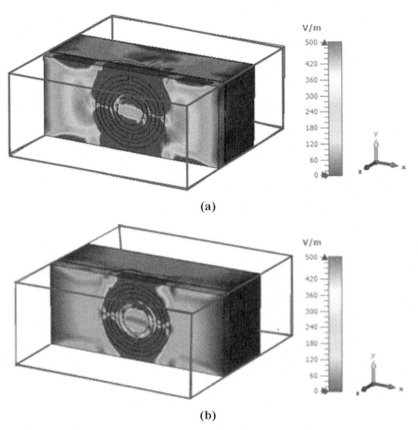

Figure 2.4. E-field distribution losses at (a) 8.34 GHz and (b) 11.59 GHz for the proposed TCSRR-based MTM structure.

the distributions are achieved at resonance frequencies of 8.34 and 11.59 GHz. The electric field strength is more concentrated in the resonator components, especially in the resonator's capacitive elements, as shown in Figure 2.4(a)–(b). At 8.34 GHz resonance frequency, the E-field distribution is effectively seen on the resonator, but at 11.59 GHz, it affects resonator parts and its surroundings due to the capacitive effect. Therefore, any small change in the sample's electrical characteristics in the sensor layer can be sensed in the proposed structure.

Figure 2.5(a)–(b) shows the **surface current distribution** simulation graphs at resonance frequencies of 8.34 and 11.59 GHz. It is seen that the surface current is accumulated more on the inner and middle inductive circles due to the strong magnetic fields. In addition, more currents are disseminated in the resonator's left and right sides, regulating the electric and **magnetic responses**, respectively. Parallel currents were caused by the electric field, whereas anti-parallel currents were caused by the magnetic response to the applied EM field. The importance of resonators in resonant states is further explained in the Figure. The resonators are affected

Liquid and Solid Material Sensing Using Metamaterial 33

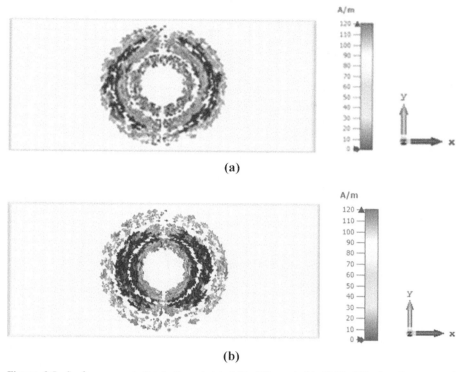

Figure 2.5. Surface current distribution at (a) 8.34 GHz and (b) 11.59 GHz for the proposed TCSRR-based MTM structure.

differently by each part of the resonators. The simulated surface current distribution for the recommended structure is a clear indicator of the presence of the electric dipole that generates the **resonance phenomena**.

2.6 Double negative (DNG) metamaterial characteristics investigation

The **effective parameters** are calculated from the results of S_{11} and S_{21} using the NRW method-based MATLAB code. The relative permittivity, relative permeability, refractive index, and impedance graphs are shown in Figure 2.6(a)–(d). The frequency range of negative effective parameters is listed in Table 2.2.

From Table 2.2, it is seen that the real part of the relative permittivity and relative permeability are negative in the frequency range of 10.98–11.57 GHz, which identifies the proposed metamaterial structure as a potential **double negative property.**

34 *Metamaterial for Microwave Applications*

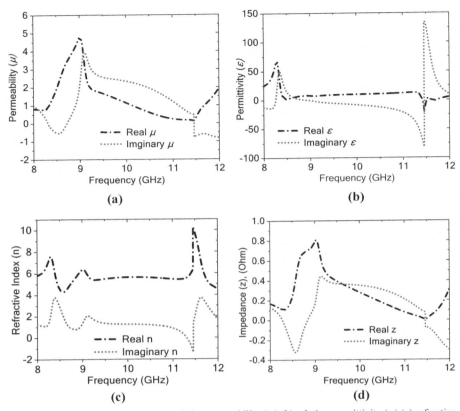

Figure 2.6. Effective parameters: (a) relative permeability (μ_r) (b) relative permittivity (ε_r) (c) refractive index (n) & (d) impedance (Z).

Table 2.2. Effective parameters for the TCSRR-based metamaterial structure.

Parameter	Frequency range (GHz)	Parameter	Frequency range (GHz)
Simulated negative real permittivity	11.42–11.51	Extracted negative real permittivity	11.44–11.57
Simulated negative real permeability	10.98–11.49	Extracted negative real permeability	11.26–11.47
Simulated double negative property	11.42–11.49	Extracted double negative property	11.44–11.47

2.7 Parametric study

Figure 2.7(a) and (b) depict the various resonator designs and **transmission responses** (S_{21}) to them. In design 1, only the 1st ring has been used on the substrate and the achieved S_{21} magnitude is –27.79 dB at 11.38 GHz. To increase the electrical length, the 2nd ring is added with the 1st ring on the substrate, and the obtained resonance frequency is 8.23 GHz with a magnitude of –15.64 dB (in design 2). In

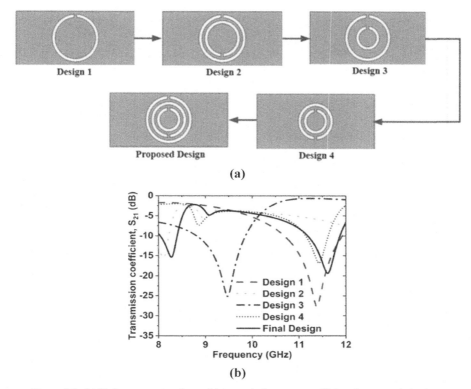

Figure 2.7. (a) Various resonator shapes (b) transmission response (S_{21}) each proposed structure.

design 3, the 3rd ring is added with the 1st ring on the substrate, and the obtained S_{21} value is –25.18 dB at 9.47 GHz. In design 4, the 2nd, and 3rd rings are with the 1st ring on the substrate and the obtained S_{21} value is –25.18 dB at 9.47 GHz. In designs 4, 2nd and 3rd rings have been used on the substrate, and the achieved magnitude of S_{21} is –16.78 dB at 11.43 GHz. In the final design the, 1st, 2nd, and 3rd rings have been used on the substrate for increasing the sensing effect and the S_{21} values are –15.37 dB at 8.34 GHz and –19.35 dB at 11.59 GHz. Since the two resonance frequencies and more sensing effects have been found from the tri circle SRR it has been selected for the proposed MTM sensor.

The resonance frequency (f_r) of the suggested sensor has been affected by the resonator width, which is shown in Figure 2.8(a). When the **sample holder** was empty, the simulation was conducted to show the resonance frequency shift. In the proposed sensor, the **resonator width** has been changed to observe the effect of the resonance frequency. There are five different widths of the resonator that have been used in this MTM-based sensor, i.e., w = 0.4, 0.5, 0.6, 0.7, and 0.8 mm. It is seen that the f_r is 8.20 and 11.21 GHz for the 0.4 mm resonator width and 8.46 and 11.79 GHz for the 0.8 mm resonator width. Since the resonator width is inversely proportional to the inductance, and inductance is inversely proportional to the f_r, hence f_r is increasing with increasing width of the resonator. So, the sensor is properly obeying

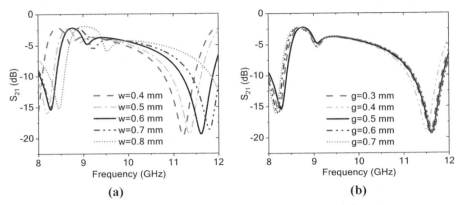

Figure 2.8. Impact of (a) width (w), (b) split gap (g) on the transmission response.

the relation of f_r and inductance. The resonator width of 0.6 mm has been selected for the suggested MTM sensor.

Figure 2.8(b) indicates the impact of the different split gaps on the resonance frequency (f_r). When the **split gap** varies from 0.3 mm to 0.7 mm with an increment of 0.1 mm, then the f_r has been seen to be 8.15 and 11.50 GHz for the 0.3 mm split gaps and 8.27 and 11.64 GHz for the 0.7 mm split gaps. Since the split gap is inversely proportional to the capacitance and capacitance is inversely proportional to the f_r, so the value f_r increases with an increasing split gap, i.e., the sensor properly obeys the relationship between f_r and capacitance. A 0.5 mm split gap was used to fix the suggested MTM sensor design.

2.8 Mathematical modelling of the TCSRR based metamaterial sensor

The waveguide port (WR90) uses the conventional model for classical EM wave propagation during the sensing measurement. Because medium variation during propagation deals with relatively small bands of radiation, studying fixed frequency problems to build mathematical models is extremely realistic. The extended waveguide linked to port 1 passes through the air inside the guided route during the measurement. The propagating wave interacts with the immediate medium changes such as substrate and sample oil. Waveguide port 2 then receives the wave that has travelled through the resonator. As a result, during wave propagation, the transmission coefficient (S_{21}) of the sensing element receives EM energy deviations utilizing distinct samples. Assuming that all time-dependent components in both electric and magnetic fields are, $e^{-j\omega t}$ using the form [30] we get,

$$\nabla \times E - j\omega\mu H = 0 \qquad (2.1)$$

$$\nabla \times H + j\omega\varepsilon E = 0 \qquad (2.2)$$

Permittivity and permeability (ε_{air}, μ_{air}) are considered general parameters with the physical structure of the propagation medium inside the waveguide. At each point of consideration, the rectangular waveguide dielectric constant assumes a uniform distribution of the dielectric constant, $\varepsilon(x) = (x_1, x_2, x_3)$. For uniform distributions, the periodicity in the **propagation direction** inside the waveguide along the extended line should be the same. Because the resonator structure is linked to the line's opening edge, a dielectric medium fluctuation occurs, and the signal goes through two distinct mediums. The substrate permittivity and permeability are denoted by (ε_s, μ_s) and the sample oil is denoted by $(\varepsilon_{oil}, \mu_{oil})$ Figure 2.9.

In general, EM waves lose their magnitude and phase velocity as they propagate through a dielectric substance. As a result, the absorbing capacity of **dielectric materials** (in our example, substrate and oil samples) characterizes energy absorption, and subsequent synthesis theoretically explains the sensing features of the suggested resonator. Furthermore, at each point of consideration, the structure has a finite size. Therefore, we assume a non-conductive permittivity component for point x_1 rather than the point x_2, x_3. As a result, the EM wave is regarded periodic before entering the resonator and sample, but quasi-periodic during propagation in the dielectric medium, according to the well-known 'Floquet-Bloch' theorem. So, for the substrate and resonator field equations $E_{as} = e^{-jas} E(s)$ and $H_{as} = e^{-jas} H(s)$, for oil sample, $E_{ao} = e^{-jas} E(o)$ and $H_{ao} = e^{-jas} H(o)$. Now substituting these values in Equations (2.1) and (2.2), we get [31]

$$\nabla_{as} \times E_{as} - j\omega\mu_s H_{as} = 0 \quad \text{(For sensing element)} \qquad (2.3)$$

$$\nabla_{as} \times H_{as} + j\omega\varepsilon_0 E_{as} = 0 \quad \text{(For sensing element)} \qquad (2.4)$$

The differential component with a complex quantity for the respective medium variation is represented by the modified '∇' operator. Due to the tangential and conical diffraction of the incident wave, the orthogonal quality of EM waves is now lost in both dielectric materials. The method of variable separation used to solve these partial differential equations is highly sophisticated and difficult to describe in general. By reducing Equations (2.3) and (2.4) to second-order ordinary **differential equations**, the simplest solution strategy is to find a second-order partial differential equation

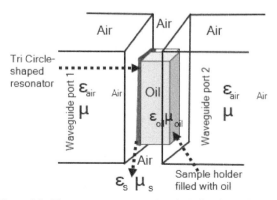

Figure 2.9. The measurement procedure is depicted as a diagram.

in 'n' variables. As a result, **Maxwell's equations** can be reduced to a simple scalar 'Helmholtz equation' [32] as follows:

$$\left(\Delta_{ao} + \omega^2 \varepsilon_{ao} \mu_{ao}\right) u_{ao} = 0 \tag{2.5}$$

$$\left(\Delta_{as} + \omega^2 \varepsilon_{as} \mu_{as}\right) u_{as} = 0 \tag{2.6}$$

where, Δ_{ao} and Δ_{as} represent second-order differential equations based on the medium and u_{ao} and u_{as} are the field components in the Z-direction. Now, discretization of Equations (2.5) and (2.6) by applying the 'finite-difference time domain' method [33, 34] gives

$$S_{z,z_n}(x,y) = MR_e \left(E_{as} H_{ao}^* - E_{ao} H_{as}^*\right) \tag{2.7}$$

where M is the S-magnitude parameter to the resonator and the oil sample. To examine the sensitivity for the realization of the sensing element, extract the S-parameter signal from another edge of the waveguide.

The transmission coefficient component of every oil sample should be jointly propagated across the measurement equipment, according to mathematical modelling. The dielectric property and concentration are constant in every arbitrary sample but changing the sample would affect these two parameter reactions S_{21}. The figure depicts the total reaction of all samples examined during the measurement. The f_r gradually changes throughout eight (8) samples implying that the magnitude variation of the transmission response (S_{21}) (dB) is as observed in the graph. When the oil concentration is high, the magnitude decreases, and when the concentration is low, the magnitude increases. We employed eight (8) oil samples in this investigation; they are, 1. olive oil, 2. olive oil (extra virgin), 3. coconut oil, 4. coconut oil (extra virgin), 5. canola oil. 6. sunflower oil, 7. clean engine oil, and 8. waste engine oil. For example, coconut oil, olive oil, and EV olive oil have high concentrations showing S_{21} as −22.48 dB, −21.77dB and −22.33 dB, respectively. On the other hand, clean engine oil and waste engine oil magnitude are illustrated as −25.95 dB and

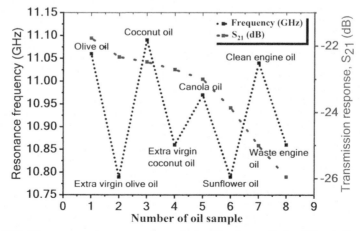

Figure 2.10. The resonance frequency and magnitude of S_{21} changes for the different oil samples.

−25.01 dB, respectively. Hence, we propose the definition of **sensitivity** of the resonator structure as

$$S = \frac{G_{\text{arg}}}{\Delta C_{oil}} \qquad (2.8)$$

where G_{arg} is an average gain variation for different oil samples and ΔC_{oil} is a concentration variation of the oil sample which is 0.28 mg/L in our case. The average **gain variation** was −23.375 dB. Therefore, the sensitivity approximately comes to −83.48 dB/(mg/L), which is quite significant as a potential prototype.

2.9 Sensing performance evaluation

2.9.1 Dielectric constant investigation of oil samples

Figure 2.11 shows the dielectric constant and loss tangent measuring setup for various **liquid under test (LUT)** samples. An open-ended coaxial probe was used to investigate the electrical properties of each sample. The N1500A **dielectric probe kit** was used to test the dielectric constant (DK) and the loss tangent (LT) in this study. In the frequency range of 50 MHz to 43.5 GHz, the probe kit was connected to a power network analyser (PNA)-L and series vector network analyser (VNA) N5224A. The dielectric probe was calibrated using air and pure water at room temperature (25°C) with well-known EM features in the frequency range of 8–12 GHz. The dielectric constant and loss tangent of the various LUTs were determined in this frequency range. Each LUT sample underwent the same procedure.

After measuring the DK of the LUT, the graphs have been prepared using the obtained measured data. It is seen that the value of DK is decreasing with increasing frequency as shown in Figure 2.12. These phenomena show that the DK is inversely proportional to f_r.

Different types of oils may be having properties depending on their constituents and natural conditions. Therefore, it was expected that an oil's electrical properties could be utilized as a feasible instrument for assuring quality and reinforcing the

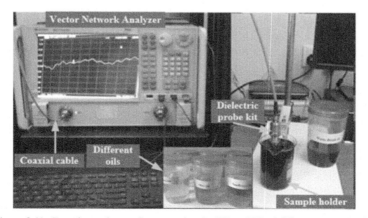

Figure 2.11. Experimental setup for measuring the DK and LT of different LUT samples.

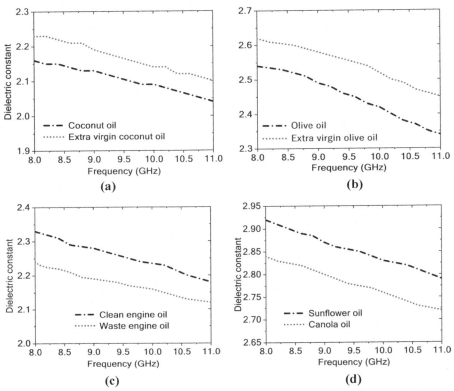

Figure 2.12. Measured dielectric constant for (a) coconut and extra virgin (EV) coconut oil (b) olive and EV olive oil (c) clean and waste engine oil and (d) sunflower and canola oil.

conception of its health benefit potential. Coconut oil is extracted from dried and old coconuts and refined with the application of high heat. Since refined coconut oil may have additional chemicals it is not recommended for external use, i.e., on the skin and hair. Extra virgin coconut oil is unrefined oil obtained by cold-pressing. **Unrefined oil** is the least processed oil, and it contains no extra additives. Unrefined oil could also be recommended for skin and hair care, in addition to dietary preferences. Firstly, the dielectric constant (DK) of these oils was measured using an open-ended coaxial dielectric probe kit in the 8–12 GHz frequency range. It is noticeable that the dielectric constants of coconut oil and extra virgin coconut oil are 2.16 and 2.24 at the 8 GHz frequency shown in Figure 2.12(a).

Olive oil is a mixture that includes both cold-pressed and processed oils, while extra-virgin olive oil is made from pure, cold-pressed olives. Extra virgin olive oil is the least processed or refined type of oil. Extra-virgin olive oil is rich in **monounsaturated fats** and contains a small number of vitamins E and K; it is also high in antioxidants, some of which have important health benefits. The dielectric constant (DK) of these oils is measured using an open-ended coaxial dielectric probe kit at the 8–12 GHz frequency range as shown in Figure 2.12(b). It is noticeable that the DKs for olive oil and extra virgin olive oil are 2.54 and 2.63.

Waste oil affects the engine due to its thickness and lack of lubricating ability. Waste oils degrade engine efficiency, reduce horsepower, reduce mileage, and shorten engine life. The dielectric constant (DK) of the clean engine and waste engine oils are measured using an open-ended coaxial dielectric probe kit at an 8–12 GHz frequency range as shown in Figure 2.12(c). It is noticeable that the DKs of clean engine oil and waste engine oil are 2.33 and 2.24.

The primary distinction between sunflower and canola oils is in the types of fat they contain. Sunflower oil is high in monounsaturated and polyunsaturated fats, which help lower cholesterol, and canola oil is high in omega-3 fatty acids, a form of polyunsaturated fat that can help lower triglycerides. The dielectric constant (DK) of these two oils was measured using an open-ended coaxial dielectric probe kit at the 8–12 GHz frequency range, shown in Figure 2.12(d). It is noticeable that the DKs of sunflower oil and canola oil are 2.92 and 2.83.

2.9.2 Loss tangent analysis of oil samples

The **loss tangent** (LT) for the LUT samples was measured using an open-ended coaxial dielectric probe kit in the 8–12 GHz frequency range as shown in Figure 2.13. The measured LTs of the coconut and **extra virgin coconut oils** are 0.28

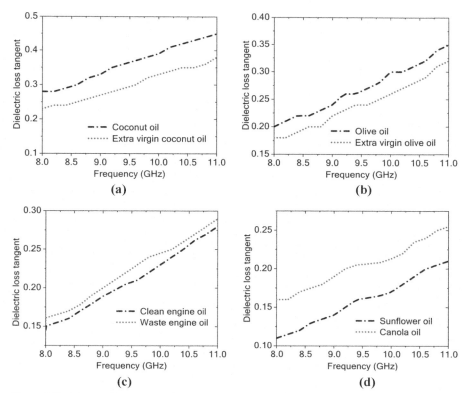

Figure 2.13. Measured loss tangent for (a) coconut and extra virgin (EV) coconut oil (b) olive and EV olive oil (c) clean and waste engine oil and (d) sunflower and canola oil.

and 0.22 at a 8 GHz frequency as presented in Figure 2.13(a). It is noticeable that the LTs for olive oil and extra virgin olive oil are 0.20 and 0.17 in the same frequency range as depicted in Figure 2.13(b). The LTs of clean and waste engine oils are 0.14 and 0.15 in Figure 2.13(c). The LTs of sunflower and canola oils are 0.11 and 0.16 at a 8 GHz frequency in Figure 2.13(d).

2.9.3 Detection and evaluation of different LUTS using TCSRR structure

Figure 2.13 depicts the measurement procedure of the proposed MTM sensor. The front and back views of the fabricated MTM sensor are indicated in Figure 2.14(a)–(b). Figure 2.14(c) shows the oil insertion process in the sample holder. To fill oils, there is a gap on the upper side of the holder, which is closed with the same material to prevent adverse effects on the environment. By using the syringe, oils have been filled in the sample holder. The sample holder is filled with the 1.625 ml **oil samples**. To prevent contamination between the samples, each oil sample was placed in a different sample holder. The MTM sensor outputs are directly influenced by the dielectric constant of oils positioned inside the sample holder. Two waveguide ports and one **extended guided wave** attached to the coaxial cable linked by the TNC female connector and N5227A PNA microwave network analyser using the PNA-L series vector network analyzer (VNA) in the 10 MHz to 67 GHz frequency range are depicted in Figure 2.14(d). The MTM sensor is attached to the waveguide, which is shown in Figure 2.14(f). A **calibration kit** Agilent N4694–60001, was used to calibrate the VNA. Firstly, the fabricated metamaterial sensor is attached to the guided opening of the X-band waveguide. The capacitive parts related to the split

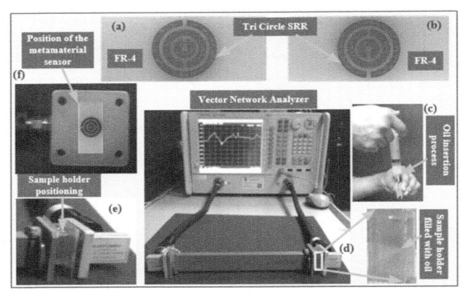

Figure 2.14. (a) Fabricated MTM sensor front view (b) back view (c) oil insertion process in the sample holder (d) VNA with waveguide structure (e) position of the sample holder (f) MTM sensor attached with the waveguide.

gaps of the resonator are assembled as sensor layers. The sample holder is placed on the suggested sensor between the waveguide and the extended guided wave, as shown in Figure 2.14(e). So, there is no direct contact between the oil sample and the **metallic layer** of the suggested sensor. This configuration was used to measure the transmission response (S_{21}) in the 8–12 GHz frequency range. For each sample, the same experimental procedure was followed.

The simulated and measured results of the transmission coefficient (S_{21}) for coconut oil and extra virgin coconut oil in the X-band frequency range are illustrated in Figure 2.15(a–b). The simulated magnitude of S_{21} for coconut oil is –20.71 dB at 10.87 GHz and –20.93 dB at 10.66 GHz for extra virgin coconut oil. Besides, the measured magnitude of S_{21} for coconut oil is –22.49 dB at 11.10 GHz and –22.76 dB at 10.85 GHz for extra virgin coconut oil. These results indicate that despite the near similarities of their dielectric behaviour, the recommended structure accurately detects various types of liquids.

The simulated and measured S_{21} results for olive oil and extra virgin olive oil operate within the X-band as shown in Figure 2.16(a)–(b). The simulated magnitude of S_{21} for olive oil is –21.07 dB at 10.82 GHz and –22.04 at 10.65 GHz for extra virgin olive oil. Besides, the measured magnitude of S_{21} for olive oil is –21.77 dB at 11.05 GHz and –22.41 dB at 10.82 GHz for **extra virgin olive oil.**

Figure 2.17(a) and (b) show the simulated and measured results of the S_{21} for clean engine and waste engine oils operating in the X-band. The simulated magnitudes of S_{21} for **clean engine oil** are –23.18 dB at 10.73 GHz and –22.39 dB at 10.88 GHz for waste engine oil. Besides, the measured magnitudes of S_{21} for clean engine oil are –25.95 dB at 10.86 GHz and –25.02 dB at 11.03 GHz for waste engine oil.

Figure 2.18(a) and (b) depict the simulated and measured results of S_{21} for sunflower oil and canola oil in the frequency spectrum of the X-band. The **simulated magnitude** of S_{21} for sunflower oil is –23.26 dB at 10.60 GHz and –22.26 dB at 10.80 GHz for canola oil. Besides, the **measured magnitude** of S_{21} for sunflower oil is –23.86 dB at 10.79 GHz and –23.01 dB at 10.97 GHz for canola oil.

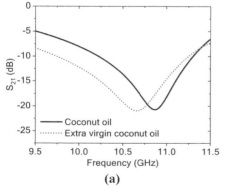
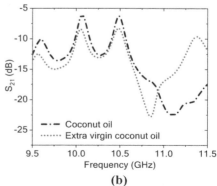

(a)   (b)

Figure 2.15. Transmission response (S_{21}) (a) simulated and (b) measured for the coconut oil and extra virgin coconut oil.

44 *Metamaterial for Microwave Applications*

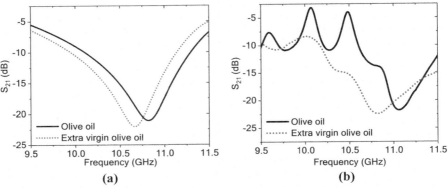

Figure 2.16. Transmission response (S_{21}) (a) simulated (b) measured for the olive oil and extra virgin olive oil.

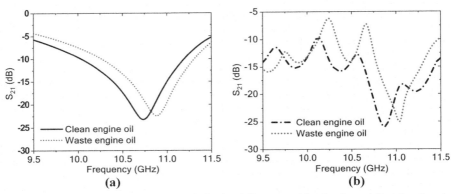

Figure 2.17. Transmission response (S_{21}) (a) simulated (b) measured for the clean engine oil and waste engine oil.

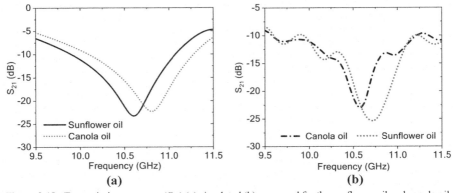

Figure 2.18. Transmission response (S_{21}) (a) simulated (b) measured for the sunflower oil and canola oil.

2.10 Design and analysis of complementary square SRR-based metamaterial

The proposed MTM sensor is very compact, and it is designed on commercially accessible **low-cost** material Rogers RT5880. The dielectric constant of the used Rogers RT5880 is 2.2, the substrate height is 1.575 mm, and 35 μm thick copper is used on the front side and there is no copper on the backside. Copper is used as a resonator, which is **annealed copper**, and its conductivity (σ) is 5.95×10^7 S/m. The front side of the sensor consists of a three-square SRR. The size of the proposed square SRR (SSRR) based MTM sensor is 8×8 mm^2, and the front and side views of the proposed MTM sensor are presented in Figure 2.19(a, b). A frequency-domain solver based on the CST microwave studio software's **Finite Integration Technique (FIT)** is used to simulate numerical unit cells. The simulation is carried out by setting the frequency range between 2–14 GHz. Table 2.3 shows the point-to-point size in mm scale of the designed square SRR-based MTM sensor.

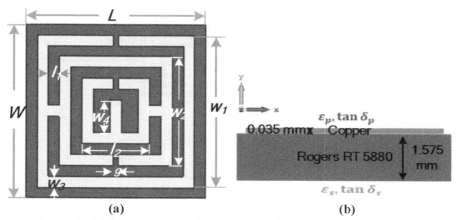

Figure 2.19. (a) Front view (b) side view of the proposed square SRR-based MTM sensor.

Table 2.3. Point-to-point size in mm scale of the designed square SRR-based MTM sensor.

Parameters	Size (mm)	Parameters	Size (mm)
L	8	w_2	5
W	8	w_3	0.5
l_1	0.5	w_4	1.5
l_2	3	g	0.25
w_1	7	-	-

Figure 2.20 shows the simulation setup of the proposed square SRR-based MTM sensor. An electromagnetic wave is incident on the Z-axis to obtain the intended outcomes from the suggested unit cell, while a perfect electric conductor (PEC) and perfect magnetic conductor (PMC) are sequentially placed along the X and Y axes. Here 'I' is the **incident wave**, 'R' is the **reflected wave** and 'T' is the transmitted wave.

46 Metamaterial for Microwave Applications

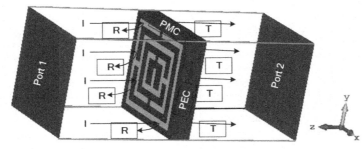

Figure 2.20. Simulation setup of the proposed square SRR-based MTM sensor.

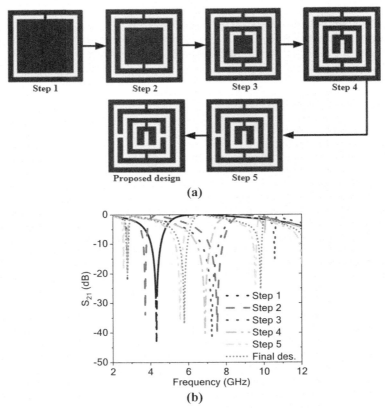

Figure 2.21. (a) Various design configurations to select the final design (b) S_{21} graph for different steps.

The proposed design is chosen by examining the distinct designs of the square SRR-based MTM structure using the trial-and-error method, as shown in Figure 2.21(a), and the graph of S_{21} for different designs is indicated in Figure 2.21(b). In step 1, one square SRR was used and found one resonance frequency.

In step 2, two square SRRs were used and found two resonance frequencies. In step 3, three square SRRs were used and found three resonance frequencies.

Table 2.4. Overall results for the different steps to select the square SRR-based MTM sensor.

Step	Resonance frequency (GHz)	Magnitude or notch (dB)	Covering band
1	4.30	–42.75	C-
2	3.72, 7.52	–36.67, –39.40	C-
3	2.77, 7.23, 10.58	–21.64, –41.72, –15.56	S-, C-, X-
4	2.78, 6.87, 13.26	–20.09, –40.10, –21.10	S-, C-, Ku-
5	2.56, 5.56, 9.53, 11.95	–18.78, –35.73, –24.35, –23.97	S-, C-, X-
Proposed Design	2.77, 5.78, 9.82, 12.29	–20.54, –36.79, –25.26, –23.62	S-, C-, X-, Ku-

In step 4, three square SRRs with one inside **stripline** were used, and found three resonance frequencies. In step 5, three square SRRs with one inside stripline and another stripline between the first and second square SRR were used, and found four resonance frequencies. In the proposed design, three square SRR, one inside stripline, and two connecting striplines are used, and found four resonance frequencies, which cover the maximum **frequency bands**. The overall results for the different designs are listed in Table 2.4.

2.11 Effective parameters extraction methods

To realize the EM properties of the MTMs, the **Nicolson–Ross–Weir (NRW) technique** [29] was used to determine three key material characteristics: permittivity (ε_r), permeability (μ_r), and refractive index (n_r). MATLAB code was used to extract effective parameters [35], and the obtained findings were compared to the CST outcomes after the execution. Based on the S_{11} & S_{21} measurements of the sample under test, the NRW process is one of the most frequent methodologies for EM metamaterial description. Impedance and wave velocity are crucial terms when measuring metamaterial characteristics. From the **normal incidence** of spreading parameter data, the NRW approach was applied to derive effective parameters [36].

The NRW method begins with the V_1 and V_2 compounds, where the scattering parameters are increased and deducted, respectively.

$$V_1 = S_{11} + S_{21} \qquad (2.9)$$

$$V_2 = S_{21} - S_{11} \qquad (2.10)$$

$$T = X \pm \sqrt{X^2 - 1} \qquad (2.11)$$

The coefficient of interface reflection is represented by T. The **S-parameters** and the unit cell can be computed to obtain effective parameters. The relative permittivity, permeability, and refractive index are all represented as follows $\varepsilon_r, \mu_r, n_r$. The number of waves and the thickness of the substrate are represented by k_0 and d

$$S_{11} = \frac{(1-Z^2)T}{1-T^2Z^2} \qquad (2.12)$$

$$S_{21} = \frac{(1-T^2)Z}{1-T^2Z^2} \qquad (2.13)$$

$$\varepsilon_r = \frac{2}{jk_0 d} \times \frac{1 - S_{21} + S_{11}}{1 + S_{21} - S_{11}} \qquad (2.14)$$

$$\mu_r = \frac{2S_{11}}{jk_0 d} + \mu_0 \qquad (2.15)$$

$$n_r = \frac{2}{jk_0 d} \sqrt{\frac{(S_{21} - 1)^2 - S_{11}^2}{(S_{21} + 1)^2 - S_{11}^2}} \qquad (2.16)$$

2.11.1 Metamaterial characteristics of the designed unit cell

The findings of the reflection coefficient (S_{11}) and transmission coefficient (S_{21}) are obtained using the CST simulator, as shown in Figure 2.22(a). The unit cell's S_{11} results show **quad resonance frequencies** of 3.18, 7.01, 10.37, and 12.77 GHz with magnitudes of −29.19, −28.78, −18.81, and −13.71 dB respectively. The unit cell's S_{21} results also show quad resonance frequencies at 2.77, 5.78, 9.82, and 12.29 GHz

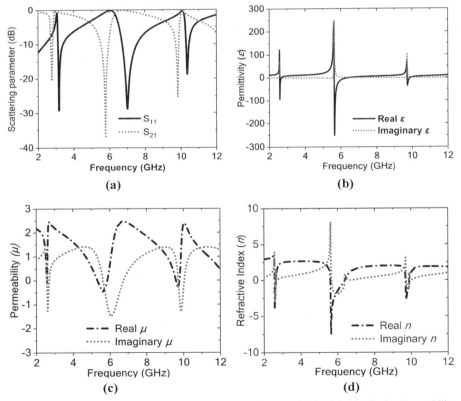

Figure 2.22. Simulated graph of (a) S-parameter (b) effective permittivity (ε_r) (c) effective Permeability (μ_r) and (d) Refractive index (n_r).

with magnitudes of –20.54, –36.79, –25.26, and –23.62 dB respectively. The –10 dB **bandwidth** of S_{21} is 30, 420, 160, and 220 MHz for the respective resonance frequencies. The effective parameters are calculated from the results of S_{11} and S_{21} using the NRW method-based MATLAB code based on equations (2.14)–(2.16), shown in Figure 2.22(b)–(d). The relative permittivity versus frequency graph is shown in Figure 2.22(b). The real value of ε_r shows four negative frequency ranges: 2.66–2.81 GHz, 5.63–6.47 GHz, 9.73–9.94 GHz, and 12.41–12.54 GHz, besides the imaginary value of ε_r shows all positive values over the whole frequency range. The relative permeability versus frequency graph is shown in Figure 2.22(c). The real value of μ_r shows four negative frequency ranges: 2.69–2.78 GHz, 5.51–5.98 GHz, 9.68–9.88 GHz, and 12.31–12.49 GHz, besides the imaginary value of μ_r also shows four negative frequency ranges: 2.71–2.82 GHz, 5.62–6.75 GHz, 9.72–10.04 GHz, and 12.40–12.60 GHz. Since the real values of both ε_r and μ_r have negative frequency ranges, hence the proposed metamaterial is of double negative features and the double negative frequency portions are 2.69–2.78 GHz, 5.63–6.47 GHz, 9.73–9.94 GHz, and 12.41–12.49 GHz. The **refractive index** versus frequency graph is shown in Figure 2.22(d). The real value of n_r shows four negative frequency ranges: 2.67–2.79 GHz, 5.62–6.37 GHz, 9.70–9.869 GHz, and 12.39–12.50 GHz, besides the imaginary value of n_r also shows four negative frequency ranges: 2.71–2.82 GHz, 5.67–6.37 GHz, 9.75–9.89 GHz, and 12.43–12.51 GHz.

2.11.2 *Surface current, E-field, and H-field analysis*

Maxwell's equations can effectively explain the link between the E-field, H-field, and surface current of the proposed metamaterial. The actual electric current generated by the incident electromagnetic field is represented by the surface current. This current generates a magnetic field, and the **fluctuating magnetic field** is the source of EMF. The following equations can be used to represent the E-field and H-field [37].

$$\nabla \times H = J + \frac{\partial D}{\partial t} \quad (2.17)$$

$$\nabla \times E = -\frac{\partial B}{\partial t} \quad (2.18)$$

$$\text{where, } \nabla = \left[\frac{\partial}{\partial x}, \frac{\partial}{\partial y}, \frac{\partial}{\partial z}\right] \quad (2.19)$$

As a result, two more equations can be used to establish the link between the electromagnetic field and metamaterial [38].

$$D(t) = \varepsilon(t) * E(t) \quad (2.20)$$

$$B(t) = \mu(t) * H(t) \quad (2.21)$$

where E and H are time-varying intensities, and D and B denote time-varying electric and magnetic flux densities, electric permittivity is represented by ε, magnetic permeability by μ, and electric current density in a medium by J. Surface current, electric, and magnetic field analysis can be used to better understand the metamaterial

50 *Metamaterial for Microwave Applications*

properties. Maxwell's equations show the **interrelationship** between these three quantities as well as the material. Figure 2.23 depicts the surface current, E-field, and H-field distribution scenarios of the suggested MTM unit cell with quad resonance frequency. Surface current flows mostly through the outer rings at the resonance frequency of 2.77 GHz in Figure 2.23(a), and the upper portions' current density is low. Surface current passes via the three outer square ring resonators and one strip resonators at 2.77, 5.78, 9.82, and 12.29 GHz. At 5.78, and 9.82 GHz resonances the current intensities are decreased, finally, the intensity is low at 12.29 GHz resonance. The E-field of the aforementioned MTM unit cell is shown in Figure 2.23(b) at four different resonance frequencies. These frequencies are 2.77, 5.78, 9.82, and 12.29 GHz. At the 2.77 GHz resonance, the E-field is almost equally apparent around the different rings. The E-field distribution is **highly concentrated** in the inner ring of the cell at 5.78 GHz resonance and marginally concentrated in the middle and outside rings. The E-field distribution is highly concentrated in the inner ring of the cell at 5.78 GHz resonance and marginally concentrated in the middle and outside rings. The lower portion of the middle ring and the upper portion of the inner ring

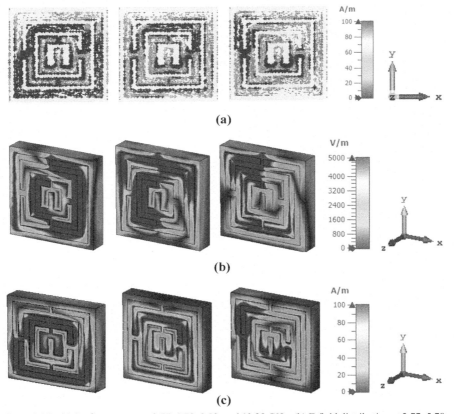

Figure 2.23. (a) Surface current at 2.77, 5.78, 9.82, and 12.29 GHz, (b) E-field distribution at 2.77, 5.78, 9.82, and 12.29 GHz, and (c) H-field distribution at 2.77, 5.78, 9.82, and 12.29 GHz.

had a large charge for the 9.82 GHz resonance frequency, whereas the outer ring and stripline had very little charge. At the 12.29 resonance, the E-field is denser in the lower part. The break or gap in the resonator ring acts as a capacitor, collecting electric charge and generating more E-fields; the E-field intensity is considerable in some places. The H-field around a wire at a reserve follows the equation $B = \mu I/2\pi r$, where µ is the permeability of free space. The H-field and E-field exhibit virtually opposite excitation for all resonance frequencies. In general, the magnetic field is created by the movement of electrical charges. As a result, as indicated by the magnetic field, maximum charge mobility raises magnetic field strength. The H-field distribution for four different resonance frequencies is shown in Figure 2.23(c). When a transverse EM wave propagates through a metamaterial for a given **frequency spectrum**, an artificial magnetic **dipole moment** is introduced in the split ring resonator. The four different resonance frequencies are 2.77, 5.78, 9.82, and 12.29 GHz, and the H-field behavior has been demonstrated for each resonance frequency. In the lower positions of the outer and middle rings, as well as the inner ring's surroundings, the H-field is strong. For the 2.77 GHz resonance, the outer square shape resonator observed moderate field intensity. For the resonance of 5.78 GHz, the field intensity is also noticeable in the middle and inner rings but is modest in the outer square SRR. For the resonance frequency of 9.82 GHz, the field intensity is very evident in the innermost ring, while it is very faint in most outer, and middle resonators. Finally, at the 12.29 GHz resonance, the intensity of the H-field is low.

2.12 Equivalent circuit analysis

Figure 2.24(a) shows the equivalent circuit model of the proposed metamaterial unit cell generated using the **ADS simulator** with lumped parts. This model is created by combining inductance (L) and capacitance (C). The resonance frequency of the metamaterial unit cell is generated using the LC circuit. Inductance is represented by an annealed copper metallic line, whereas capacitance is represented by a split ring gap [39]. Each split ring resonator is a LC series circuit with a specific resonance frequency. The length and gap of the ring resonator can be adjusted to tune the resonance frequency

$$f = \left(1 / 2\pi\sqrt{LC} \right)^{-1} \tag{2.22}$$

The total **inductance** (L) can be determined using the transmission line principle

$$L = 0.01 \times \mu_0 \left\{ \frac{2(d+g+h)^2}{(2w+g+h)^2} + \frac{\sqrt{(2w+g+h)^2 + t^2}}{(d+g+h)} \right\} t \tag{2.23}$$

The total capacitance (C) was also calculated using the equation below

$$C = \varepsilon_0 \left[\frac{2w+g+h}{2\pi(d+h)^2} \ln\left\{ \frac{2(d+g+h)}{(a-l)} \right\} \right] t \tag{2.24}$$

where, $\varepsilon_0 = 8.854 \times 10^{-12} \frac{F}{m}$, $\mu_0 = 4\pi \times 10^{-7} \frac{H}{m}$, w = strip width, h = thickness of the substrate, t = copper strip thickness, d = split gap, and l = length of the strip.

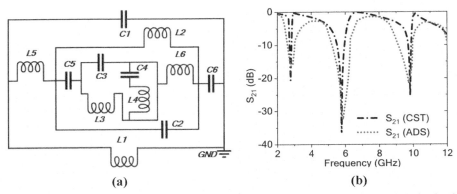

Figure 2.24. (a) Equivalent circuit of the optimized square SRR unit cell (b) CST and ADS simulated graph of transmission coefficient (S_{21}).

The used inductances and capacitances in the equivalent circuit are denoted by the *L1, L2, L3, L4, L5, L6, C1, C2, C3, C4, C5,* and *C6*. ADS results are analyzed to validate the CST result. The outer square SRR forms the inductance *L1* and capacitance *C1*, the second square SRR forms *L2* and *C2*, the third square SRR forms *L3* and *C3*. The inner stripline provides *L4*, and *C4*, where the coupling capacitor *C5* and inductor *L5* between the first and second outer square SRR, and the *L6*, and *C6* are the coupling inductances and capacitance between the second and third square SRR. It's critical to accept both internal and exterior inductance in this equivalent circuit to achieve total inductance; the values of the inductance and capacitance components are calculated using the ADS software with the required response of the S_{21}. The component values in ADS are tuned in such a way that the resonances of S_{21} produced from the CST and HFSS are identical. We tweaked the inductor *L3* and capacitor *C4* to adapt the first resonance frequency to 2.77 GHz. The inductor *L4* and capacitors *C6* have been modified once the second resonance of 5.78 GHz has been addressed. The third resonance frequency of 9.82 GHz is adjusted by tuning the inductor *L5* and capacitor *C5*. The fourth resonance frequency of 12.29 GHz is adjusted by **tuning** the inductor *L6* and capacitor *C4*. Finally, in Figure 2.24(b), the CST and ADS simulated transmission coefficients are shown in Figure 2.24(b), and these three results are nearly identical.

2.13 Parametric analysis

The transmission coefficient (S_{21}) and resonance frequency (f_r) are altering in response to changes in design specifications, i.e., split gap (g), resonator width (w), substrate material, and **metamaterial array.** These analyses are described below.

2.13.1 Effect of change of split gap

Figure 2.25 shows the impact of the split gap on the transmission resonance frequency and magnitude. To check the overall result, five alternative split gaps of 0.15, 0.25, 0.40, 0.50, and 0.60 mm are used. For each split gap, the designed structure is

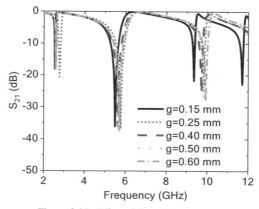

Figure 2.25. Effect of the split gaps on S_{21}.

Table 2.5. The overall results for the change of split gap.

Split gap (g) mm	Resonances (GHz)	Notches (dB)	Covering band
0.15	2.54, 5.48, 9.35, 11.71	−18.06, −36.05, −21.71, −22.83	S-, C-, X-
0.25	2.77, 5.78, 9.82, 12.29	−20.54, −36.79, −25.26, −23.62	S-, C-, X-, Ku-
0.40	2.83, 5.84, 9.93, 12.48	−21.05, −37.12, −24.89, −23.67	S-, C-, X-, Ku-
0.50	2.87, 5.92, 9.99, 12,59	−21.38, −36.87, −25.03, −24.06	S-, C-, X-, Ku-
0.60	2.95, 6.01, 10.08, 12.69	−21.88, −37.03, −24.86, −23.77	S-, C-, X-, Ku-

simulated and data is collected. When the simulation was completed we plotted the graph. The main effect of establishing a split gap in a metamaterial resonator patch is to change the intended unit cell's resonance frequency. Because the split gap of the **resonator patch** introduces a capacitance, it has a direct effect on the metamaterial's resonance frequency. The overall performance is listed in Table 2.5. In this analysis, a 0.25 mm split gap is selected for the proposed structure.

2.13.2 Effect of change of resonator width

Figure 2.26 shows the influence of the resonator width on transmission resonance and magnitude. There are five different resonator widths used to check the overall result; these gaps are 0.30, 0.40, 0.50, 0.60, and 0.70 mm. For each resonator width, the designed structure is simulated and data is collected. When the simulation is completed we plot the graph. One of the **key effects** of inserting a resonator in a MTM resonator patch is that it changes the resonance of the proposed unit cell. Because the resonator width introduces an inductance, it has a direct effect on the metamaterial's resonances. The **overall performance** is listed in Table 2.6. In this analysis, a 0.50 mm resonator width is selected for the proposed structure.

54 Metamaterial for Microwave Applications

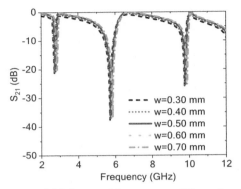

Figure 2.26. Impact of the resonator widths on S_{21}.

Table 2.6. The overall results for the change of resonator widths.

Resonator width mm	Resonance frequency (GHz)	Magnitude (dB)	Covering band
0.30	2.70, 5.70, 9.75, 12.22	−21.34, −37.59, −26.05, −24.41	S-, C-, X-, Ku-
0.40	2.72, 5.73, 9.77, 12.24	−20.94, −37.19, −25.66, −24.01	S-, C-, X-, Ku-
0.50	2.77, 5.78, 9.82, 12.29	−20.54, −36.9, −25.26, −23.62	S-, C-, X-, Ku-
0.60	2.81, 5.82, 9.85, 12.33	−20.44, −36.69, −25.16, −23.51	S-, C-, X-, Ku-
0.70	2.85, 5.86, 9.90, 12.37	−20.42, −36.68, −25.15, −23.50	S-, C-, X-, Ku-

2.13.3 Effect of change of substrate material

Figure 2.27 shows the S_{21} diagram for distinct substrate materials. The consequences of changing the dielectric material on the metamaterial performance were also investigated in this study. In this study, three types of Rogers (RO4350B, RT5880, and RT6202) with FR-4 were analyzed. FR-4 with a thickness (*th*) of 1.6 mm, an electric permittivity (ε) of 4.3, and a dielectric loss (δ) of 0.025 was utilized in the

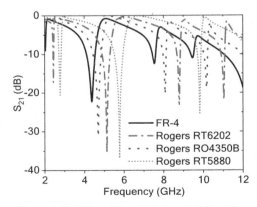

Figure 2.27. Effect of the substrate materials on S_{21}.

Table 2.7. The overall results for the different substrate materials.

Substrate material	Resonances (GHz)	Magnitude (dB)	Covering band
FR-4	4.30	−42.75	C-
Rogers RO4350B	3.72, 7.52	−36.67, −39.40	C-
Rogers RT5880	2.77, 5.78, 9.82, 12.29	−20.54, −36.79, −25.26, −23.62	S-, C-, X-, Ku-
Rogers RT6202	2.78, 6.87, 13.26	−20.09, −40.10, −21.10	S-, C-, Ku-

first test. In the second test **Rogers RO4350B** with th = 1.524 mm, ε = 3.66, and δ = 0.0037 was utilized. In the third test Rogers RT5880 with th = 1.575 mm, ε = 2.2, and δ = 0.0009 was utilized. In the last test Rogers RT6202 with th = 1.5 mm, ε = 2.2, and δ = 0.0009 was utilized. The overall results for the used substrate material are mentioned in Table 2.7.

2.13.4 Metamaterial structure fabrication and experimental results analysis

To validate the simulation results, the suggested unit cell has been fabricated and measured. When the unit cell is measured, it is positioned in between two waveguide ports. These waveguide ports are attached to the **vector network analyser** manufactured by Agilent and referred to as the N5227A. The prototype of the constructed unit cell and 16 × 16 array MTM are shown in Figure 2.28(a–b). The experimental setup for the **unit cell** and array MTM is illustrated in Figure 2.29(a)–(b). Figure 2.30(a) is a representation of the transmission coefficient of the square SRR unit cell, showing both the simulated and measured values. The simulated resonances are 2.77, 5.78, 9.82, and 12.29 GHz, whereas the experimental resonances are 2.73, 5.88, 9.87, and 12.43 GHz. Both these resonances cover the S, C, X, and Ku bands. The simulated notches are −20.54, −36.79, −25.26, and −23.62 dB, whereas the measured notches are −16.04, −29.53, −20.36, −19.66 dB. From Figure 2.30(a) it is seen that there is a slight distinction between the measured and simulated results. There might be a small number of reasons for this occurrence. First, an inaccuracy in the calibration of the Agilent N5227A vector network analyzer utilizing the Agilent N4694 60,001 Ecal produced a variance in findings between those that were measured and simulated. In addition, it might be because of very minor mistakes made during the manufacturing process of the suggested substrate layer. The mutual resonance effect of the two **waveguide ports'** transmitting and receiving ends always affects the readings and causes tiny variations in both sets of data. Moreover, this effect will always have an influence. Last but not least, the permittivity of the substrate material is a crucial factor in the outcomes. To measure the array performance, we designed the array and the fabricated prototype had a size of 128 × 128 mm^2. The measurement setup of the array prototype is shown in Figure 2.29(b), and it consists of two **horn antennas** that are utilized to evaluate the performance of the array. The prototype of the array is placed in the middle of two horn antennas that are 40 cm apart from one another for this assessment setup. The 16 × 16 array measurement result is depicted in Figure 2.30(b). Comparing the S, C, X, and Ku-band to the unit cell, it can be observed from the array graph that there are some noises present in the S, C, X, and Ku bands. In addition, the

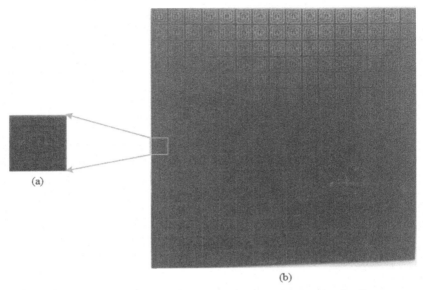

Figure 2.28. Fabricated **square SRR** MTM (a) Unit cell (b) 16 × 16–unit cell array.

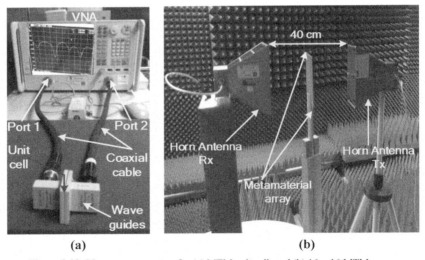

Figure 2.29. Measurement setup for (a) MTM unit cell, and (b) 16 × 16 MTM array.

findings of the array show that there is a minor variation in the resonance frequencies. Nevertheless, there is a clear and convincing parallel between the findings of the array and the results of the unit cell. The seemingly insignificant disparity can be the result of a manufacturing mistake or the mutual connection that exists between the cells of the array. Despite this, the output of the array achieves the S, C, X, and Ku band properties that are envisioned.

Liquid and Solid Material Sensing Using Metamaterial 57

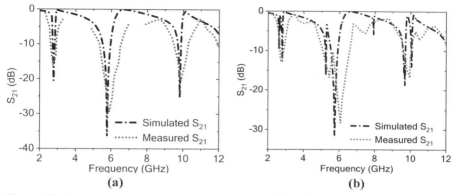

Figure 2.30. Simulated and measured S_{21} graph for (a) square SRR MTM unit cell and (b) 16 × 16 MTM array.

2.13.5 *Effective medium ratio (EMR) analysis*

In the MTM study, the effective medium ratio (EMR) is a crucial factor for examining the structure. The EMR shows how tiny and effective the metamaterial is [40, 41]. A high EMR reflects the MTM design's perfection and fulfilment criterion. If the EMR value is less than 4, the condition of the MTM subwavelength is not met. Because the metamaterial's frequency and size are inversely related, the MTM's modest size results in a high resonance frequency that is incompatible with low-frequency devices. As a result, when developing metamaterials, the EMR must be properly considered. The EMR is the ratio of the wavelength to the unit cell size, as calculated by Eq. (2.25).

$$EMR = \frac{\lambda}{L} \tag{2.25}$$

λ = wavelength at the lower resonance frequency and L = length of the square SRR MTM unit cell.

The EMR of the designed structure is 13.54, since λ = 108.30 mm at 2.77 GHz, and L = 8 mm.

2.14 Materials and thickness sensing using the proposed sensor

Metamaterial sensing has recently been examined on a large scale. The sensing use of different bands has been examined utilizing the proposed MTM sensor, even though this topic is of tremendous interest to researchers. Microwave sensors may be used to determine the moisture content, density, structure and shape of materials, and even **chemical reactions**, by measuring the characteristics of materials based on microwave contact. Microwave sensors have several benefits over traditional sensors, including quick & exact measurements, non-destructive nature, full automation, and the ability to be manufactured online or in a lab. Figure 2.31(a) shows the simulation setup of the designed square SRR-based MTM sensor model for materials and thicknesses of materials' characterizations. Figure 2.31(b) represents

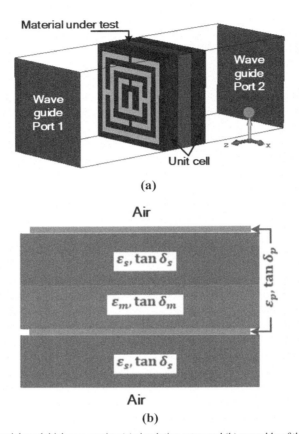

Figure 2.31. Materials and thickness sensing (a) simulation setup and (b) assembly of the different layers.

the layer arrangement of the sensor. There are five layers in the arrangement, such as the substrate, copper resonator, material under test (MUT) layer and copper patch. MUT layers are **sandwiched** between two MTM structures and placed in between two waveguide ports. The dielectric constant of the sensing MUTs was changed to investigate the sensing performance. This change has a significant impact on the microstrip line's capacitance, resulting in a shift at the resonance frequency.

Figure 2.32 shows the whole procedure of MUT sensing using compact MTM structures. Firstly, the square SRR-based MTM sensor has been designed and the prototype has been **fabricated,** then the MUT sample is prepared, after which the MUT is sandwiched by the MTM sensor, and finally, the sandwich arrangement is placed in two waveguide ports and measured by the VNA.

To test the sensing performance, we have taken three MUTs such as FR-4, Rogers RO4350B, and RT5880, with different FR-4 thicknesses of 1, 1.5, and 2.3 mm. The permittivity (ε_m), loss tangent (*tan tan* δ_m) and FR-4 thickness (t_m) are 4.3, 0.0025, and 1.5. The ε_m, *tan tan* δ_m and t_m of Rogers RO4350B are 3.66, 0.0037, and 1.524. In the case of **Rogers RT5880**, ε_m = 2.2, *tan tan* δ_m = 0.0009, and

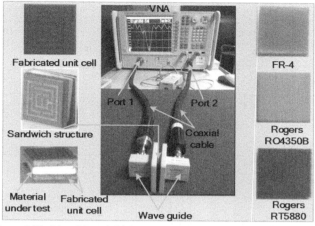

Figure 2.32. Materials and **thickness sensing** laboratory measurement setup.

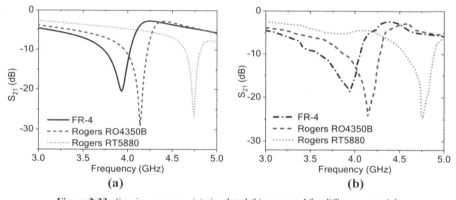

Figure 2.33. Sensing response (a) simulated (b) measured for different materials.

t_m = 1.575. When the experimental setup is complete, the scattering parameters for each MUT are measured, and graphs are created using the measured data. Figure 2.33 shows how the sensor behaviour changes with the dielectric constant changes. The change in mutual and coupling capacitance of the patch caused these modifications. It demonstrates that as the **sensor layer** dielectric constant grew, the resonance frequency-shifted towards the lower frequency, demonstrating a linear relationship between frequency and dielectric constant. The overall sensor's performance for different material characterizations is listed in Table 2.8.

From Table 2.8, it is seen that the **resonances shifted** at 210 MHz between FR-4 and RO4350B, 600 MHz between FR-4 and RO4350B, and 810 MHz between FR-4 and RT5880 due to the permittivity change.

The suggested MTM-based sensor construction can also be utilized to detect material thickness. To test the thickness sensing performance, three FR-4 thicknesses, 1, 1.5, and 2.3 mm are taken, measured, and then graphs are created using the measured

Table 2.8. Sensor's performance for different material characterizations.

MUT	Operating frequency (GHz)	Resonance frequency (GHz)		Notch (dB)		Resonance frequency shift (MHz)
		Sim.	Meas.	Sim.	Meas.	
FR-4	3–5	3.93	3.94	−20.52	−18.62	210 between FR-4 and RO4350B
Rogers RO4350B		4.14	4.14	−29.04	−25.24	600 between RO4350B and RT5880
Rogers RT5880		4.74	4.76	−26.76	−24.61	810 between RT5880 and FR-4

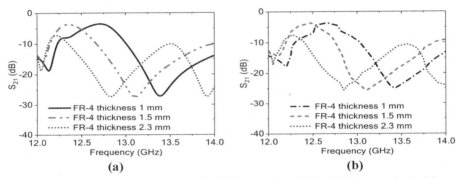

Figure 2.34. Sensing results (a) simulated and (b) measured for different thicknesses of materials.

data. Figure 2.34 shows how the sensor behavior changes for the dielectric constant changes. These adjustments were triggered by a change in the patch's mutual and coupling capacitance. It demonstrates that as the sensor layer dielectric constant grew, the resonance frequency shifted towards the lower frequency, demonstrating a linear relationship between frequency and dielectric constant. Figure 2.34 depicts that the sensor layer thickness change is inversely proportional to the resonance frequency shift. The overall sensor's performance for different thickness characterizations is listed in Table 2.9. From Table 2.9, it is seen that the resonances are shifted at 290 MHz between 1 and 1.5 mm thickness, 270 MHz between 1.5 and 2.3 mm thickness, and 560 MHz between 1 and 2.3 mm FR-4 thickness due to the permittivity change.

Table 2.9. Sensor's performance for different thicknesses of **material characterizations**.

FR-4 thickness (mm)	Operating frequency (GHz)	Resonance frequency (GHz)		Notch sim. (dB)		Resonance frequency shift (MHz)
		Sim.	Meas.	Sim.	Meas.	
1	10–14	10.91, 13.39	10.93, 13.42	−29.17, −27.15	−27.43, −25.05	290 between 1 and 1.5 mm thickness
1.5		10.85, 13.10	10.84, 13.09	−28.97, −27.47	−28.05, −25.79	270 between 1.5 and 2.3 mm thickness
2.3		10.84, 12.83	10.84, 12.85	−30.73, −27.32	−30.14, −25.64	560 between and 2.3 mm thickness

2.14.1 Quality factor and sensitivity analysis

The quality factor is the ratio of the resonance frequency to the **+3 dB bandwidth**. It is a critical consideration when investigating the sensing capability of a metamaterial sensor. The fact that most MTM sensors have a low-quality factor and considerable measurement errors is a major source of concern, which limits their use for several purposes. The formula of quality-factor is, $Q = \frac{f_r}{\Delta f}$ [42], here f_r = resonance frequency and Δf = +3 dB bandwidth. The difference between two independent variable values at which the dependent variable equals 70.7 percent in relation to the minimal transmission value is the +3 dB bandwidth. The maximum quality factor has been found in the FR-4 material sensing. Using the above formula, Q-factor has been calculated, and its value is 325.

Exhaustive sensitivity research is carried out to evaluate the influence that different substrate thicknesses and materials have on the sensitivity of the square SRR-based sensor. The extracted sensitivity S(%) is defined as follows [43]:

$$S(\%) = \frac{f_0 - f}{f(\varepsilon_r - 1)} \times 100 \tag{2.26}$$

where, f_0 is the initial frequency, i.e., when MUT is considered as air, f is the frequency when the MUT is changed one after another and sandwiched between two MTM sensors and ε_r is the materials permittivity value. The value of f_0 is 5.45 and the value of f is 3.93, 4.14, and 4.74 GHz for the MUT FR-4, RO4350B, and RT5880 respectively. So, the sensitivity of the sensor is 11.72, 12.48, and 11.9 for the above-mentioned MUT. For a change in MUT permittivity value of 0.92, the average sensitivity changes by 12.03%. Again, in the thickness sensing, for a change of 0.5 mm in the MUT thickness, the average sensitivity changes by 11.53%. Table 10 depicts the overall sensing performance comparison between the proposed square SRR-based MTM sensor and some published MTM sensors.

2.15 Summary

In this chapter, metamaterial-based microwave sensing has been discussed with the required analysis. A **tri circle** SRR shaped metamaterial sensor is effectively designed and fabricated such that it can be utilized for the detection of various liquids in the X-band frequency. The proposed sensor is shown to be capable of distinguishing between various oil samples with ease. The dielectric constant depends on the oil concentration and frequency. The dielectric constants of the coconut oil and extra virgin coconut oil are 2.16 and 2.24, although the dielectric values are very close between these two oils; even after that, the resonance frequency has shifted to 210 MHz. To summarize, the proposed sensor is well suited for detecting different oils. The recommended structure is highly sensitive; the sensitivity is –83 dB (mg/L) which can be used in a variety of fields, such as detection of various liquids, **microfluidic sensors**, and industrial applications. The sensor's structure was successfully intended for real-time, long-lasting, and specific recognition of various samples. As a result, the proposed sensor could be very useful in a variety of industrial, liquid, and chemical detection applications. Another designed compact

MTM sensor can thoroughly distinguish the materials and thickness of the materials. The optimized cell's dimensions are 8 × 8 mm^2, and obtained resonances are 2.77, 5.78, 9.82, 12.29 with notches of –20.54, –36.79, –25.26, –23.62 dB, respectively. Three different materials and thicknesses of materials are used to examine the sensing performance of the fabricated sensor. The EMR, quality factor, and sensitivity of the sensor are 13.54, 325, and 12.03% change for a permittivity value change of 0.092 and a 11.53% change for MUT thickness change of 0.5 mm respectively. The resonances are shifted due to permittivity and thickness changes of the materials, and it is inversely proportional to the permittivity and thickness of the materials for the change in mutual and coupling capacitance of the copper resonator. The materials detection sensor works in the S band whereas the thickness detection sensor performs in the X band. Resonances are shifted at 210 MHz between FR-4 and RO4350B, 600 MHz between FR-4 and RO4350B, and 810 MHz between FR-4 and RT5880. Also, resonances are shifted at 290, 270 and 560 MHz between 1 and 1.5 mm thickness of the FR-4, 1.5 and 2.3 mm thickness of the FR-4, and 1 and 2.3 mm thickness of the FR-4, respectively. All the laboratory results are identical to the simulation outcomes. Because the proposed sensor is small and **cost-effective**, it has a high sensitivity and good **quality factor**, hence this sensor can be employed in a variety of industries for sensing various materials and their thicknesses.

References

[1] Itoh, T. and C. Caloz. 2005. *Electromagnetic Metamaterials: Transmission Line Theory and Microwave Applications*. John Wiley & Sons.
[2] Argyros, A., A. Tuniz, S. Fleming and B. Kuhlmey. 2015. Drawn metamaterials. pp. 29–54. In: *Optofluidics, Sensors and Actuators in Microstructured Optical Fibers*, ed: Elsevier.
[3] Awai, I., I. Matsuda, M. Hotta, A. Sanada and H. Kubo. 2006. Microwave applications of single negative materials. *Journal of the European Ceramic Society*, 26: 1811–1815.
[4] Veselago, V. G. 1967. Electrodynamics of substances with simultaneously negative electrical and magnetic permeabilities. *Usp. fiz. nauk*, 92: 517–526.
[5] Pendry, J. B., A. J. Holden, D. J. Robbins and W. Stewart. 1999. Magnetism from conductors and enhanced nonlinear phenomena. *IEEE Transactions on Microwave Theory and Techniques*, 47: 2075–2084.
[6] Shelby, R. A., D. R. Smith and S. Schultz. 2001. Experimental verification of a negative index of refraction. *Science*, 292: 77–79.
[7] Martın, F., J. Bonache, F. a. Falcone, M. Sorolla and R. Marqués. 2003. Split ring resonator-based left-handed coplanar waveguide. *Applied Physics Letters*, 83: 4652–4654.
[8] Martin, F. 2015. *Artificial Transmission Lines for RF and Microwave Applications*. John Wiley & Sons.
[9] Baena, J. D., J. Bonache, F. Martín, R. M. Sillero, F. Falcone, T. Lopetegi et al. 2005. Equivalent-circuit models for split-ring resonators and complementary split-ring resonators coupled to planar transmission lines. *IEEE Transactions on Microwave Theory and Techniques*, 53: 1451–1461.
[10] La Spada, L. and L. Vegni. 2018. Electromagnetic nanoparticles for sensing and medical diagnostic applications. *Materials*, 11: 603.
[11] Liberal, I. and N. Engheta. 2017. Near-zero refractive index photonics. *Nature Photonics*, 11: 149–158.
[12] Chen, T., S. Li and H. Sun. 2012. Metamaterials application in sensing. *Sensors*, 12: 2742–2765.
[13] Bakır, M., M. Karaaslan, F. Dinçer, K. Delihacioglu and C. Sabah. 2016. Tunable perfect metamaterial absorber and sensor applications. *Journal of Materials Science: Materials in Electronics*, 27: 12091–12099.

[14] Altintaş, O., M. Aksoy, E. Ünal and M. Karaaslan. 2019. Chemical liquid and transformer oil condition sensor based on metamaterial-inspired labyrinth resonator. *Journal of The Electrochemical Society,* 166: B482.
[15] Akgol, O., M. Karaaslan, E. Unal and C. Sabah. 2017. Implementation of a perfect metamaterial absorber into multi-functional sensor applications. *Modern Physics Letters B,* 31: 1750176.
[16] Bakır, M., Ş. Dalgaç, M. Karaaslan, F. Karadağ, O. Akgol, E. Unal et al. 2019. A comprehensive study on fuel adulteration sensing by using triple ring resonator type metamaterial. *Journal of the Electrochemical Society,* 166: B1044.
[17] Wu, J., P. Wang, X. Huang, F. Rao, X. Chen, Z. Shen et al. 2018. Design and validation of liquid permittivity sensor based on RCRR microstrip metamaterial. *Sensors and Actuators A: Physical,* 280: 222–227.
[18] Salim, A., M. U. Memon and S. Lim. 2018. Simultaneous detection of two chemicals using a TE20-mode substrate-integrated waveguide resonator. *Sensors,* 18: 811.
[19] Ferrer González, P. J. 2015. Multifunctional Metamaterial Designs for Antenna Applications. Doctoral thesis, Universitat Politecnica de Catalunya.
[20] Nyfors, E. 2000. Industrial microwave sensors—A review. *Subsurface Sensing Technologies and Applications,* 1: 23–43.
[21] Muñoz-Enano, J., P. Vélez, M. Gil and F. Martin. 2020. Planar microwave resonant sensors: A review and recent developments. *Applied Sciences,* 10: 2615.
[22] Sturm, G. S., M. D. Verweij, A. I. Stankiewicz and G. D. Stefanidis. 2014. Microwaves and microreactors: Design challenges and remedies. *Chemical Engineering Journal,* 243: 147–158.
[23] Berube, M. 1985. *The American Heritage Dictionary: Second College Edition*: Houghton Mifflin.
[24] Yang, J. J., M. Huang, H. Tang, J. Zeng and L. Dong. 2013. Metamaterial sensors. *International Journal of Antennas and Propagation,* vol. 2013.
[25] Chuma, E. L., Y. Iano, G. Fontgalland and L. L. B. Roger. 2018. Microwave sensor for liquid dielectric characterization based on metamaterial complementary split ring resonator. *IEEE Sensors Journal,* 18: 9978–9983.
[26] Abdulkarim, Y. I., L. Deng, H. Luo, S. Huang, M. Karaaslan, O. Altıntaş et al. 2020. Design and study of a metamaterial based sensor for the application of liquid chemicals detection. *Journal of Materials Research and Technology,* 9: 10291–10304.
[27] Su, L., J. Mata-Contreras, P. Vélez and F. Martín. 2017. A review of sensing strategies for microwave sensors based on metamaterial-inspired resonators: dielectric characterization, displacement, and angular velocity measurements for health diagnosis, telecommunication, and space applications. *International Journal of Antennas and Propagation,* vol. 2017.
[28] Hossain, M., M. Faruque, M. Islam and S. Islam. 2017. An effective medium ratio obeying metaatom for multiband applications. *Bulletin of the Polish Academy of Sciences. Technical Sciences,* vol. 65.
[29] Islam, M. T., M. Moniruzzaman, T. Alam, M. Samsuzzaman, Q. A. Razouqi and A. F. Almutairi. 2021. Realization of frequency hopping characteristics of an epsilon negative metamaterial with high effective medium ratio for multiband microwave applications. *Scientific Reports,* 11: 1–23.
[30] Sadiku, M. N. and S. V. Kulkarni. 2009. *Principles of Electromagnetics*: Oxford University Press.
[31] Boyer, C., E. Kalnins and W. Miller. 1976. Symmetry and separation of variables for the Helmholtz and Laplace equations. *Nagoya Mathematical Journal,* 60: 35–80.
[32] Kozlov, A. I., L. P. Ligthart and A. Logvin. 2007. *Mathematical and Physical Modelling of Microwave Scattering and Polarimetric Remote Sensing: Monitoring The Earth's Environment Using Polarimetric Radar: Formulation and Potential Applications* vol. 3: Springer Science & Business Media.
[33] Haffa, S., D. Hollmann and W. Wiesbeck. 1992. The finite difference method for S-parameter calculation of arbitrary three-dimensional structures. *IEEE Transactions on Microwave Theory and Techniques,* 40: 1602–1610.
[34] Chew, W. C. 2021. Lectures on theory of microwave and optical waveguides. *arXiv preprint arXiv:2107.09672.*
[35] CST AG, D. 2018. Germany. CST STUDIO SUITE. Accessed: 2018, ed.

[36] Luukkonen, O., S. I. Maslovski and S. A. Tretyakov. 2011. A stepwise Nicolson–Ross–Weir-based material parameter extraction method. *IEEE Antennas and Wireless Propagation Letters*, 10: 1295–1298.

[37] Wartak, M. S., K. L. Tsakmakidis and O. Hess. 2011. Introduction to metamaterials. *Physics in Canada*, 67: 30–34.

[38] Hasan, M. M., M. R. I. Faruque and M. T. Islam. 2018. Beam steering of eye shape metamaterial design on dispersive media by FDTD method. *International Journal of Numerical Modelling: Electronic Networks, Devices and Fields*, 31: e2319.

[39] Paul, C. R. 2011. *Inductance: Loop and Partial*. John Wiley & Sons.

[40] Islam, M. S., M. Samsuzzaman, G. K. Beng, N. Misran, N. Amin and M. T. Islam. 2020. A gap coupled hexagonal split ring resonator based metamaterial for S-band and X-band microwave applications. *IEEE Access*, 8: 68239–68253.

[41] Islam, M. R., M. Samsuzzaman, N. Misran, G. K. Beng and M. T. Islam. 2020. A tri-band left-handed meta-atom enabled designed with high effective medium ratio for microwave based applications. *Results in Physics*, 17: 103032.

[42] Omer, A. E., G. Shaker, S. Safavi-Naeini, H. Kokabi, G. Alquié, F. Deshours et al. 2020. Low-cost portable microwave sensor for non-invasive monitoring of blood glucose level: Novel design utilizing a four-cell CSRR hexagonal configuration. *Scientific Reports*, 10: 1–20.

[43] Abdolrazzaghi, M., M. Daneshmand and A. K. Iyer. 2018. Strongly enhanced sensitivity in planar microwave sensors based on metamaterial coupling. *IEEE Transactions on Microwave Theory and Techniques*, 66: 1843–1855.

Chapter 3

Metamaterial for Future Generation Wireless Communications

Samir Salem Al-Bawri[1,*] and *Mohammad Tariqul Islam*[2]

3.1 Introduction

The tremendous development of the upcoming **fifth-generation (5G)** (in both millimeter-wave, mm-wave, and sub 6 GHz bands), as well as **sixth generation (6G)** communication technology, has been a prime requirement in recent years in order to meet the demand for lower energy consumption, low latency, low cost, higher data rate, and for supported data transmission within a large number of user terminals which deal with numerous applications outside of the cellular industry and mobile communications [1, 2]. Due to the scarcity of spectrum resources, 5G wireless communication system research is now focused mostly on the mm-wave/sub 6 GHz bands. Furthermore, due to its innovative spectrum and large bandwidth, the frequency range has gotten a lot of attention around the world [3]. However, the researchers' key problem (as an example) is improving isolation between the adjacent elements of the suggested massive/multiple-input multiple-output (mMIMO/MIMO) **antennas** and circuit designs as well as reducing the size, as intercoupling would affect the total efficiency and overall system performance. Many ways of achieving high isolation between every two adjacent elements of MIMO antenna have been described in the literature. Hence, a wide range of MTM-based structure options are presented in order to emphasize the enhancement of performance metrics that are uncommon and rarely explored in the reported state of the art.

---

[1] Space Science Centre, Institute of Climate Change, Universiti Kebangsaan Malaysia (UKM), Bangi, Selangor, Malaysia.
[2] Department of Electrical, Electronic and Systems Engineering, Universiti Kebangsaan Malaysia, Bangi, Selangor, Malaysia.
Email: tariqul@ukm.edu.my
* Corresponding author: samir@ukm.edu.my

Besides, high throughput provided by 5G and beyond enables efficient spectrum utilization because frequencies may be reused over short distances. Line-of-sight (LOS) and non-line-of-sight (NLOS) signal propagation, as well as a massive antenna array, perform a significant role in 5G (mmWave/Sub 6 GHz) and 6G performances as demonstrated in Figure 3.1. Massive MIMO antennas loaded with filled metamaterials will be the key to producing highly directed beams that overcome path losses and NLOS circumstances. Moreover, the Dynamic Metasurface Metamaterial Antenna (DMA) is an interesting radiating technology for wireless 5G/6G systems, which allows to dynamically control the **massive MIMO** antenna elements and produce a reduction in power consumption and cost compared with conventional antennas (see Figure 3.2).

In fact, metamaterials are structures that have been purposely designed to produce electromagnetic properties which are not created in nature [4]. **Left-handed** metamaterials are those that have both negative permittivity and negative permeability ($\epsilon < 0$) ($\mu < 0$) (LHMs). The scientific, technological, and engineering sectors have all done extensive research on LHMs [5]. They can also be defined as compound structures that are engineered to modify the electromagnetic properties of the material.

Metamaterials can be categorized based on structures of the targeted wavelength. These structures primarily have electromagnetic characteristics due to interference effects from the unit cells between dispersed fields that create frequent band gaps. The electromagnetic characteristics of that group are obtained primarily through a homogenization process and hence have effective electromagnetic properties. It should also be mentioned that metamaterials enable the emulation of electromagnetic properties of a known material or the desired electromagnetic properties that are not readily accessible by default. However, their limited electrical size, as opposed to the operative wavelength, is the secret to applying such resonators to the synthesis of metamaterial-induced planar circuits. Consequently, these resonators are virtually lumped components that give rise to the miniaturization capacity of

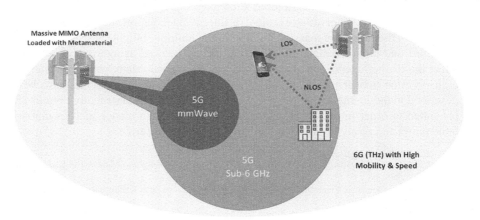

Figure 3.1. Evolution of 5G and 6G wireless communication.

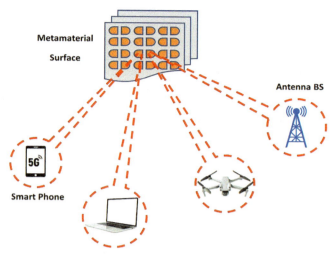

Figure 3.2. Metasurface mMIMO antenna of multi-user wireless communication.

planar microwave structures and circuits. Single negative metamaterial (SNG) [6], double-negative metamaterials (DNG) [7], **near-zero index (NZI)** of permeability (μ) and permittivity (ε), along with a refractive index (NZRI) [8–10] characteristics and electromagnetic bandgap (EGB) [11, 12] are the most common types of metamaterial structures.

Various kinds of metamaterials are also utilized to improve the designed antenna [13, 14]. Furthermore, the previously suggested decoupling approaches are difficult to build and operate on tiny MIMO parts. Unlike conventional antenna arrays, this chapter will explain a compact small **split-ring resonator (SRR)** which can be used to reduce the coupling between antenna arrays. The use of the metamaterial structure within the 5G and beyond antennas design also results in a noticeable increase in gain as well as good radiation characteristics [6, 7]. In this chapter, we introduce several unique metamaterial unit cells based SNG, DNG, NZRI of refractive index (n), (μ), and (ε) for simultaneous improvement of the overall performance of 5G antennas such as isolation and gain. The details of a few examples of metamaterial structures are explained and applied on antenna design to explain the validity of the proposed design. The accuracy is also investigated by experimental results.

3.2 Metamaterial's background

In nature, all materials have positive **permeability** (μ) and **permittivity** (ε). The materials which have positive permittivity and permeability will be called the right-hand material (RHM). As for materials with passive permittivity and permeability, it can be achieved in a section of synthetic materials called the left-hand material (LHM). The inclusion of small inhomogeneities to enforce efficient macroscopic behavior gives these substances their features rather than their composition. Metamaterials are used to denote synthetic materials, which possess electromagnetic properties that are not common in nature. It has also made its way into the primary electromagnetic stream. The properties of electromagnetic waves of the composite media interact with impurities, producing electricity and magnetism, and thus affect the permittivity and permeability of the compound medium [15].

Metamaterials can be categorized into two groups from the unit cell's point of view: The first group refers to structures of the order of the directed wavelength, which include so-called electromagnetic bandgaps (EBGs). These structures primarily have electromagnetic characteristics due to interference effects from the unit cells between dispersed fields that create frequent band gaps. As a homogenous means for incident electromagnetic waves, metamaterials have unit cells much smaller than the directed wavelength. The electromagnetic characteristics of that group are obtained primarily through a homogenization process and hence have effective electromagnetic properties. It should also be mentioned that metamaterials enable the emulation of desired electromagnetic properties of a known material that are not readily accessible by default. However, as opposed to the operative wavelength, their limited electrical size is the secret to applying such resonators to the synthesis of metamaterial-induced planar circuits. Consequently, these resonators are virtually lumped components that give rise to the miniaturization capacity of planar microwave structures and circuits.

The second group, in June 2002, the Caloz Group proposed the exploration of SRR using the LHM wire medium and its similar circuits [16, 17]. The major aim is to build a transmission line with the series ability and shunting inductance, which leads to the corresponding circuit topology such as T-type or Ś for the line structure. SRRs can be activated in the magnetic frequency as comparable media for the periodic and circular metal rings; they have resonant permeability. In the resonant frequency range, the SRR structure shows negative values for electromagnetic properties.

The observation of left-handed materials was noted in 1967 by Veselago who examined the wave propagation of the electromagnetic plane theoretically in a lossless medium at a certain frequency that concurrently had negative actual permittivity and permeability [18]. Furthermore, materials generally have two exceptional factors: permeability and permittivity. These factors influence how the natural material interacts with electromagnetic (EM) radiation, including microwaves, light, and radio waves. A material that has both positive permeability and permittivity simultaneously is called a right-handed material. In the literature, right-handed materials are referred to as **double-positive materials** (DPS). The mid finger specifies the direction of propagation of the wave when it is typically put in both fingers when the magmatic field (H) and electric field (E) directions are characterized respectively by the index finger and thumb on the right hand. Furthermore, energy flow or wave propagation in RHMs is characterized by Poynting vector $Pav = 0.5 \, Re \, [E \times H*]$ and the phase changes are indicated by phase constant $k = w \sqrt{\mu} \times \sqrt{\varepsilon}$ both being on the same axis. In all known natural materials, the spread of electromagnetic waves follows the **Right-Hand** rule with positive refractive indicators [19], as shown in Figure 3.3.

A left-handed material is, in contrast, a material with simultaneous negative permeability and permittivity. Left-handed materials are often referred to as **double negatives** (DNGs). LHM, as Sihvola defines, is an engineered material, which does not occur in nature and obtains its material properties not directly from its composition

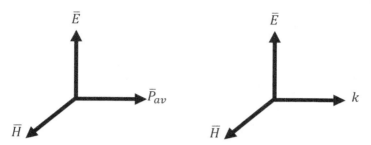

Figure 3.3. Wave propagation/energy flow in the right-handed medium.

material [20]. Considering the influence that the LHM has on the **refractive index** (n) characterized by equations (3.1) to (3.4):

$$n = \sqrt{\epsilon r} \times \sqrt{\mu r} \qquad (3.1)$$

where, μr and ϵr are considered as the relative permeability and permittivity of the material. In the case when both of them are negative equation (3.1) takes the following forms,

$$n = -\sqrt{\epsilon r} \times -\sqrt{\mu r} \qquad (3.2)$$

$$n = j\sqrt{\epsilon r} \times j\sqrt{\mu r} \qquad (3.3)$$

Hence n will be negative and is presented as follows:

$$n = \sqrt{\epsilon r} \times \mu r \qquad (3.4)$$

Due to a lack of investigational verification about LHM materials of the Vassalage's theory, Pendry et al. have demonstrated and verified in theoretical and experimental situations that with only one negative value of the achieved real parts of the material parameters [21]. MTM can be applied to several applications such as cloaks of microwave invisibility, revolutionary electronics, imperceptible submarines, lenses with negative refractive-index, antennas with compact and efficient properties, microwave components, and waveguides to improve the overall performance.

Besides, the spreading **electromagnetic waves** experience the compound structure as a consistent medium with efficient permittivity and permeability because the unit cells of an effective medium are smaller compared with the directed wavelength. The efficient parameters of the substance are derived from the geometry and may vary from them. This is the main feature of powerful media that allows us to develop artificial material with exotic electromagnetic properties that do not exist easily in nature [22].

3.3 Metamaterial based mm-wave

The **mm-wave** at 28 GHz frequency band has drawn a lot of interest from the scientific community around the world because of the expected wide range and huge data transfer as well as lower atmospheric attenuation, which is considered as one of the non-ignorable and extremely crucial challenges in mm-wave systems [23]. Metamaterial has been used to improve the overall system performance based on unique characteristics.

3.4 Metamaterial particle design

The goal of particle-level design is to create geometric dimensions for metamaterial particles that will execute the material properties defined by the proposed design's system level. The S-parameter retrieval and full-wave simulation techniques are used. A rapid design has the advantage of requiring just a small number of simulations, regardless of the number of particles or material properties.

Metamaterial unit cell design

Figure 3.4 shows a MTM unit cell schematic perspective as well as its geometrical design characteristics (a). A low profile of a hexagonal-shaped structure and square split ring resonator (SSRR) are combined by a 0.3 mm wide slab to form the unit cell. The MTM unit cell is built on a low-loss Rogers 5880 substrate with a 0.0009 loss tangent, a dielectric constant (ε) of 2.2, and a 0.79 mm thickness. Two simulation setups were used to test the unit cell operating principle, as shown in Figure 3.4(b) in the y-direction and Figure 3.4(c) in the x-direction. It depicts the suggested metamaterial design's simulated electromagnetic wave propagation when it was put between two waveguide ports. The boundary condition of perfect **electric conductor** (PEC) is used on the x- and y-axes, although the perfect **magnetic conductor** (PMC)

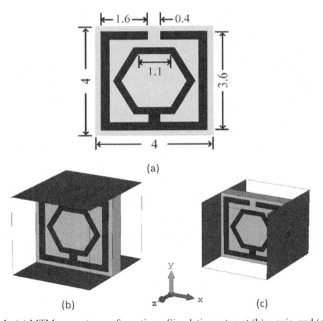

Figure 3.4. (a) MTM geometry configurations; Simulation setup at (b) y-axis, and (c) x-axis.

boundary condition is used on the z-axis. This one is made in a vertical x-axis on top of the chosen Rogers substrate, within a 3 mm gap between each pair of units. For simulations based on Computer Simulation Technology, a finite-difference time-domain solution was utilized (CST).

Two metallic rings divide the planned MTM unit cell. Prior to adding the inner hexagonal structure conductor, the initial outer square ring conductor was split by 0.4 mm on the upper arm as shown in Figure 3.5a. As demonstrated in Figure 3.5, a 0.4 mm wide metallic strip is used to join the two rings together (c). The metal strip lines will act as inductors, although the gaps in the inner and outer strip line connectors will provide capacitive properties. At the y and x-axes, S21 transmission coefficients show that the resonance frequency-shifted directly toward the lower frequency band. In contrast, Figures 3.5 and 3.6 describe all the suggested MTM design steps.

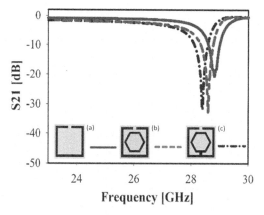

Figure 3.5. S21 at y-axis. (a) no inner line connector, (b) included inner line connector, (c) suggested MTM.

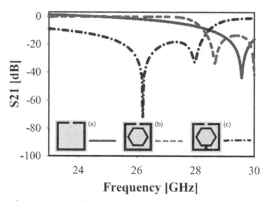

Figure 3.6. S21 at x-axis. (a) no inner line connector, (b) included inner line connector, (c) suggested MTM.

3.4.1 Principle of metamaterial working

The **unit cell** of MTM behavior and the magnetic field and electric field zones in physical phenomena are studied using the noticeable surface current distribution at various frequency bands. Figures 3.7(a) and 3.7(b) illustrate the **surface current** distributions of the suggested MTM at 27 and 28 GHz. The surface current distribution is represented by arrows, while the surface current density is represented by color. At 27 GHz, there is a detectable surface current. On the top edges of the outer square-shaped parts, nevertheless, the surface current is further intense. Furthermore, the overall structure of the MTM unit cell has been shown with a noticeable perturbing surface current. Although, once the **current flows**, opposite side orientations of the observed MTM-structure etching strips current distribution are seen nullifying the current and generating a stopband.

Figures 3.8 and 3.9 show the observed **S-parameters** findings for the y and x directions. The y-axis shows that the resonance frequency band (26.4–29.1 GHz) is both a component of the Ka-frequency band and covers the 5G spectrum. The reached stopband operational band is attributed to the resonator's split outer square-shaped structure integrated with the internal hexagonal structure. However, above 23 GHz, a considerable stopband resonance is detected in the x-direction.

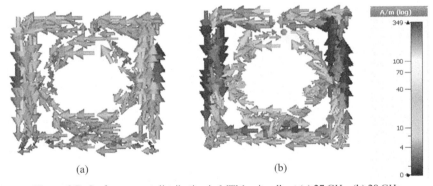

Figure 3.7. Surface current distribution in MTM unit cells at (a) 27 GHz, (b) 28 GHz.

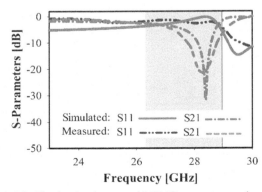

Figure 3.8. Simulated and measured MTM S-parameters on the y-axis.

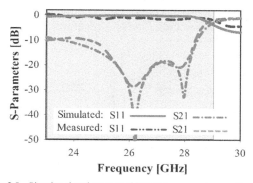

Figure 3.9. Simulated and measured MTM S-parameters on the x-axis.

3.4.2 Metamaterial properties extraction

Responses to electric fields with dispersive permittivity or responses to magnetic fields with dispersive permeability result from the resonances. Utilizing the standard obtained incidences data of scattering parameters, a **robust approach** is used to extract the effective MTM parameters [24]. The transmission and refraction coefficients of the designed MTM unit cell are first examined using simulations in the frequency band from 23 GHz to 30 GHz.

By applying the Microwave Network Analyzer (Agilent N5227A PNA) with two waveguides to co-axial adapters, the S-parameters of the unit cell are obtained. The needed frequency range was covered by a SAR-1834031432-KF-S2-DR wave guide in the range of (18 GHz–40 GHz), while the MTM prototype was used to measure the distance between two similar waveguides in the directions of x the and y axes, as indicated in Figure 3.10. CST (**Computer Simulation Technology**), an electromagnetic full-wave software with proven precision, is used to simulate and adjust the S-parameters. The designed MTM is simulated in both the y and x dimensions, with periodic structural boundaries imposed in z-y and z-x directions. Moreover, the resonant behavior is supplemented by resonant property phenomena, which mostly occur from attempting to extract the characteristics of spatially material at the interest frequency band. Figures 3.11

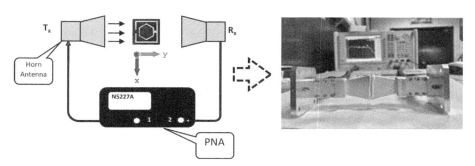

Figure 3.10. Metamaterial experimentation setup.

74 *Metamaterial for Microwave Applications*

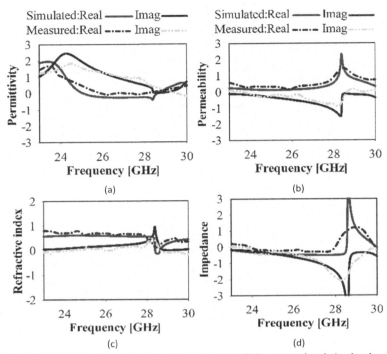

Figure 3.11. At a y-axis of 1 × 1 unit cell, Metamaterial (MTM) measured and simulated outcomes: (a) ε, (b) μ, (c) n, (d) impedance (z).

and 3.12 show the proposed metamaterial effective parameters. For various MTM unit cell configurations, these characteristics include effective **imaginary and real values** of the unique properties such as permeability, refractive index, permittivity, and impedance. Hence, the negative real value indexed zone is marked with a light-yellow tint for **single-negative MTM** (SNG), NZRI near zero, Epsilon (ENZ), and Mu (MNZ) MTM.

As demonstrated in Figure 3.10 (c), extensive NZRI real values are displayed in the region of (23–30 GHz) for wave propagation along the y-axis, whereas ENZ and MNZ sections are obtained with more than 6 GHz; however, 5.1 GHz BW is as seen in Figure 3.10(b). In wave propagation along the x-axis, the NZRI property achieves a **bandwidth** of more than 3 GHz, as shown in Figure 3.10(c). Electromagnetic cloaking and high gain antenna design can both benefit from this frequency spectrum. In addition, as shown in Figure 3.12(b), the MNZ property exists for the achieved frequency range of 27.6 to 28.9 GHz, but the lower band of the KA frequency band is also included by the suggested MTM in **wave propagation** along the x-axis.

3.5 Metamaterial based Sub 6 GHz

Along with mmWave technology, 5G solutions in **sub-6 GHz** bands will be available for mobile and base station applications. The electric and magnetic responses are frequently defined by constitutive properties like permittivity and permeability,

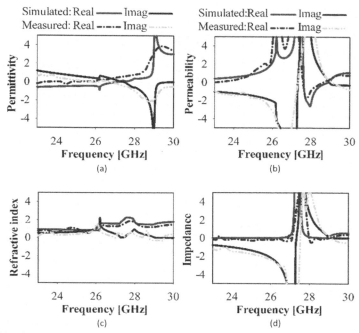

Figure 3.12. MTM, 1 × 1 unit cell x-axis, measured and simulated achieved results: (a) ε, (b) μ, (c) n, (d) impedance (z).

which can be modified by changing the geometrical dimensions of metamaterial particles.

3.5.1 Metamaterial example

The proposed unit cell structure is shown in front view in Figure 3.13(a). It is made up of numerous couples of symmetrical C-shaped structure split-ring resonators that are combined and covered to permit wideband operations of the unit cell. The unit cell was created on a FR4 substrate with a dielectric constant and a thickness of 4.6 and 1.6 mm, respectively. Two simulation setups were used to verify the working concept of the unit cell. The structure was first placed in the x-direction between two wave guiding ports, as shown in Figure 3.13(b), and subsequently in the y-direction, as shown in Figure 3.13(c).

The metamaterial simulation was carried out in the frequency span of 2–20 GHz. Negative electric permeability and permittivity are common material properties utilized to characterize how the materials polarize in the occurrence of **magnetic and electric fields**, resulting in a negative refractive index.

Figure 3.14(a, b) shows the results of the attained simulated (S21) **transmission coefficient** and (S11) **reflection coefficient** in the x-direction and y-direction, respectively. The frequency resonance band in the span of 2–8.6 GHz that is a portion of **5G** (lower band) and UWB applications, is shown in the x-direction; moreover, many bands in the frequency span of 9.8–10, 13.5–15, and 17–19.6 GHz are shown

76 *Metamaterial for Microwave Applications*

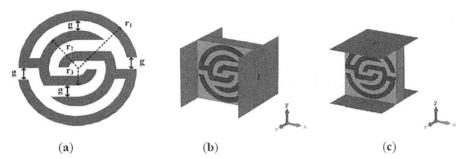

Figure 3.13. (a) MTM configuration with r1 = 2.9 mm, g = 0.5 mm, r2 = 1.9 mm, r3 = 0.9 mm; (b) x-axis set-up; (c) y-axis set-up.

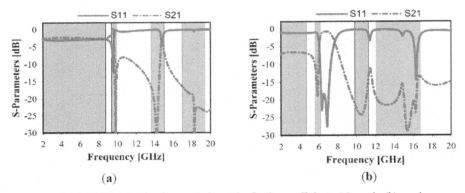

Figure 3.14. MTM simulated transmission and reflection coefficients: (a) x-axis, (b) y-axis.

as well. The effective reason of the acquired multiband range is thought to be the outer and inner split-ring-shaped resonators, with the blue dotted line indicating the higher resonance at 10 GHz. Furthermore, in the y-axis case of the MTM unit cell, where the S21 is in the frequency range of 12–16.9, 9.8–11.2, 5.5–6.5, and 2–4.6 GHz, various bands are evaluated.

Figures 3.15–3.17 show the relative simulated imaginary part and real value part of efficient results for ε, μ, n, and z for various unit cell array topologies. The negative real value indexed region for DNG (double-negative MTM) or SNG (single-negative MTM) is shown in dark brown in these pictures; however, the light brown color denotes a NZRI and both (MNZ/ENZ) near-zero MTMs.

In the wave propagation along the x-axis, the wide real value of the double-negative metamaterial refractive index is seen in the frequency span of 2–8.5 GHz, as illustrated in Figure 3.15(a), although a narrow section of NZRI is attained with a realized BW of 255 MHz. However, as demonstrated in Figure 3.15(b), a more significant BW than 2 GHz can be produced with a NZRI feature on the y-axis along with wave propagation; this could be useful for EM cloaking and antenna systems with a high gain. Figure 3.16 shows true negative real and imaginary values of (μ) and (ε) for frequency spans of 2.6 to 8.8 GHz and 2 to 6.4 GHz, respectively, for the 1 × 1 unit cell. As a result, the unit cell takes on the properties of a

Metamaterial for Future Generation Wireless Communications 77

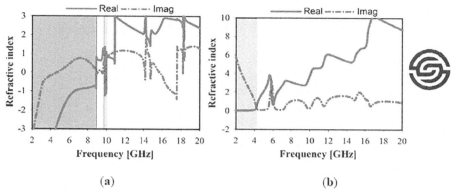

Figure 3.15. MTM simulated (n) of 1 × 1 unit cell: (a) x-direction; (b) y-direction.

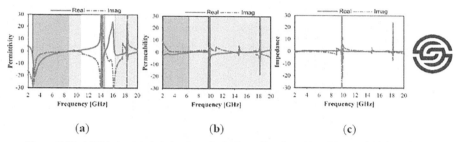

Figure 3.16. MTM simulated attained results in the x-direction: (a) ε; (b) μ; and (c) impedance.

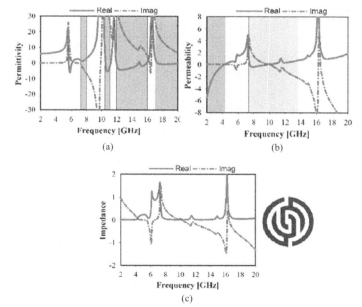

Figure 3.17. MTM simulated result in y-direction: (a) ε; (b) μ and (c) impedance.

78 *Metamaterial for Microwave Applications*

double-negative metamaterial (DNG). Furthermore, the suggested metamaterial only covers the **X and KU bands** in the x-direction of wave propagation with the realization of the MNZ property.

Figure 3.17 shows numerous bands with negative actual SNG permeability, although SNG permittivity in the frequency range of 2–4.3 GHz is observed, which is suitable for 5G (sub 6 GHz) and **S-band** applications and it can also induce NZRI in the same range. Wide MNZ, on the other hand, is clearly visible in the range of 7.2–13.8 GHz for the real value effective permeability magnitude.

The relative real value refractive index in the case of simulated results for various MTMs of the 2 × 2, 1 × 2, and 1 × 1 array architectures is shown in Figure 3.18. NZRI has been demonstrated in the x-direction over a very **wide frequency range** of

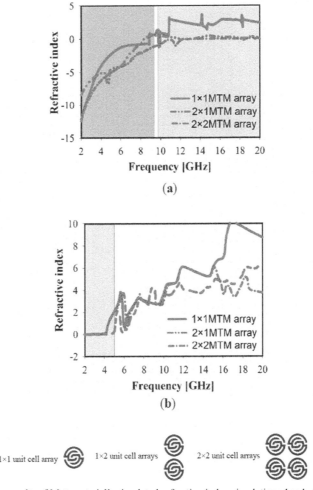

Figure 3.18. The results of Metamaterial's simulated refractive index simulations 1 × 1, 1 × 2 and 2 × 2 array MTM structures: (a) x-direction; (b) y-direction.

10–20 GHz, utilizing a 2 × 2 or 1 × 2 MTM unit cell array that contains the resonance spans in the KU and X bands, as opposed to a non-indexed characteristic in the case of the 1 × 1 MTM unit cell. In y-axis wave propagation, the general effective NZRI bandwidth increases from 2.38 GHz for the 1 × 1 and 1 × 2 **array MTM structures** to 3.1 GHz for the array MTM structure with 2 × 2 unit cells.

3.6 Analysis, design and simulation of 5G metamaterial

The length and width of several **symmetric split rings** and the **gap distances** between them were modified through numerous numerical simulations to design the proposed metamaterial (MTM), which is intended for use in both sub-6 GHz and mm-wave applications. These parameters have a significant influence on the resonance phenomena of the MTM. The proposed MTM design configuration is shown in Figure 3.19, which is etched on a Regress 5880 substrate with a 1.57 mm thickness and an epsilon of 2.2.

Additionally, to achieve the desired results, the proposed MTM strip lines lengths and widths are adjusted and changed accordingly. As a result, many design phases are required to reach the final design of the proposed MTM unit cell. The first step is to create a **copper ring** with two **split gaps** on opposing sides, as shown in design 1 of Figures 3.20 and 3.21 for Sub 6 GHz and mm-waves, respectively.

Figures 3.21(a) and (b) demonstrate how the single ring induces a **resonance** of transmission coefficients (**S21**) at 5.2 GHz and 28.6 GHz (b). Prior to adding the third split circle, another split ring is added that contains two splits at the rotating circles, as seen in Figures 3.21(a) and (b). This generates an additional resonance and shifts it toward a **lower frequency**, as seen in Figure 3.22. Through **field analysis** (magnetic and electric) and a study of the current distribution of the MTM at a specific frequency of 5.2 GHz and 28.6 GHz, where resonance for design 1 occurs, the impact of various design steps is further examined. As seen in design 1, the outermost ring's vertical side is susceptible to the magnetic field. When electromagnetic waves strike

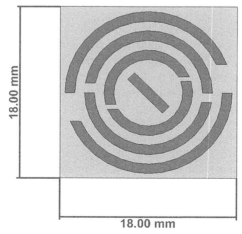

Figure 3.19. Sub-6 GHz and mm-wave MTM configuration.

80 *Metamaterial for Microwave Applications*

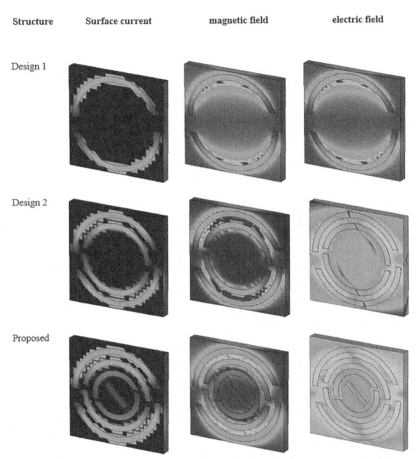

Figure 3.20. Scattering parameters for Sub 6 GHz MTM in different steps: (a) Surface current, (b) magnetic field, (c) electric field at 5.2 GHz.

the split ring, magnetic dipoles are produced, which in turn produces current dipoles with current flowing both in the clockwise and counter clockwise directions through these two edges of the outer ring. A horizontally oriented spreading electric field creates an **electric dipole**. As a result, there is a connection between magnetic and electric dipoles, and electromagnetic resonances of S21 are clearly observed. Two additional rings are added, which not only results in an additional resonance but also changes the resonance's 5.2 GHz bandwidth. This is due to the mutual inductance and co-planar capacitance that were attained.

Figures 3.23 and 3.24 for sub-6 GHz and mm-wave, respectively, show the outcome of scattering parameters using CST related to permittivity, permeability, and refractive index. Figure 3.23 shows that the actual component of permittivity is negative between 5 and 6 GHz, and the obtained bandwidths are 1 GHz (a) due to the realized transmission coefficients occurring at 5.2 GHz with a wide bandwidth. The permeability plot is shown in Figure 3.23(b) indicates wide bandwidth for

Metamaterial for Future Generation Wireless Communications 81

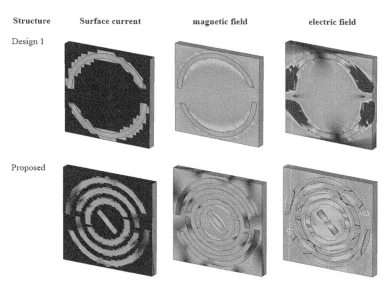

Figure 3.21. Scattering parameters for mm-wave MTM in different steps: (a) Surface current, (b) magnetic field, (c) electric field at 28.6 GHz.

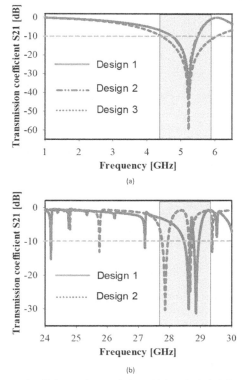

Figure 3.22. MTM Coefficients of transmission (S21) at (a) Sub 6 GHz, (b) mm-wave.

82 *Metamaterial for Microwave Applications*

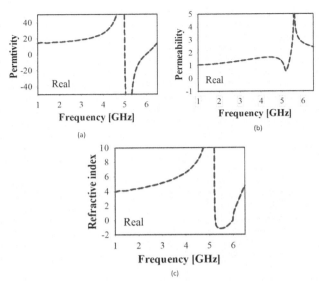

Figure 3.23. Sub 6 GHz MTM characteristics (a) permittivity (b) permeability (c) refractive index.

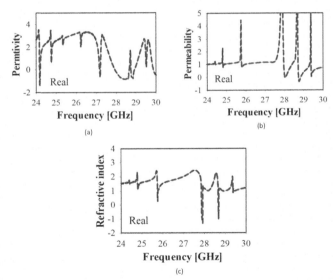

Figure 3.24. MM-Wave MTM characteristics (a) permittivity (b) permeability (c) refractive index.

near zero permeability with a magnitude less than 1.2 in the area of 1–5.5 GHz. Additionally, as shown in Figure 3.23(c), the real component of the refractive index exhibits negative values in the vicinity of frequencies with low permittivity and near zero magnitudes in the range of frequencies between 5 and 6 GHz. With both permeability and refractive index close to zero, the suggested MTM exhibits **epsilon**

negative (ENG) behaviour, making it appropriate for antenna applications for gain improvement.

Furthermore, a wide near zero epsilon negative (ENG) is achieved in the mm-wave part with more than 3 GHz bandwidth, as shown in Figure 3.24(a), whereas near zero **mu negative (MNG)** is also attained with a very wide bandwidth in the frequency span of 24–30 GHz. However, three **negative real values** of the refractive index have been realized, as demonstrated in Figure 3.24(c).

3.7 Emerging metamaterial applications

Significant improvements have been noticed in applying metamaterials to many **prototypes** and designs, such as antennas. For instance, an antenna size can be reduced by loading it with metamaterials. Moreover, gain, bandwidth, and multiband characteristics can be generated and enhanced due to the use of metamaterial. The metamaterials are utilized as distinct functions for the antennas depending on the technical needs of the built antenna.

3.7.1 Antennas - gain improving

Gain is one of the most critical features of the planar antennas to meet the requirements of active transmission and reception. Therefore, it must be overcome due to its effect on the antenna's performance when used in various applications. Recently, metamaterials have been used in the design of antennas because the metamaterials used are synthetic magnetic conductors (AMCs) or artificial magnetic materials (AMMs). The metamaterial unit cells are applied as an environmental plane for the antenna, such as arranging unit cell arrays surrounding the radioactive elements or using substrates above or below radioactive elements. Furthermore, metamaterials can also be used for loading the antennas to improve the antenna gain [7]. The MTM unit cell-based CPW antenna is shown in Figure 3.25, whereas, at 3.5 GHz, the proposed antenna's highest realized gain rises from 3.4 dBi (without metamaterial) to 6.47 dBi (with metamaterial). The MTM layers serve as a superstrate and, unlike

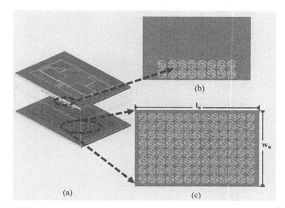

Figure 3.25. MTM CPW antenna (a) 3D-view, (b) back view, (c) separator unit cell layer in suspension.

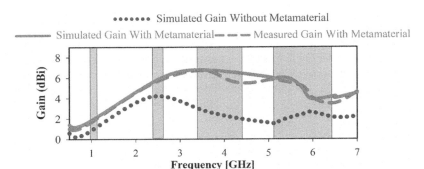

Figure 3.26. Gain and efficiency of the proposed antenna.

reflectors, use solid metal that may be positioned in close proximity to the radiating elements, resulting in a 3 dB improvement as demonstrated in Figure 3.26.

3.7.2 Miniaturization of the size of antennas

The increasing demand for the multi-function device has led to finding a method for the size reduction of the antenna. Many technical solutions have been used in designing pressurized antennas such as high dielectric substrates, shortening pins and introduction of some disturbances in the antenna structure. Consequently, many engineers have used the metamaterial as a **defective ground structure** to reduce the size of the antenna. In this case, the metamaterials have unusual unit cell properties at the resonant frequency of the antenna [25], and also using the metamaterial unit cell comprises a newly redesigned circular slot ring resonator and interdigital capacitor. CSRR has been used in [26], and wideband **miniaturized antennas** result. Microstrip antenna dimensions with an additional ring resonator (CSRR) are lower than those of conventional built-in microstrip antennas as shown in Figure 3.27.

Figure 3.27. CSRR loaded antenna.

3.7.3 Antenna's bandwidth enhancement

Metamaterials are used to improve the antenna frequency bandwidth in addition to the advantages described above. Metamaterials are used as antenna components or as a superstrate positioned above the radiation surface to accomplish this objective. In addition to reducing the size, the application of MTMs as the DGS of the antenna increases the antenna bandwidth, depending on the case [27]. The use of metamaterials has increased the bandwidth in antenna architecture. Various metamaterial structures with different methods of application are chosen to obtain the best antenna bandwidth, depending on the technical specifications of the constructed antenna. A UWB antenna structure loaded with MTM is shown in Figure 3.28. It is constructed from FR-4 material and has an overall dimension of $22.5 \times 14 \times 1.6$ mm^3. However, as shown in Figure 3.29, the reflection coefficients (S11) of the simulated and fabricated prototype demonstrate the enlargement in bandwidth. It illustrates how the observed bandwidths for both scenarios, which range from 3.08 GHz to 14.1 GHz and satisfy the requirements for **ultra-wideband**, are similar with respect

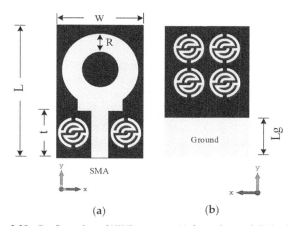

Figure 3.28. Configuration of UWB antenna: (a) front view and (b) back view.

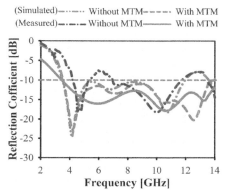

Figure 3.29. Simulated and measured reflection coefficient (S11).

to the reflection coefficients (S11). The antenna without MTM, however, exhibits minor peaks beyond the 10 dB thresholds at higher frequencies, which prevents broad performance. When MTM was introduced, the peaks could be suppressed, resulting in continuous functioning across the whole bandwidth.

3.7.4 Metamaterials for multiband generation

Multiband antennas are required to combine multiple functions on a single apparatus, which is in high demand. The use of metamaterials in antenna construction is an enticing development to minimize sizes, increase power gains, and increase bandwidth and layout antennas with multiple frequency bands [28]. They can be used to design multiple-frequency antennas that are smaller than conventional. MTM embraces negative refraction index and symmetrical pairs in the cell structures of the units at resonant frequencies. This allows the design of multiband antennas using MTM as radioactive components or loaded components.

3.7.5 Metasurface

In the past decade, metamaterials have shifted to a field with established and commercialized applications from a simple theoretical notion. Three-dimensional metamaterials can be expanded to a two-dimensional design on a surface or interface by electrically placing tiny spreaders or holes. The **metasurface** comprises metamaterials with two-dimensional or surface components. Like metamaterials, their reaction may be characterized by their electrical and magnetic polarization. They are sometimes called metafilms in literature [29]. These fascinating metamaterial forms are thinner than 3D metamaterials, as they are 2D structures. As a result, fewer bulky constructions are available. Metasurfaces can be utilized instead of metamaterials in various applications. It has the benefit of taking up less physical area than complete three-dimensional metamaterial structures, allowing for the creation of structures that are less lossy.

Due to their specific permittivity and permeability values, metamaterials can control the diffusion of light, whilst the waves are manipulated on one very thin layer. Therefore, the two-dimensions of metasurfaces reduce the antenna size and allow for small structures. Metasurfaces provide a very promising option to confine electromagnetic waves [30]. Their flat construction makes it easy for metasurfaces to be manufactured using flat instruments [31]. Therefore, as two-dimensional materials, metasurfaces may be readily incorporated into other instruments, making them a major feature of nano-photonic circuits and allowing them to be part of the photonics "chip laboratory" [32].

The negative metamaterial index is due to the resonance of each meta-atom. This characteristic naturally disperses the metamaterials and hence limits the bandwidth of these materials. It was demonstrated in [33] that the dispersion properties might be tailored using extremely thin metasurfaces with deep underwave distance marks in two-layered fishnet constructions. The scientific community has demonstrated a strong recent interest in this field because of the many benefits given by meta-surfaces over metamaterials. This method has brought significant growth in the underlying physics that regulates the conduction and possible uses of metasurfaces. The metasurface is described as a periodic (or aperiodic) structure, whereas compared to operational wavelengths, the thickness and periodicity of the different elements (scatterers) are tiny, including a range (symmetric or asymmetric) of resonant subwavelength scatterers that govern the electromagnetic reaction of the surface [34]. For a metasurface, as the surface is thin, the fields and the polarization must equal the average of the surface area. The homogenization technique in [35] was utilized to construct leak-wave antennas with a leaky-wave with an electronic beam scan that is modulated sinusoidally, i.e., fractal metasurfaces [36], whereas non-linear metasurfaces were used to shield and minimize interference on a common platform [37].

When a wave in available space impacts a surface, the effective surface characteristics, which may be defined as the impedance of reflection and transmission, dominate. Frequency selectivity surfaces (FSS), due to their capacity to selectively transmit, or reflect waves of different frequencies or angles of incidence, are widely researched and utilized on filters, **absorbers**, random **antennas** and more. Furthermore, a typical **frequency selective surface (FSS)** is a periodic structure having each individual element resonant at the resonance frequency. FSSs typically have a periodicity equal to half the wavelength of the resonant frequency [38]. The tiny size of the unit cell likewise permits the metasurfaces to have a consistent reaction in the lighting electromagnetic wave angle variations. Such a construction was presented in [39], where a wire grid on the other side of the dielectric backs a layer of patches. For some applications, it is necessary not only to absorb regular and angular incidents of leaky spatial waves to carefully absorb surrounding currents on the metal body of the system. The surface wave absorber absorbs spreading waves.

3.8 Summary of existing research related to DNG metamaterial

Unique techniques have been used in [40] to minimize the antenna size (metamaterial and fractal), and SRR structure being used. The metamaterial coating consists of multiple parallel rings based on fractal techniques and when the overlying material is loaded, the final antenna size is heavily reduced by (40%) at 4.5 GHz and another resonant frequency of 2.5 GHz occurs. On the other hand, the best **axial ratio (AR)** is obtained with a **reconfigurable** qualification by making openings, whereby they are used in the surface current, resulting in circular polarization. Hence, when using metamaterials (DNG) miniaturization of the antenna size can be achieved while achieving circular polarization simultaneously.

Miniaturization of a rectangular microstrip antenna operates at 2.4 GHz, about 65% compared to a conventional dielectric microstrip antenna using insulating magnetic substrates. The proposed magnetic substrate was constructed or designed using metamaterial, which can reduce the antenna size dimensions by increasing the main parameters of the substrate. This substrate uses a medium-loaded plate (SRR). When the antenna is minimized, the magnetic transmittance can be improved to compensate for the loss of bandwidth. The bandwidth of the miniaturized antenna is unchanged because the designed 2.4 GHz frequency increases its magnetic permeability [41].

Microstrip antennas are one of the types most compatible with microwave applications due to their lightweight and low cost. With the increasing demand for miniature antennas in recent years, antenna miniaturization has an important role in wireless communications. Here, a miniature rectangular correction antenna with two cells in configuration was designed and manufactured as a double negative medium (DNG) with air as a double positive medium (DPS) as an antenna substrate after being placed under the microstrip. After examining and analyzing the resonance structure of the proposed metamaterial unit cell, a double echo sub-wavelength is created. For this resonant frequency, the basic metamaterial parameters, permittivity and transmittance are negative, the dimensions of the proposed antenna are smaller than the wavelength of the conventional rectifying antenna, by about 54.6%, and thus miniaturization occurs [42]. In [43], phased array antennas have been investigated based on a Circular Split Ring Resonator Metamaterial CSRR presented for the 5G framework. This proposal aims to enhance antenna efficiency concerning gain and bandwidth based on circular divided ring CSRR antennas, in the frequency of 28 GHz. However, a narrow bandwidth influences the antenna microstrip. The metamaterial resonators will solve this issue, and the antenna's return loss is below −10 dB. Using CSRR on the antenna increases guidance and gains.

In the presented work in [44], a high gain resonator antenna is equipped with the antenna patch for the microscope and a cell for the metamaterial unit. A unitary cell that acts as a dual-band reflector is implanted in 2 closed ring resonators. The unit cell has high reflectivity as a double negative (DNG) metamaterial where both the effective permeability and permittivity are negative, resonant at 5.9 GHz and 9.9 GHz, with characteristic 3 dB stopbands between 2 and 4.5 GHz. By using some of the algorithms, it has been seen enhancement of bandwidth from 160 MHz to 305 MHz at 5.7 GHz and 270 MHz to 755 MHz at 10.3 GHz with a gain increment of 4 dBi and 3.1 dBi, respectively is possible. The proposed antenna's distinguishing characteristic is its compact size coupled with a unit cell to achieve a high gain at both frequencies. Authors in [45] explained new cells and array arrangements, with a larger bandwidth, in the main component of the C- and X-bands with double-negative characteristics (DNG). The proposed metamaterial can exhibit a negative region with a bandwidth greater than 3.6 GHz. For computing, the Computer Simulation Technology (CST) Microwave Studio uses the finite integration technique (FIT), and the experimental setup for calculating output is implemented within two waveguide ports. At 0-degree and 90-degree rotation angles, the observed data agrees well with the simulated data of the unit cell. In the microwave spectra, the built unit cell has a negative refractive index from 3.482 to 7.096 GHz (bandwidth 3.61 GHz), 7.876 to 10.047 GHz (bandwidth 2.171 GHz), and 11.594 to 14 GHz (bandwidth 2.406 GHz). Finally, compared to other metamaterials in the microwave frequency spectrum, the proposed double-negative metamaterial unit cell has a better capacity for broader bandwidths in the case of negative refractive index applications.

However, in [46], good work was done on the design and analysis of the substrate (DNG) based on the split-ring resonator (SRR) and its merging with the antenna, using the Nicolson-Ross-Wier method. The extracted parameters show that the slab's relative transmittance is close to the trope without affecting the combined air resonance frequency bandwidth. Thus, the results of the experiments specifically indicate that the additional parchment filling rate increases by 7.6 dB near the resonant frequency with a deflection of less than 1 dB. Currently, high gain is the main requirement for wireless communication applications. In [47], the provided work has been done on designing a rectangular microstrip antenna which has become increasingly popular due to its distinct advantages; it can also be easily inserted into radio equipment to meet the requirements of 5G mobile network applications. A high gain and directional antenna are adopted, operating at a frequency of 3.5 GHz with a bandwidth of 2.28% and a gain of 7.43 dB. The proposed microstrip antenna can be miniaturized in several approaches, most notably the use of high dielectric fixed substrates, short circuit or resistance loading techniques, and splitting techniques, which affect the antenna performance. A small new strip antenna with a normal structure and striking miniature effects is designed by symmetric prospecting of ten ringed resonator material units on a rectangular antenna radiation patch. While maintaining the antenna gain and orientation, loading the metamaterial to form the symmetric circular resonator effectively reduces the microstrip antenna's total size. The average antenna size has been reduced from 40 × 45 mm to 35 × 40 mm, which is a 22.2 percent decrease over the same microstrip antenna without any piercing buildup. Antenna miniaturization is affected by metamaterial loading.

An extremely miniaturized triple-band antenna loaded with a new adapted metamaterial cell with two circular slot loop resonators (MDCsRR) is presented in [48]. It contains a resonator in which the frequency slot is low and compact. The concept for the proposed patch antenna feature is to add series power to reduce the frequency of resonance at half the wavelength, thereby decreasing the electrical dimensions of the proposed patch antenna. The suggested microstrip patch is loaded in series to limit the resonating frequency by an interdigital condenser. The new low-frequency portable metamaterial unit is also used to load the proposed patch antenna into a multi-mode operation. The model transmission line is used to analyze the resonance band's passenger band and **stop band** characteristics. The antenna suggested offers triple band operation at 3.28 dBi at the first frequency of 3.2 GHz, 2.76 dBi for the 2nd frequency of 5.4 GHz, 3.1 dBi for the 3rd frequency of 5.8 GHz, which are considered important measurements. In comparison with the traditional patch antenna, the band widths of the proposed electric antenna are reduced to around 68.83%. The antenna has a broad spectrum of **WiMax and WLAN band** implementations.

The article in [49] presented a WLAN/WIMAX triple band patch antenna based on the complementary ring resonator. In this study, a standard patch antenna with a 3.6 GHz band width working in the intermediate WiMAX band (3.2 GHz to 3.8 GHz) is loaded into a ground plane with an appropriate rectangular CSRR, enabling the lower WLAN band to be excited (2.4 GHz to 2.484 GHz). Two CSRRs of adequate sizes are put on the ground to resonate on the high-frequency bands of WiMAX (5.25 GHz to 5.85 GHz). However, a new triple-band simple microstrip antenna design using the metamaterial concept is presented in [50]. The multi-metamaterial unit cells were considered key to the multi resonance response. This was achieved by etching one rectangular split-ring resonator (SRR) and two circular unit cells in the ground plane of a conventional patch operating at 3.56 GHz. The circular unit cells resonate for the upper band of the Wi-MAX at 5.6 GHz, whereas the designed rectangular cell was created to generate resonance for the lower band of The WLAN at 2.45 GHz. The WLAN's/WiMAX's operating bands are covered by the triple resonances which are achieved by the proposed antenna with quite an enhanced performance. A detailed parametric analysis of the location of the proposed unit cells is presented, and the most appropriate locations are selected to be the unit cell locations for an improved overall performance. Good reliability between measurement and simulation confirms the ability of the proposed antenna to achieve an enhanced gain at the three frequencies.

3.9 Summary

In this chapter, we introduced the design of the instant metamaterial for 5G applications in terms of mmWave and sub 6 GHz bands. According to the results, the designed low-profile MTM unit cell at 28 GHz has a wide frequency range at a near zero index for permeability, refractive index, and permittivity properties. Furthermore, x and y directions of wave propagation have been applied with a compact square SRR metamaterial unit cell which was addressed and developed to statistically validate several bands, including the lower 5G band at 3.5 GHz, whereas DNG and NZRI characteristics revealed negative real values of permittivity, refractive index, and permeability throughout numerous bands. The suggested designs can be applied to the 5G and beyond MIMO antennae designs to improve isolation, envelope correlation coefficient (ECC), gain, efficiency, and overall performance.

References

[1] Andrews, J. G., S. Buzzi, W. Choi, S. V. Hanly and A. Lozano. 2014. What will 5G be? *IEEE Journal on Selected Areas in Communications,* 32(6): 1065–1082.
[2] Attiah, M. L., A. Isa, Z. Zakaria, M. Abdulhameed, M. K. Mohsen et al. 2019. A survey of mmWave user association mechanisms and spectrum sharing approaches: an overview, open issues and challenges, future research trends. *Wireless Networks,* pp. 1–28.
[3] Rappaport, T. S., Y. Xing, G. R. MacCartney, A. F. Molisch, E. Mellios et al. 2017. Overview of millimeter wave communications for fifth-generation (5G) wireless networks—With a focus on propagation models. *IEEE Transactions on Antennas and Propagation,* 65(12): 6213–6230.
[4] Sindreu, M. D., J. Naqui, J. Selga, P. VØlez, J. Bonache et al. 2013. Composite right-/left-handed transmission line metamaterials. pp. 1–25. *In:* Wiley Encyclopedia of Electrical and Electronics Engineering. Hoboken, NJ, USA: Wiley, Apr. 2013.
[5] Lai, A., T. Itoh and C. Caloz. 2004. Composite right/left-handed transmission line metamaterials. *IEEE Microw. Mag.,* 5(3): 34–50, Sep. 2004.
[6] Al-Bawri, S. S., H. Hwang Goh, M. S. Islam, H. Y. Wong, M. F. Jamlos et al. 2020. Compact ultra-wideband monopole antenna loaded with metamaterial. *Sensors,* 20: 796.
[7] Al-Bawri, S. S., M. S. Islam, H. Y. Wong, M. F. Jamlos, A. Narbudowicz et al. 2020. Metamaterial cell-based superstrate towards bandwidth and gain enhancement of quad-band CPW-fed antenna for wireless applications. *Sensors,* 20: 457.
[8] Al-Bawri, S. S., M. T. Islam, T. Shabbir, G. Muhammad, M. S. Islam et al. 2020. Hexagonal shaped near zero index (NZI) metamaterial based MIMO antenna for millimeter-wave application. *IEEE Access,* 8: 181003–181013.
[9] Shabbir, T., M. T. Islam, S. S. Al-Bawri, R. W. Aldhaheri, K. H. Alharbi et al. 2020. 16-Port non-planar MIMO antenna system with near-zero-index (NZI) metamaterial decoupling structure for 5G applications. *IEEE Access.*

[10] Shabbir, T., R. Saleem, S. S. Al-Bawri, M. F. Shafique and M. T. Islam et al. 2020. Eight-port metamaterial loaded UWB-MIMO antenna system for 3D system-in-package applications. *IEEE Access*.
[11] Ebrahimpouri, M., E. Rajo-Iglesias, Z. Sipus and O. Quevedo-Teruel. 2018. Cost-effective gap waveguide technology based on glide-symmetric holey EBG structures. *IEEE Trans. Microw. Theory Tech.*, 66(2): 927–934, Feb. 2018, doi: 10.1109/TMTT.2017.2764091.
[12] Assimonis, S. D., T. V. Yioultsis and C. S. Antonopoulos. 2012. Design and optimization of uniplanar EBG structures for low profile antenna applications and mutual coupling reduction. *IEEE Transactions on Antennas and Propagation*, 60(10): 4944–4949.
[13] Zhang, Q.-L., Y.-T. Jin, J.-Q. Feng, X. Lv and L.-M. Si et al. 2015. Mutual coupling reduction of microstrip antenna array using metamaterial absorber. pp. 1–3. *In: 2015 IEEE MTT-S International Microwave Workshop Series on Advanced Materials and Processes for RF and THz Applications (IMWS-AMP)*, IEEE.
[14] Farahani, M., J. Pourahmadazar, M. Akbari, M. Nedil, A. R. Sebak et al. 2017. Mutual coupling reduction in millimeter-wave MIMO antenna array using a metamaterial polarization-rotator wall. *IEEE Antennas and Wireless Propagation Letters*, 16: 2324–2327.
[15] Brodie, G. 2019. Energy transfer from electromagnetic fields to materials. *Electromagn. Fields Waves*, Jan. 2019.
[16] Caloz, C. and T. Itoh. 2004. Transmission line approach of left-handed (LH) materials and microstrip implementation of an artificial LH transmission line. *IEEE Trans. Antennas Propagation*, 52(5): 1159–1166, May 2004.
[17] Caloz, C., A. Sanada and T. Itoh. 2004. A novel composite right-/left-handed coupled-line directional coupler with arbitrary coupling level and broad bandwidth. *IEEE Transactions on Microwave Theory and Techniques*, 52: 980–992.
[18] Limaye, A. 2021. Size reduction of microstrip antennas using left-handed materials realized by complementary split-ring resonators. Accessed: Jun. 01, 2021.
[19] Pashaeiadl, H., M. Naserpour and C. J. Zapata-Rodríguez. 2018. Scattering of electromagnetic waves by a graphene-coated thin cylinder of left-handed metamaterial. *Optik (Stuttg)*, 159: 123–132, Apr. 2018.
[20] Sihvola, A. 2007. Metamaterials in electromagnetics. *Metamaterials*, 1(1): 2–11, Mar. 2007.
[21] Goussetis, G., A. P. Feresidis and A. R. Harvey. 2010. Experimental realisation of electromagnetic metamaterials. *J. Mod. Opt.*, 57(1): 1–16, Jan. 2010.
[22] Pendry, J. B., A. J. Holden, D. J. Robbins and W. J. Stewart. 1999. Magnetism from conductors and enhanced nonlinear phenomena. *IEEE Trans. Microw. Theory Tech.*, 47(11): 2075–2084.
[23] Shayea, I., T. A. Rahman, M. H. Azmi and M. R. Islam. 2018. Real measurement study for rain rate and rain attenuation conducted over 26 GHz microwave 5G link system in Malaysia. *IEEE Access*, 6: 19044–19064.
[24] Chen, X., T. M. Grzegorczyk, B.-I. Wu, J. Pacheco Jr and J. A. Kong et al. 2004. Robust method to retrieve the constitutive effective parameters of metamaterials. *Physical Review E*, 70(1): 016608.
[25] Varamini, G., A. Keshtkar and M. Naser-Moghadasi. 2018. Miniaturization of microstrip loop antenna for wireless applications based on metamaterial metasurface. *AEU-International Journal of Electronics and Communications*, 83: 32–39.
[26] Al-Bawri, S. S. and M. F. Jamlos. 2018. Design of low-profile patch antenna incorporated with double negative metamaterial structure. pp. 45–48. *In: 2018 IEEE International RF and Microwave Conference (RFM)*.
[27] Rahman, M. M., M. S. Islam, M. T. Islam, S. S. Al-Bawri and W. H. Yong et al. 2022. Metamaterial-based compact antenna with defected ground structure for 5G and beyond. *Computers, Materials and Continua*, 72: 2383–2399.
[28] Rajkumar, R. and K. Usha Kiran. 2016. A compact metamaterial multiband antenna for WLAN/WiMAX/ITU band applications. *AEU - Int. J. Electron. Commun.*, 70(5): 599–604, May 2016.
[29] Kuester, E. F., M. A. Mohamed, M. Piket-May and C. L. Holloway. 2003. Averaged transition conditions for electromagnetic fields at a metafilm. *IEEE Trans. Antennas Propag.*, 51(10 I): 2641–2651, Oct. 2003.
[30] Quevedo-Teruel, O., H. Chen, A. Diaz-Rubio, G. Gok and A. Grbic. 2019. Roadmap on metasurfaces. *J. Opt.*, 21(7): 073002, Jul. 2019.

[31] Yoon, G., I. Kim and J. Rho. 2016. Challenges in fabrication towards realization of practical metamaterials. *Microelectron. Eng.*, 163: 7–20, Sep. 2016.
[32] AV, K., B. A and S. VM. 2013. Planar photonics with metasurfaces. *Science*, 339(6125): 12320091–12320096, Mar. 2013.
[33] Jiang, Z. H., S. Yun, L. Lin, J. A. Bossard, D. H. Werner et al. 2013. Tailoring dispersion for broadband low-loss optical metamaterials using deep-subwavelength inclusions. *Sci. Reports* 31, 3(1): 1–9, Mar. 2013.
[34] Moon, S. W., Y. Kim, G. Yoon and J. Rho. 2020. Recent progress on ultrathin metalenses for flat optics. *iScience*, 23(12): 101877, Dec. 2020.
[35] Minatti, G., F. Caminita, E. Martini, M. Sabbadini, S. Maci et al. 2016. Synthesis of modulated-metasurface antennas with amplitude, phase, and polarization control. *IEEE Trans. Antennas Propag.*, 64(9): 3907–3919, Sep. 2016.
[36] Moeini, S. 2019. Homogenization of fractal metasurface based on extension of babinet-booker's principle. *IEEE Antennas Wirel. Propag. Lett.*, 18(5): 1061–1065, May 2019.
[37] Achouri, K., G. D. Bernasconi, J. Butet and O. J. F. Martin. 2018. Homogenization and scattering analysis of second-harmonic generation in nonlinear metasurfaces. *IEEE Trans. Antennas Propag.*, 66(11): 6061–6075, Nov. 2018.
[38] Wilton, D. 1997. R, Book Review: Frequency Selective Surfaces: Analysis and Design by John C. Vardaxoglou, Research Studies Press, Taunton, England.
[39] Sarabandi, K. and N. Behdad. 2007. A frequency selective surface with miniaturized elements. *IEEE Trans. Antennas Propag.*, 55(5): 1239–1245, May 2007.
[40] Varamini, G., A. Keshtkar and M. Naser-Moghadasi. 2018. Miniaturization of microstrip loop antenna for wireless applications based on metamaterial metasurface. *AEU - Int. J. Electron. Commun.*, 83: 32–39, Jan. 2018.
[41] Farzami, F., K. Forooraghi and M. Norooziarab. 2011. Miniaturization of a microstrip antenna using a compact and thin magneto-dielectric substrate. *IEEE Antennas Wirel. Propag. Lett.*, 10: 1540–1542.
[42] Jahromi, A. G. and F. Mohajeri. 2014. Design and fabrication of a miniaturized microstrip antenna loaded by DNG metamaterial. *International Journal of Electronics and Communication Engineering*, 8(5): 802–806.
[43] Essid, C. and A. Samet. 2018. A design of phased array antenna with metamaterial circular SRR for 5G applications. *In: IEEE International Symposium on Personal, Indoor and Mobile Radio Communications, PIMRC*, Feb. 2018, 2017-October: 1–4.
[44] Rajak, N., N. Chattoraj and R. Mark. 2019. Metamaterial cell inspired high gain multiband antenna for wireless applications. *AEU - Int. J. Electron. Commun.*, 109: 23–30, Sep. 2019.
[45] Hasan, M. M., M. R. I. Faruque, S. S. Islam and M. T. Islam. 2016. A new compact double-negative miniaturized metamaterial for wideband operation. *Materials (Basel)*, 9(10): 1–12.
[46] Gangwar, D., S. Das and R. L. Yadava. 2017. Gain enhancement of microstrip patch antenna loaded with split ring resonator based relative permeability near zero as superstrate. *Wirel. Pers. Commun.*, 96(2): 2389–2399, Sep. 2017.
[47] Li, R., Q. Zhang, Y. Kuang, X. Chen, Z. Xiao et al. 2019. Design of a miniaturized antenna based on split ring resonators for 5G wireless communications. *2019 Cross Strait Quad-Regional Radio Sci. Wirel. Technol. Conf. CSQRWC 2019 - Proc.*, pp. 1–4.
[48] Singh, A. K., M. P. Abegaonkar and S. K. Koul. 2018. Miniaturized multiband microstrip patch antenna using metamaterial loading for wireless application. *Prog. Electromagn. Res. C*, 83: 71–82.
[49] Hamad, E. K. I. and A. Abdelaziz. 2019. Metamaterial superstrate microstrip patch antenna for 5G wireless communication based on the theory of characteristic modes. *J. Electr. Eng.*, 70(3): 187–197, Jun. 2019.
[50] Ali, W., E. Hamad, M. Bassiuny and M. Hamdallah. 2017. Complementary split ring resonator based triple band microstrip antenna for WLAN/WiMAX applications. *Radioengineering*, 26(1): 78–84, Apr. 2017.

Chapter 4

Metamaterial Structure Exploration for Wireless Communications

Mohammad Tariqul Islam* and *Md. Moniruzzaman*

4.1 Introduction

Metamaterial (MTM) is the latest research area in the dimension of complex and unconventional media. The term metamaterial is initiated in the scientific community in the research of complex media to explore surprising and unusual properties of artificial materials that are beyond natural materials [1]. The metamaterials can be classified as single negative metamaterials in which either the real part of the permittivity is negative or permeability is negative [2], or double negative metamaterials that exhibit both negative permeability and permittivity in the frequency of interest. Double negative metamaterials bear some other names such as left-handed materials, negative index materials, or backward wave media due to particular properties they possess [1]. The double negative (DNG) property of the metamaterial was first investigated by Veselgo, a Russian scientist [3], and the first application of a DNG metamaterial was proposed by Pendry as a perfect lens [4]. Some usages of metamaterials include a planar and metallic waveguide, microstrip antennas, monopoles, absorbers, dipoles, and so forth [5–12]. Invisibility cloaks [13, 14], hyper lenses [15], electromagnetic transformation devices [16], and optical nano-circuits [17] are some new devices that can be implemented with metamaterials. Moreover, various miniaturized microwave components including filters, phase shifters and couplers can be devised using the concepts of composite right- and left-handed transmission lines employing forward and backward wave

---

Department of Electrical, Electronic and Systems Engineering, Faculty of Engineering and Built Environment, Universiti Kebangsaan Malaysia, Bangi 43600, Selangor, Malaysia.
Email: p99997@siswa.ukm.edu.my
* Corresponding author: tariqul@ukm.edu.my

propagation [18–20]. Demand for metamaterials is increasing day by day due to their ability to improve the performance of microwave devices. Moreover, cutting edge manufacturing technology inspires the development of compact metamaterials at low costs. It is expected that the global market of metamaterials will grow rapidly due to their necessity for applications in communication antenna, such as sensing, displays, solar panels, windscreens, radio imaging, medical imaging, radar systems in self-driving cars, acoustics and radars [21]. Moreover, in various industries such as automotive, aerospace, defence, security, aviation, and medical, there is a growing demand for durable and lightweight metamaterials.

A metamaterial can be considered as a macroscopic composite made with a periodic or non periodic structure with a size smaller than or equal to the sub wavelength so that it can be regarded as an effective medium. A periodic structure metamaterial is like a homogeneous medium whereas, a non-periodic structure is an unhomogeneous medium. In microwave frequencies, it can be fabricated on a printed circuit board whose properties mainly depend on the various architectures of the fabricated design and substrate materials such as FR-4, Rogers, F4B. Since properties of the metamaterial can be controlled by architectural variation, it provides the flexibility to explore new properties that are not available in natural materials, indicating added advantages offered by metamaterials [22]. Electric permittivity (ε) and magnetic permeability (μ) are the two important parameters that define the material properties. In addition, refractive index is another improtant parameter that can be defined as $n = \sqrt{\mu_r \varepsilon_r}$, where μ_r is the relative permeabilty and ε_r is the relative permittivity of the material.

In the recent work, many MTM structures are presented by various authors such as L-shaped resonator-based MTM by Tamim et al. [23], modified H shaped by Hossain et al. [24], RLC resonator based MTM by Ahamed et al. [25], ring-shaped resonator by Hoque et al. [26]. The overall dimension of the unit cell in each of the above-mentioned works is comparatively large. Since in the modern wireless communication systems, compact designs are required to include metamaterials with miniaturized devices, new metamaterials with compact designs may be a research goal from this perspective. On the other hand, metamaterial superstrates have been utilized by Saravanan et al. [27], Pandey et al. [28] and Ojo et al. [29] to improve the antenna gain. Though they have achieved increasing the antenna gain significantly with this method, further research may still be conducted to utilize metamaterials with antennas in a more efficient way for further improvement of the antenna performance. Moreover, most of the metamaterials in available literature are designed for specific applications. So, additional research work can be done to design them for utilisation in multipurpose applications. It is observed that compactness in MTM unit cells has been achieved through asymmetrical structural designs [30] but harmonic distortion and frequency shifts occur in the case of the arrays of MTM unit cells. These frequency shifts and distortions are due to mutual coupling between the array elements due to a lack of structural symmetry. Symmetrically structured MTMs may help to eliminate harmonic distortions and frequency shifts but they also increase the overall size of the unit cell. So, a new technique for the structural design of the metamaterial requires exploration so that significant compactness may be attained. In addition to this, different mechanisms of frequency tuning have been exercised by the researcher to control the metamaterial resonance frequencies. Electrical frequency tuning by varying the CMOS gate voltage [31], coupling method [32], temperature change [33] and varactor diode based tuning [34] are some approaches that are used in metamaterials. However, each of these approaches require either a specific manufacturing technology, or control circuit, or specific controlling property (such as temperature). Thus, in case of frequency tuning, complexity arises in design, as well as measurement preparation. Considering these factors, research work can be continued to explore a new frequency tuning technique that will be simple but provides a flexible way to adjust the resonance frequency.

In this chapter, five different metamaterials are presented that possess five different structural symmetries and all can be used for wireless communication system applications. The metamaterials are designed and simulated in CST microwave studio-2019 suite. The metamaterial properties are analyzed, and their performances are also studied through experiments. The major findings of these studies are also included in this chapter.

4.2 Metamaterial with asymmetrical resonator

In this section metamaterials with different structures are explored. The first metamaterial presented here is an asymmetrical structure as shown in Figure 4.1. The metamaterial is designed over a FR4 substrate with a dielectric constant and loss tangent of 4.3 and .02 respectively. The overall dimensions of the MTM are set as $8 \times 8 \times 1.5$ mm^3 [30]. The design consists of multiple split rings that are optimized through numerical simulations. To achieve a low frequency coverage the rings are interconnected and thus the MTM provides resonance that fall into the S, C and X bands. This metamaterial shows transmission coefficient resonances of 2.24 GHz, 4.77 GHz, 5.94 GHz, 8.94 and 10.84 GHz with magnitudes of –20.4 dB, –24 dB, –20 dB, 21.5 dB and 15 dB respectively. Whereas reflection coefficients exhibit their minima at 2.5 GHz, 5.2 GHz, 6.6 GHz, 9.9 GHz and 11.4 GHz with magnitudes of –23 dB, –15 dB, –24 dB, –15 dB and –42 dB respectively. Figure 4.2(a) shows the transmission coefficient (S_{21}) and reflection coefficient (S_{11}) spectra for the unit cell of this metamaterial. Since the frequency of every $|S_{21}|$ minimum is always lower than the $|S_{11}|$ minimum frequency so, every resonance can be considered as an electrical one in the unit cell [35]. The metamaterial exhibits negative permittivity due to electrical resonances caused by the resonator and it is observed in Figure 4.2(b) that negative permittivity occurs in the frequency ranges of 2.22 GHz–2.48 GHz, 4.68 GHz–5.2 GHz, 5.85 GHz–6.4 GHz, 8.66 GHz–9.4 GHz, 9.7 GHz–10 GHz and 10.63 GHz–11.14 GHz. The compactness of the metamaterial is determined by the effective medium ratio (EMR) which can be defined as the ratio of the wavelength at the lowest resonance frequency to the metamaterial unit cell length. The value of the EMR is 16.74 which is achieved by setting various splits and ringed interconnections at different places, optimized through numerical simulation. Since the structure is asymmetric in case of the arrays of unit cells the resonance phenomena is greatly

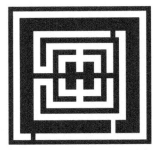

Figure 4.1. Coupled ring split ring resonator based MTM.

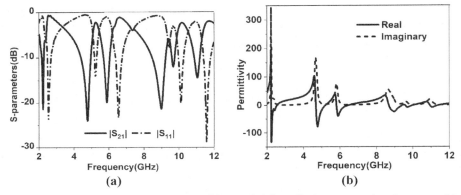

Figure 4.2. (a) Scattering parameter spectra of the coupled ring split ring resonator based metamaterial (a) unit cell, (b) permittivity plot of the unit cell.

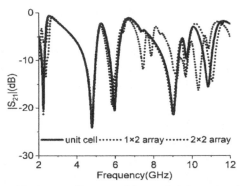

Figure 4.3. Array performance of the coupled ring metamaterial.

affected due to the mutual coupling effect between the unit cells of the array. The transmission spectra for different arrays are presented in Figure 4.3 that shows the resonance deflections in the array. As the array size increases the distortion in the resonance patterns also increases. This distortion is more pronounced in the high frequencies compared to lower frequencies as shown in Figure 4.3.

4.3 Metamaterial with single axis symmetric resonator with applications

4.3.1 Design of MTM unit cell

It has already been discussed that an asymmetrical structural design provides compactness but in the case of arrays the resonances shift, and these distortions form the unit cell response and increase with an increasing number of cells in an array. The question is how to maintain similar responses between unit cells and arrays. So other designs with changing structural symmetry have been explored and thus, the performance of a single axis symmetric metamaterial has been investigated. In Figure 4.4, a metamaterial named, inductively tuned modified split ring resonator is presented. As shown in Figure 4.4 this metamaterial is initiated on a FR-4 substrate with dimensions of 8 × 8 mm^2, thickness of 1.5 mm, permittivity of 4.4, and a loss tangent of 0.02 [36]. The resonator is placed at one side of the substrate with copper strips of 0.035 mm thickness. The resonator comprises four split rings with three square-shaped ones of the same width and a circular one. All the square-shaped rings have an extended rectangle at each corner that increases their surface area. In the outer ring, two split gaps are connected with two shunt copper paths parallel to the split gaps that modify the resonance frequencies. The two innermost rings are interconnected and the structure is symmetric with the vertical axis. Figure 4.5 shows the design steps and the corresponding S_{21} response is presented in Figure 4.6(a). It is observed from Figure 4.6(a) for this metamaterial the resonance frequency shifts significantly when the inductive shunt patch is connected to the outer ring. From the extracted data it is found in Step 1 of the design evolution, the metamaterial shows resonances at 4.9 GHz and 11.15 GHz. Whereas, with the inclusion of the second ring in Step 2, a total of four resonances at 5.95 GHz, 11.3 GHz, 13.8 GHz and 17.5 GHz are obtained. In Step 3, resonances are found at 5.65 GHz, 7.2 GHz, 11.4 GHz, 13.8 GHz and 17.4 GHz. Step 4 impacts less but when the two innermost rings are interconnected in Step 5, the resonances are shifted significantly at 5.5 GHz, 6.2 GHz, 11.2 GHz, 14.4 GHz, and 17.5 GHz. In the final design a drastic shift of the resonances towards the lower frequency is noticed and S_{21} resonances are found at 2.38 GHz, 4.24 GHz, 5.98 GHz, 9.55 GHz, 12.1 GHz, and 14.34 GHz that are included in the S, C, X and Ku bands [36].

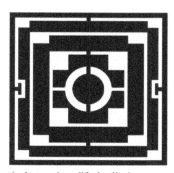

Figure 4.4. Inductively tuned modified split rings resonator based MTM.

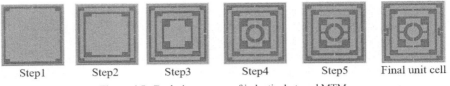

Figure 4.5. Evolution process of inductively tuned MTM.

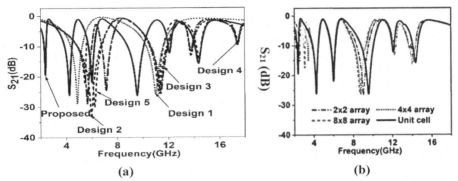

Figure 4.6. (a) S_{21} response for different design steps toward final unit cell of inductively tuned metamaterial unit cell, (b) array performance of the inductively tuned metamaterial.

4.3.2 Result analysis of the metamaterial

The array performances of this metamaterial are observed by taking different array sizes of the unit cells. It is observed that the resonance frequencies shift in the high and very low frequency range. Within 4 GHz to 8 GHz the resonances are almost unchanged. However, in comparison to the asymmetric structure, this shift in resonances is less, and, thus due to the single axis symmetry, the array response is improved significantly. The MTM properties are explored through the study of relative permittivity, permeability and refractive index plots as depicted in Figure 4.7(a)–(c). As shown in Figure 4.7(a), negative permittivity is obtained at frequencies of 2.381 GHz, 4.28 GHz, 6.06 GHz, 9.68 GHz, 12.2 GHz and 14.53 GHz. On the other hand, near zero permeability is obtained in the vicinity of frequenies 2.38 GHz, 4.28 GHz, 6.06 GHz, 9.75 GHz, 12.15 GHz, and 14.47 GHz as depicted in Figure 4.7(b) [36]. Additionaly, a near zero refractive index is also noticed in the Figure 4.7(c). Thus this MTM shows ENG properties with near zero permeability and refractive index. Figure 4.7(d) exhibits the measured S_{21} and a close similarity is observed compared with the simulated result. The calculated EMR value is 15.75 that indicates the compactness of this metamaterial that is achieved through the inductively tuned method.

4.3.3 Application of metamaterial in antenna for performance enhancement

As discussed above, the proposed MTM-1 exhibits negative permittivity along with near-zero permeability and refractive index. These properties have immense applications in the field of antennas for improving performances, especially the NZI

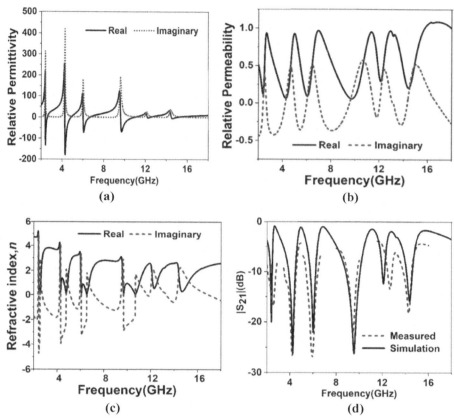

Figure 4.7. (a) Permittivity plot (b) Permeability plot (c) Refractive index plot (d) measured S_{21} in comparison of simulation.

property that can be used to boost antenna gain [37, 38] and directivity [39]. To observe the effect of the metamaterial on antennas, a modified log periodic patch antenna has been designed named tapered log periodic dipole (TLPD) antenna [40] using FR4 substrate. The TLPD antenna consists of 9 arms on each side of the substrate of dimensions 161.08 mm × 110.8 mm all arms being connected with tapered copper. This antenna exhibits a wideband reflection coefficient but the gain is very low which is insufficient for various applications such as microwave-based imaging sensors and energy harvesting. To increase the gain of the TLPD antenna, a metamaterial array is used as a superstrate over it. The array contains 21 × 14 unit cells of this MTM unit cell and it is placed above the TLPD antenna at a distance of 30 mm, optimized by numerous simulations. The antenna and MTM array are separated by a small polystyrene block. The reflection coefficient measurement process is shown in Figure 4.8(a) and gain is measured using the satimo near the field measurement set up as shown in Figure 4.8(b). The measured reflection coefficient is depicted in Figure 4.9 that exhibits the same bandwidth for TLPD antenna both with and without MTM. However, as expressed in Figure 4.10, due to the MTM-1

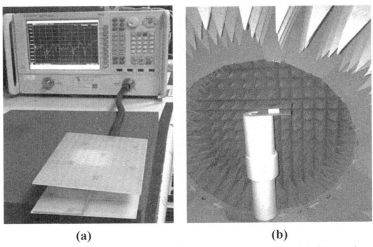

Figure 4.8. (a) Reflection coefficient measurement of antenna with metamaterial, (b) experimental setup for gain measurement using the Satimo near field measurement system.

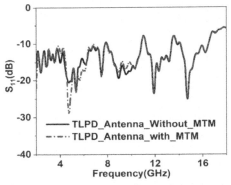

Figure 4.9. Reflection coefficient of the TLPD antenna with and without metamaterial array.

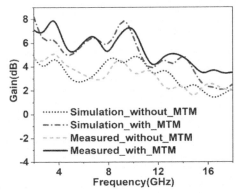

Figure 4.10. Gain of TLPD antenna with and without metamaterial superstrate.

superstrate, a significant gain enhancement is achieved. Almost 3.5 dBi more gain is obtained resulting in a maximum gain of over 8 dBi. A higher gain over the frequency range 2–18 GHz with an average gain of 5 dBi (~ 3 dBi without MTM-1 array) is also observed with an average enhancement of about 73%. From Figure 4.10, it is also observed that, above 10 GHz antenna gain with and without MTM decreases significantly. It is due to the effect of the FR4 substrate's high dielectric loss that is very dominant in the high frequency (above 10 GHz) range.

4.3.4 Application of the metamaterial as an absorber

Another important feature of the discussed inductively tuned MTM is that when a copper back plane is used it shows good absorption properties. As revealed in Figure 4.11, this metamaterial with a copper back shows a near-unity absorption around 11.73 GHz and 3.98 GHz. In addition, at 5.5 GHz and 13.47 GHz absorption is more than 90%. High absorption is achieved due to the value of the impedance of this metamaterial being closer to the free space impedance. Two lower valued absorption maxima are obtained at 2.38 and 8.5 GHz. This metamaterial provides narrow band absorption having high Q factors with values of 28.4, 34.4, 17.6, 23.0, and 32.0 at 3.98 GHz, 5.5 GHz, 8.48 GHz, 11.73 GHz and 13.4 GHz respectively. The high selectivity makes this absorber suitable for sensing and detection applications [41]. The peak absorption mechanism underlies the current distributions at the top and bottom layers of this metamaterial absorber. Though for brevity reasons current distribution images are not included here it is observed that at near-unity absorption frequencies, front and back layer currents are antiparallel, and thus a current loop is constructed which initiates a magnetic dipole resonance [42]. This magnetic field coupling with the incident field causes maximum absorption. Thus, inductively tuned metamaterial discussed here can be suitbale for dual purposes. Firstly, as a metamterial superstrate it can be used as a gain enhancer of the multi band antenna and with a copper backplane it can be used as a good absorber suitable for sensing and detection applications.

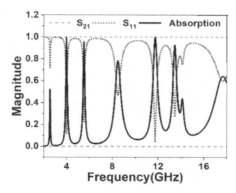

Figure 4.11. Absorption property of the inductively tuned metamaterial with full copper backplane.

4.4 Two axes symmetric metamaterial and it's application
4.4.1 MTM design and result analysis

In section 3.3, metamaterial with single axis symmetry has been discussed. The presented metamaterial discussed in this section suffers from two major limitations. One is the frequency shift in the unit cell arrays. The second one is at high frequencies (above 10 GHz), when themetamaterial loaded antenna shows a decreasing gain compared with that at lower frequenicies. To eliminate these problems, a new two axes symmetrical structure metamaterial has been presented in Figure 4.12. This metamaterial is named gap coupled symmetric split ring resonator based metamaterial that is designed over a 1.6 mm thick FR4 substrate. An overall unit cell dimension of 10 × 10.5 mm² is selected for this design. Three split rings are designed for the resonating patch as depicted in Figure 4.12. Among these SRRs, square-shaped rings are used for the two outer SRRs and the rest are circular. Near the split gaps of the outermost ring, two pairs of metallic stubs are utilized for its interconnection with the nearer ring as depicted in Figure 4.12. In addition to this, in the innermost ring, two shunt inductive paths are introduced near the split gaps. The dimensions of the rings, position of the split gaps as well as coupling metallic stubs are so chosen that the entire structure becomes symmetric about the horizontal and vertical axes. This symmetric structure is selected for the elimination of harmonics as well as the coupling effect that originates when several unit cells are employed to form an array. Moreover, the coupling near the split gaps helps to shift the resonances towards the lower frequencies which helps to avoid losses at high frequencies. The reflection as well as the transmission coefficient plots of this metamaterial are depicted in Figure 4.13(a). From this figure, it is found that resonances of the transmission spectra are obtained at frequencies of 2.78 GHz, 7.7 GHz, and 10.16 GHz. Moreover, negative permittivity and near zero permeability are also achieved from this design (as depicted in Figure 4.13(b) and (c)). As presented in Figure 4.13(d), it is observed that the refractive index undergoes a positive to negative transition with near zero properties in the vicinity of the resonances of the transmission and reflection spectra. The EMR value of this metamaterial is 10.7 which also indicates significantly compact dimensions of this metamaterial [43].

Figure 4.12. Gap coupled metamaterial.

Metamaterial Structure Exploration for Wireless Communications 103

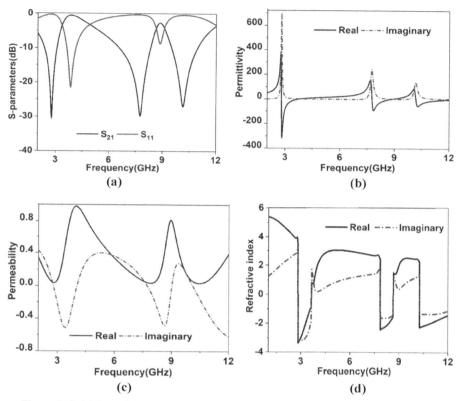

Figure 4.13. (a) Scattering parameters plot (b) permittivity (c) permeability (d) refractive index.

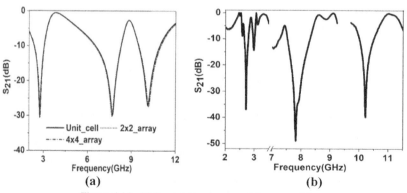

Figure 4.14. (a) S_{21} response in array, (b) measured S_{21}.

The array performances are investigated for different arrays and it is observed the unit cell and its array exhibits similar results as demonstrated in Figure 4.14(a). The similarity is due to the full symmetric structure of the MTM unit cell design. The measured result is displayed in Figure 4.14(b). The measured resonances of S_{21} are observed at 2.72 GHz, 7.81 GHz, and 10.22 GHz having magnitudes of –36.88 dB, –48.89 dB, and –39.78 dB, respectively. It is noteworthy to mention that harmonics and noises prevail in the measured graph at low as well as mid-frequency regions. Moreover, a small deviation is observed between the measured and simulated transmission coefficients. Despite this small inaccuracy, both results match well with each other, and multiband resonances fall into the S, C, and X bands.

4.4.2 Antenna gain enhancement using MTM superstrate

An array of 4 × 4 elements of this metamaterial with an overall dimension of 41 mm × 40 mm is used as a superstrate of the monopole antenna. This array's elements provide the same designed antenna area that covers the total surface of the antenna. Thus, it interacts with the most emitted radiations of the antenna, which ultimately helps to improve the antenna gain. Figure 4.15(a) shows the top view of a monopole antenna whereas the bottom view of the antenna is depicted in Figure 4.15(b). As shown in Figure 4.15(c), an array of this metamaterial is placed as a superstrate at the ground side of the antenna at a spacing of 30 mm. The space is selected by parametric study targeting that position of this metamaterial array which will not affect the bandwidth of the antenna much rather it will increase the gain. The monopole antenna exhibits a bandwidth extended from 2.5 GHz to 4.24 GHz and exhibits an average gain of 2 dBi with a maximum of 2.95 dBi. The measured values of the S_{11} for antenna with and without metamaterial are depicted in Figure 4.16 in comparison with the outcomes of the simulation. From Figure 4.16, it is noticed that measured S_{11} results of the antennas with and without metamaterial exhibit close similarity. The measured –10 dB impedance bandwidth of the test antenna without MTM extends from 2.56 GHz to 4.2 GHz with a resonance peak at 3.5 GHz having a magnitude of –17 dB. When the MTM is used as the superstrate –10 dB bandwidth is observed extending from 2.58 GHz to 4.1 GHz with a resonance peak of –24 dB at 3.67 GHz. In both cases, the bandwidth is slightly less than the simulation result, but the deviations are less than 5%. Gain data is plotted in Figure 4.17 to compare measured gain with an antenna without the metamaterial and simulation results. A comparison of the gain for the antenna with and without the MTM superstrate

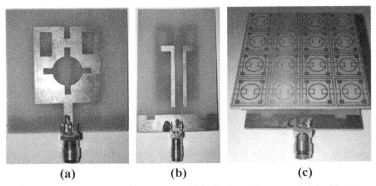

(a) (b) (c)

Figure 4.15. (a) Front view of the antenna, (b) back view, (c) metamaterial with antenna.

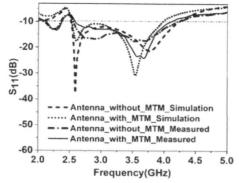

Figure 4.16. Simulated and measured S_{11} of the monopole antenna with and without metamaterial superstrate.

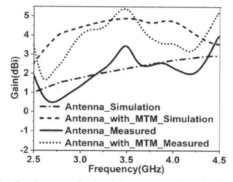

Figure 4.17. Simulated and measured gain of the antenna with and without MTM superstrate.

indicates a high gain with the MTM superstrate. The MTM cover provides a maximum gain of around 4.95 dBi, whereas a bare antenna offers a maximum gain of around 2.5 dBi, indicating a 95% gain enhancement. It is noteworthy to mention that, measured antenna gain with and without metamaterial is slightly varied compared to the simulation result. This mismatching is due to fabrication tolerances of the antenna and calibration errors associated with the Satimo near field measurement system. Measured antenna gain is less compared to the simulated gain except in the frequency range of 3.2 GHz–3.8 GHz where the measured gain is well above the simulated gain. However, comparing the measured gain between the antenna with and without metamaterial it is observed that the metamaterial exhibits its dominant impact on the antenna and helps to boost up the antenna gain.

4.5 Mirror symmetric resonator based tuned metamaterial

4.5.1 Design of the metamaterial

From the above discussions, it is evident that the two axis symmetric structure provides better array performances and a FR4 substrate is good for designing the

106 *Metamaterial for Microwave Applications*

metamaterial working within or around 10 GHz. So, for the applications above 10 GHz frequencies we need to use low loss substrate materials. So, another metamaterial structure can be presented that is structurally different compared to metamaterials in the earlier sections. This metamaterial named mirror symmetric resonator based tuned metamaterial is designed on the Rogers RT5880 substrate having a thickness of 1.57, a dielectric constant of 2.2, and a loss tangent of 0.001. The selected dimensions of the substrate are 10 × 10 mm^2 which makes the unit cell small enough compared to the wavelength in our target frequency range from 4–18 GHz covering the C, X, and Ku-bands so that an effective response of the metamaterial can be realized [44]. The unit cell of the metamaterial is depicted in Figure 4.18. The patch of this metamaterial is a two-fold mirror-symmetric structure that can be divided into four equal parts. Each part contains two square split-ring resonators that are connected by two metal strips. The quartiles are then interconnected by square shaped metal strips placed at the centre of the total structure. Four tuning metallic stubs are extended towards the two perpendicular axes from the bisecting point on four sides of this squared metal strip. The length of these metallic stubs plays a vital role in changing the resonance frequencies. These resonance frequencies can be changed by modifying the length of these stubs. The scattering parameter plots of this metamaterial are shown in Figure 4.19 that exhibits four resonances

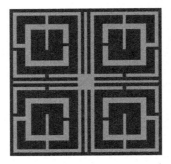

Figure 4.18. Unit cell of mirror symmetric resonator-based metamaterial.

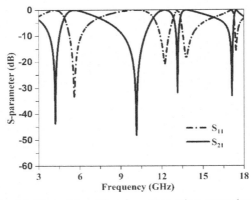

Figure 4.19. Simulated S_{21} and S_{11} of the mirror symmetric resonator based metamaterial.

of the transmission coefficient that occur at 4.20 GHz, 10.14 GHz, 13.15 GHz, and 17.1 GHz covering the C, X and Ku bands. It is also noticed that this metamaterial exhibits ENG properties with near zero index permeability and refractive index with an EMR value of 7.17 [45].

4.5.2 Frequency tuning of the metamaterial

The scattering parameters of the proposed MTM can be adjusted as per the application requirement by using four tuning stubs extended in horizontal and vertical directions from the centre of the structure, as shown in Figure 4.20. Lengths, l of these stubs are changed simultaneously from a maximum length of 5 mm to a half-length of 2.5 mm with an equal distance of 0.5 mm to observe the response of the MTM where l is measured from the centre of the resonator structure.

Resonances of the transmission coefficients obtained from these changes in stub lengths are depicted in Figure 4.21(a)–(c). Investigating Figure 4.21(a), a nominal shift in resonance frequency occurs in the C band in which shifts are within 4.21 GHz–4.32 GHz for changes in l with a maximum shift of 2.61%. The tuning effect in this band is observable in bandwidth, where it changes within 0.7 GHz–1.45 GHz with a maximum percentage change of 107%. In the X band resonance frequency shifts are more pronounced where a nearly linear shift ranges from 10.58 GHz–11.5 GHz, as shown in Figure 4.21(b). The maximum bandwidth of 1.2 GHz with a central frequency of 10.58 GHz is exhibited in this band, whereas the minimum bandwidth is 0.11 GHz with resonance at 11.7 GHz. In Figure 4.21(c), responses of the MTM in the Ku band are depicted in which dual resonances are observed for $l = 4$ mm and 4.5 mm. In this band, the resonances are spread over a large frequency region extending from 13.8 GHz to 17.4 GHz with a minimum bandwidth of 0.16 GHz and a maximum of 0.46 GHz with resonances at 16.85 GHz and 16.9 GHz, respectively for $l = 2.5$ mm and 5 mm, respectively. It is noteworthy to mention that the maximum length, $l = 5$ mm, contributes to the maximum bandwidth of the C band resonance and minimum bandwidth of the X and Ku band resonances. At this length, horizontal metal stubs are adjacent to the PEC boundary,

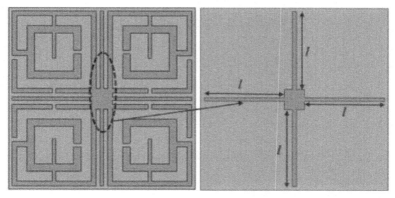

Figure 4.20. Tuning metallic stubs of the metamaterial unit cell.

108 *Metamaterial for Microwave Applications*

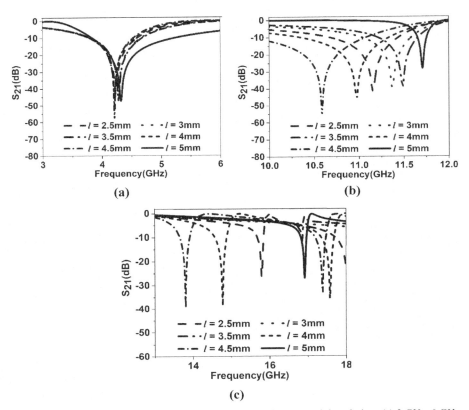

Figure 4.21. Transmission coefficient (S21) for variation of tuning stub length, l at: (a) 3 GHz–6 GHz (b) 10 GHz–12 GHz (c) 13 GHz–18 GHz.

Table 4.1. Performance comparison of the mirror symmetric resonator based metamaterial for different stub lengths in terms of resonance frequency and bandwidth.

Length (mm)	Resonance frequency (GHz)	Bandwidth (GHz)	Length (mm)	Resonance Frequency (GHz)	Bandwidth (GHz)
2.5	4.28, 11.5, 16.85	0.83, 0.46, 0.04	3	4.28, 11.37	0.8, 0.57
3.5	4.23, 11.15, 15.8	0.8, 0.77, 0.16	4	4.21, 10.98, 14.8, 17.6	0.75, 0.96, 0.3, 0.37
4.5	4.21, 10.58, 13.8, 17.4	0.7, 1.2, 0.2, 0.3	5	4.32, 11.7, 16.9	1.45, 0.11, 0.16

whereas vertical stubs are closest to the PMC boundaries. S_{21} responses are shown in Table 4.1 (a) 4–9 GHz (b) 10 GHz–12 GHz (c) 13 GHz–18 GHz.

4.6 Rotating symmetric resonator based metamaterial absorber

4.6.1 Metamaterial absorber (MMA) design

This metamaterial contains three layers (i) a substrate layer, (ii) a resonator patch top layer (iii) a metallic bottom layer. The patch and metallic backplane are placed on two opposite sides of the FR4 substrate that bears a thickness of 1.6 mm with a dielectric constant of 4.3, and a loss tangent of .02. Figure 4.22(a) shows the front view of this metamaterial absorber whereas the backview is depicted in Figure 4.22(b). Since a perfect absorber reflection, as well as transmission, should be zero, in the proposed MTM a full copper backplane is used to ensure zero transmission. On the other hand, the resonating patch is optimized through the performed numerical simulation in the CST microwave studio-2019 suite to acquire zero reflection at our target frequencies of 2.4 GHz and 5 GHz. The resonating patch includes three 0.035 mm thick copper rings over the substrate. Four tuning metallic stubs are produced from each of the two outer rings that have the functionality of frequency modification, whereas the innermost split rings work along with the metallic stubs to obtain maximum absorption in the desired frequencies. The total structure is not only two axes symmetric but also rotating symmetric as the unit cell of this metamterial absorber is depicted in Figure 4.22. The thickness of the bottom copper layer is a crucial factor as it determines whether this layer can create a complete obstacle for the transmitted electromagnetic wave or not. This can be determined by calculating the skin depth using equation (1) [46].

$$\text{Skin depth, } \delta = \sqrt{\frac{\rho}{\pi \mu f}} \qquad (4.1)$$

The calculated skin depth at 2.4 GHz is 1.347 μm as the resistivity of the copper, $\rho = 1.72 \times 10^{-8}$ Ω-meter and permeability, $\mu = 1$. Absorptivity can be expressed as, $A(\omega) = 1 - T(\omega) - R(\omega)$. In the proposed structure, the bottom copper thickness is 0.035 mm which allows a complete omission of the transmission wave spectra, $T(\omega)$. Thus, absorption depends on the reflected wave and can be calculated as absorptivity, $A(\omega) = 1 - R(\omega)$, where reflection spectra, $R(\omega) = $ (Reflection Coefficient, S_{11})^2. Figure 4.23 shows the reflection, transmission and absorption spectra for this metamaterial absorber. It provides two absorption peaks at 2.4 GHz and 5 GHz with near-perfect absorption having absorption peaks of 99.2% and 99.99% [47].

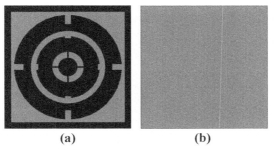

Figure 4.22. Metamaterial absorber (a) front view (resonator), (b) back view (copper back plane).

110 *Metamaterial for Microwave Applications*

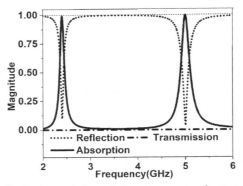

Figure 4.23. Reflection, transmission and absorption spectra of metamaterial absorber.

4.6.2 *Frequency tuning of the MMA*

As specified, this includes eight metallic stubs in the two outer rings that have the functionality of controlling the frequency of resonance of S_{11}. Thus, peak absorption frequencies can be adjusted by changing the length of these metallic stubs. The stubs are placed orthogonal to the horizontal and vertical axes so that the total structure maintains an axis symmetry as shown in Figure 4.24. All the outer ring metallic stubs having a length, *l1* are changed simultaneously from 0 mm to 2 mm keeping all other ring parameters constant to observe the effect on absorption characteristics. The absorption spectra for the change in *l1* are presented in Figure 4.25(a)–(b). In Figure 4.25(a), the absorption spectrum indicates that there is a linear shift of the

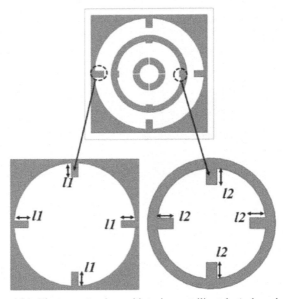

Figure 4.24. The two outer rings with tuning metallic stubs (enlarged view).

Metamaterial Structure Exploration for Wireless Communications 111

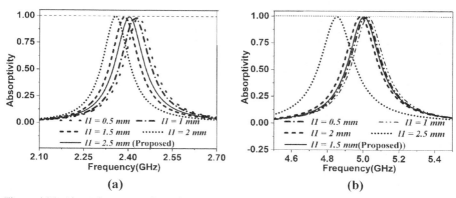

Figure 4.25. Absorption spectra for various values of l1 (a) for frequency interval 2.1 GHz–2.7 GHz, (b) 4.5 GHz–5.5 GHz.

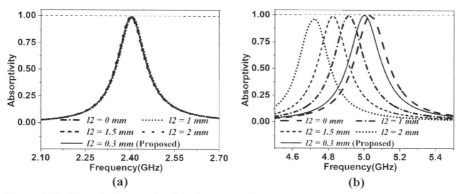

Figure 4.26. Absorption spectra for changing values of second ring metallic stubs length, l2 (a) for frequency interval 2.1 GHz–2.7 GHz, (b) for frequency interval 4.5 GHz–5.5 GHz.

Table 4.2. Comparison of the response of MTM-4 for different stub lengths.

Outer ring metallic stub, *l1*			middle ring metallic stub, *l2*		
Length (mm)	Frequency (GHz)	Absorption (%)	Length (mm)	Frequency (GHz)	Absorption (%)
0.5	2.434, 5.01	98.7, 99.75	0	2.4, 5.04	99, 99.86
1	2.424, 5.02	98.88, 99.8	0.3	2.4, 5	99.2, 99.9
1.5	2.4, 5	99.2, 99.9	1	2.41, 4.9	98.9, 99.4
2	2.38, 4.98	99.35, 99.99	1.5	2.4, 4.82	99.1, 99.2
2.5	2.36, 4.85	99.6, 99.6	2	2.4, 4.72	99, 96

first peak absorption frequencies from 2.44 GHz for *l1* = 0.5 mm to 2.36 GHz for *l1* = 2.5 mm. On other hand, absorption peak resonance at 5 GHz is almost unaffected due to the change of *l1* from 0.5 mm up to 2 mm as shown in Figure 4.25(b). A drastic change in resonance at 5 GHz is observed for *l1* = 2.5 GHz and due to this stub length change the peak resonance frequency shifts to 4.84 GHz. This abrupt change in response around 5 GHz happens because, for ll = 2.5 mm, the distance between the second metallic ring and metallic stubs is only 0.1 mm, Thus, a strong coplanar capacitance is formed between these metallic stubs and the second circular ring due to this small gap as capacitance, $C = \varepsilon \frac{A}{d}$; where A is the area of the metallic stubs and d is the gap distance.

4.6.3 Metamaterial property analysis of the MMA

The derived relative permittivity and permeability are depicted in Figure 4.27 and Figure 4.28. By investigation of the obtained results, it is revealed that the MMA exhibits single negative behaviour as permittivity shows its negative values from 2 GHz–2.4 GHz, 2.88 GHz–5 GHz, 5.88 GHz–6 GHz, whereas negative permeability exists in the frequency bands extending from 2.4 GHz–2.88 GHz and 5 GHz–5.88 GHz. This result indicates that the MMA exhibits both electrical and magnetic resonances near the peak absorption frequencies. Matching of the absorber impedance with the free space impedance is very essential for a perfect absorber. Since the free space impedance is 377Ω, a mismatch between these two impedances causes reflection, which eventually decreases the absorption. The normalized impedance, Z of a medium can be calculated by using the relation, $Z = \sqrt{\frac{\mu_r}{\varepsilon_r}}$. The calculated

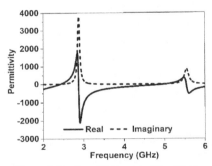

Figure 4.27. Permittivity plot of the MMA.

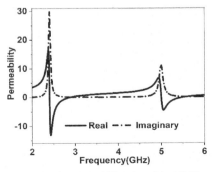

Figure 4.28. Permeability plot of the MMA.

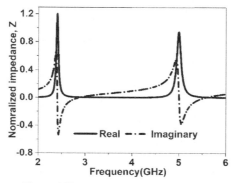

Figure 4.29. Impedance graph of the MMA.

normalized impedance is plotted in Figure 4.29 which indicates that the real part of the impedance nearly approaches unity whereas the imaginary part approaches zero at the target frequencies. By analysing the data, it is found that $Z = 1.2 + j\ 0.033$ at 2.4 GHz, whereas at 5 GHz, $Z = 0.94 + j\ 0.01$, where j is the imaginary operator. Z is related to S_{11} according to the relation [48], $S_{11} = \dfrac{Z - Z_0}{Z + Z_0}$, where Z_0 is the free space impedance that gives $S_{11} = 0.09$ at 2.4 GHz and $S_{11} = 0.03$ at 5 GHz providing absorptions of 99.2% and 99.9%, respectively.

4.6.4 Power and current distribution analysis of MMA

The power analysis is performed to understand the absorption behaviour of the MMA. In the simulation method, a port is assigned as a source that transmits 0.5 watt power, and a portion of this power is reflected and the remaining portion is received by the MMA as expressed in Figure 4.30. The reflected power spectrum exhibits two notches of nearly zero power at 2.4 GHz and 5 GHz. Due to the copper backplane, no power is transmitted indicating that the received power is absorbed by the MMA which is very plausible from Figure 4.31. Though the absorber contains conducting metallic layers as well as a substrate, the power received by this metamaterial is solely absorbed by the substrate layer and the metallic layer loss is almost zero due to its high conductivity as depicted in Figure 4.31. The power loss in the substrate material is due to the high dielectric constants of the FR4 substrate [49], which is the key factor of the power absorption mechanism.

For understanding the absorption phenomena, surface currents have been studied at 2.4 GHz and 5 GHz. Figure 4.32(a) presents the surface current distribution at 2.4 GHz and it is noticed from this Figure that the surface current at the top resonator is mostly concentrated around the vertical metallic stubs connected with the outer ring indicating the contribution of this ring to the absorption peak obtained at 2.4 GHz. The current at the backplane is distributed all over the copper but mostly localized below the top resonator localized current. The top resonator localized current flows in a clockwise direction whereas the bottom copper plane current flows in the anticlockwise direction. Thus, two antiparallel surface currents form a circulating

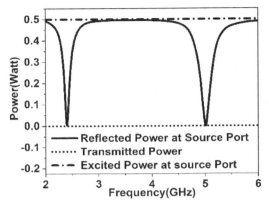

Figure 4.30. Power at different ports.

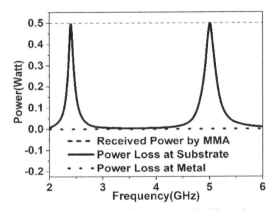

Figure 4.31. Power spectra for absorber with different losses.

current loop that triggers the magnetic resonance that is essential for obtaining peak absorption at this frequency [49, 50]. Similarly, as expressed in Figure 4.32(b), an opposite flowing current is observed between the middle ring of the resonator and the copper back at the bottom. Thus, peak absorption at 5 GHz is solely contributed by the electromagnetic resonance offered by the middle ring. It is noteworthy to mention that the innermost ring also contributes a small amount of current field which also helps to increase the absorption.

The power distribution pattern in the substrate material can be examined by using the density of power loss at the substrate material. Figure 4.33(a) shows the power loss density at the resonator side and copper backside at 2.4 GHz whereas, 4.33(b) exhibits the same for 5 GHz. A close investigation of the image in Figure 4.33(a) reveals that on the front resonator side the power loss is concentrated below the outer ring of the resonator whereas on the backside power is more widely spread but it covers the area under the influence of the outer ring. In the central region, power loss density is almost zero indicating that at 2.4 GHz, this region makes no contribution to the absorption. On the other hand, power is concentrated mostly

Metamaterial Structure Exploration for Wireless Communications 115

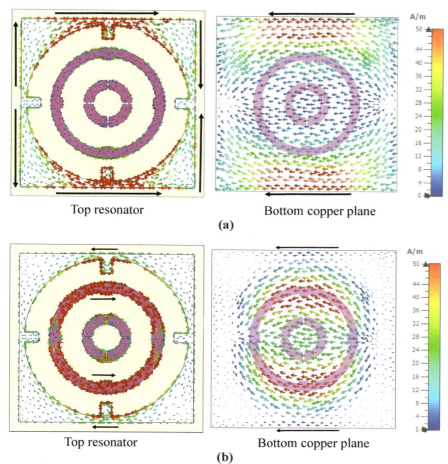

Figure 4.32. Current distribution in the metamaterial absorber at (a) 2.4 GHz, (b) 5 GHz.

below the middle ring loss at 5 GHz as shown in Figure 4.33(b). The innermost ring and the horizontal metallic stubs of the outer ring also contribute significantly to the power loss at this frequency. On the bottom side where a full copper back is used, power is more widely spread covering an area influenced mostly by the middle ring and partly by the innermost ring and the outermost ring horizontal stubs. In the central region the power loss is almost zero. The power loss is also nullified in the peripheral region of the substrate at 5 GHz, as depicted in the back view of Figure 4.33(b). So, from the power loss distribution pattern it can be concluded that the outer ring acts for the absorption at 2.4 GHz, whereas the middle ring along with the innermost ring and horizontal metallic stubs of the outermost ring make a combined contribution to the absorption at 5 GHz with a major contribution by the middle ring, and thus near-unity absorption is achieved at this frequency.

116 *Metamaterial for Microwave Applications*

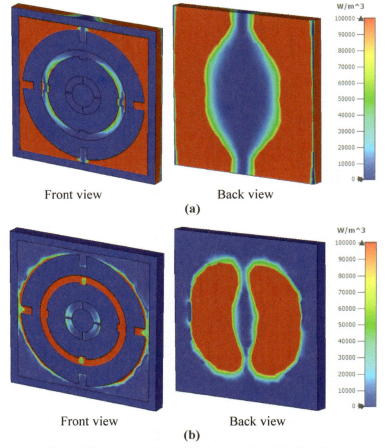

Figure 4.33. Power loss density in substrate (a) 2.4 GHz, (b) 5 GHz.

4.6.5 *Angular stability and polarization insensitivity study of MMA*

High angular stability in the absorption profile is essential for a MMA to be a perfect absorber. This performance depends on the thickness of the MMA that characterizes the structural profile. The effect of the incident angular change is studied using transverse electric (TE) or transverse magnetic (TM) waves. Figure 4.34(a) shows the absorption spectra for the TE mode of the incident wave. An investigation of the absorption spectra indicates that it exhibits high stability observed with the maximum absorption at the same frequencies without a deviated bandwidth. A similar, performance is also noticed for the TM mode signal with an incident angle variation that is displayed in Figure 4.34(b). This high stability in incident angle variation is due to the symmetric MMA cell structure that is essential to achieve the same response at the same frequencies for the TE, and TM modes of propagation [51, 52]. The MMA performance as a perfect absorber is also further studied by considering the cross-polarization effect. Since the incident wave is polarized and if we consider

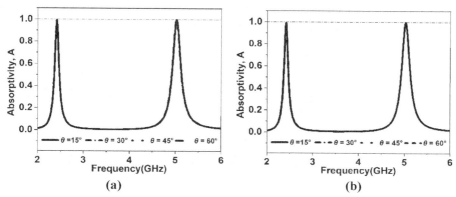

Figure 4.34. Absorption for various incident angles (a) TE mode, (b) TM mode.

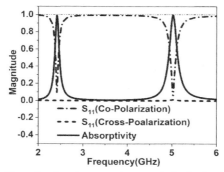

Figure 4.35. Absorption based on co and cross polarized reflection coefficient.

the electric field to be parallel to the X-axis, then the co-polarization reflectance from X to X can be defined as $r_{xx} = E_{xr}/E_{xi}$ and cross-polarization reflectance from X to Y can be defined as $r_{yx} = E_{yr}/E_{xi}$, where E denotes the electric field and subscripts r and i symbolize reflection and incidence of the electromagnetic waves. Then, absorptivity can be calculated considering the co and cross polarized reflectances using the relation, $A(\omega) = 1 - r_{xx}^2 - r_{yx}^2$. The absorption spectrum considering co and cross polarization effects is shown in Figure 4.35 which indicates that the cross polarization is essentially zero at our frequencies of interest (2.4 GHz and 5 GHz). Thus, the cross-polarization effect is nullified and the absorber exhibits maximum absorption of 99.2% and 99.9% at the target frequencies.

4.6.6 Experimental result of the MMA

The measurement setup is presented in Figure 4.36 which consists of a vector network analyzer (VNA) and waveguide adapter. An MMA array is placed at the gap's position of the waveguide adapter as depicted in Figure 4.36. The S_{11} is measured using the VNA and the measured result is illustrated in Figure 4.37. Since transmission is zero due to the full copper back of the MMA the transmitted signal is

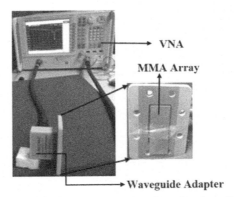

Figure 4.36. Measurement set up to determine absorption.

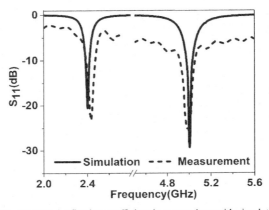

Figure 4.37. Measured reflection coefficient in comparison with simulation result.

not measured. The absorption is calculated From the measured S_{11} and the results are compared with the simulation and plotted in Figure 4.38.

In a comparison of the measured results with simulated ones, it is observed that measured S_{11} exhibits a deviation in frequency as well as magnitude. Due to this deviation, the same shift in the magnitudes of the peak absorption and its frequency is observed. Scrutinizing Figure 4.38, it is found that peak absorption occurred at 2.44 GHz with 99.5% absorption and 4.99 GHz with 99.4%. Thus frequency shifts of 1.25% and 0.25% happen from the target frequency of 2.4 GHz and 5 GHz. At 2.4 GHz and 5 GHz, the measured absorption is about 94% and 99.3%, respectively. Contrary to this more than 80% of the absorption bandwidth is 0.11 GHz extending from 2.37 GHz to 2.48 GHz for the first absorption peak, and 0.5 GHz for the second absorption peak extending from 4.7 GHz to 5.2 GHz. This absorption bandwidth is higher as compared with the simulated one. It can be concluded that the MMA exhibits a good performance in the Wi-Fi frequencies with near-unity absorptions. Thus, this MMA can be a good candidate for wireless applications especially for microwave shielding of Wi-Fi singals.

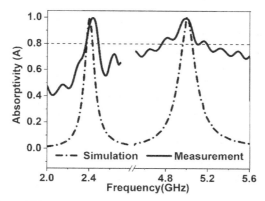

Figure 4.38. Meausred absorption in comaparison with simulation.

4.7 Summary

In this chapter, a number of metamateial structures have been explored for wireless communications' applications. It is observed that an asysmmetrical structural design provides compact dimensions with a high EMR of 16.74. This can be utilized for multiband wireless communications. But, in the case of unitcell arrays this MTM exhibits resonance frequency shifts. The above discussed single axis symmetric metamaterial exhibits multiresonances with negative permittivity, near-zero permeability, and near-zero refractive. It also shows a high EMR value of 15.75 indicating the compactness of the designed structure. Another unique characteristic of this design is that with the copper backplane the resonating structure exhibits high absorptions of 99.6%, 95.7%, 99.9%, and 92.7% with Q-factors of 28.4, 34.4, 23, and 32 at 3.98 GHz, 5.5 GHz, 11.73 GHz, and 13.47 GHz respectively. Thus, this metamaterial also exhibits about 73% gain enhancement of a multiband antenna and as a superstrate for mulit applications. On the other hand, the presented two axes symmetric metamaterial provides three S_{21} resonances of 2.78 GHz, 7.7 GHz, and 10.16 GHz with the reduction of the mutual coupling effect in the array elements. Moreover, it exhibits negative permittivity, near-zero permeability, and refractive index within its compact dimensions and moderate EMR of 10.7. It helps to increase the gain of a monopole antenna by 95% when used as a superstrate.

The presented mirror symmetric resonator based metamaterial provides excellent cross coupling reductions when an array is formed containing several unit cells. It provides resonances that occur at 4.2 GHz, 13.15 GHz, 17.1 GHz, and 10.14 GHz covering the C, X, and Ku bands. The ENG property with near-zero refractive index and permeability is obtained through the unique compact design and a maintained high EMR value of 7.17. The resonance frequencies of this metamaterial can be easily modulated by varying the length of the four tuning metallic structures that make this metamaterial structure suitable for frequency selective applications for wireless communication systems. Finally, a rotating symmetric resonator based metamaterial acts as an absorber that is designed to target the absorption of the Wi-Fi band signals at 2.4 GHz and 5 GHz. About 99.2% and 99.9% absorption at these two targeted frequencies are achieved through a simplistic metamaterial with a symmetrical structure to avoid any effects of cross polarization and incident angle variation. The distinguishing feature of this metamaterial absorber is that the peak absorption frequencies can be tuned through length variation of the tuning metallic stubs connected with the two outer rings. Two different sets of metallic stubs assist in adjusting the two maximum absorption frequencies almost independently. These variety of metamaterials can be utilized for various applications in wireless communications.

References

[1] Bilotti, F. and L. Sevgi. 2012. Metamaterials: Definitions, properties, applications, and FDTD-based modeling and simulation. *International Journal of RF and Microwave Computer-Aided Engineering,* 22: 422–438.

[2] Alù, A., N. Engheta, A. Erentok and R. W. Ziolkowski. 2007. Single-negative, double-negative, and low-index metamaterials and their electromagnetic applications. *IEEE Antennas and Propagation Magazine,* 49: 23–36.

[3] Martin, F. 2006. Special issue on Microwave metamaterials: Theory, fabrication and applications. *EuMA Proc,* vol. 2.

[4] Pendry, J. B. 2000. Negative refraction makes a perfect lens. *Physical review letters,* 85: 3966.

[5] Engheta, N. and R. W. Ziolkowski. 2006. *Metamaterials: Physics and Engineering Explorations.* John Wiley & Sons.

[6] Erentok, A. and R. W. Ziolkowski. 2008. Metamaterial-inspired efficient electrically small antennas. *IEEE Transactions on Antennas and Propagation,* 56: 691–707.

[7] Bilotti, F., A. Alù, N. Engheta and L. Vegni. 2005. Anomolous properties of scattering from cavities partially loaded with double-negative or single-negative materials. *Departmental Papers (ESE),* p. 210.

[8] Alù, A., F. Bilotti, N. Engheta and L. Vegni. 2007. Subwavelength, compact, resonant patch antennas loaded with metamaterials. *IEEE Transactions on Antennas and Propagation,* 55: 13–25.

[9] Bilotti, F., L. Nucci and L. Vegni. 2006. An SRR based microwave absorber. *Microwave and Optical Technology Letters,* 48: 2171–2175.

[10] Alù, A., F. Bilotti, N. Engheta and L. Vegni. 2007. Subwavelength planar leaky-wave components with metamaterial bilayers. *IEEE Transactions on Antennas and Propagation,* 55: 882–891.

[11] Alici, K. B., F. Bilotti, L. Vegni and E. Ozbay. 2010. Experimental verification of metamaterial based subwavelength microwave absorbers. *Journal of Applied Physics,* 108: 083113.

[12] Bilotti, F., A. Toscano, K. B. Alici, E. Ozbay, L. Vegni et al. 2010. Design of miniaturized narrowband absorbers based on resonant-magnetic inclusions. *IEEE Transactions on Electromagnetic Compatibility,* 53: 63–72.

[13] Cai, W., U. K. Chettiar, A. V. Kildishev and V. M. Shalaev. 2007. Optical cloaking with metamaterials. *Nature Photonics,* 1: 224–227.

[14] Silveirinha, M. G., A. Alù and N. Engheta. 2007. Parallel-plate metamaterials for cloaking structures. *Physical Review E,* 75: 036603.

[15] Zhang, W., H. Chen and H. O. Moser. 2011. Subwavelength imaging in a cylindrical hyperlens based on S-string resonators. *Applied Physics Letters,* 98: 073501.

[16] Pendry, J. B., D. Schurig and D. R. Smith. 2006. Controlling electromagnetic fields. *Science,* 312: 1780–1782.

[17] Engheta, N. 2007. Circuits with light at nanoscales: optical nanocircuits inspired by metamaterials. *Science,* 317: 1698–1702.

[18] Martin, F., F. Falcone, J. Bonache, R. Marqués, M. Sorolla et al. 2003. Miniaturized coplanar waveguide stop band filters based on multiple tuned split ring resonators. *IEEE Microwave and Wireless Components Letters,* 13: 511–513.

[19] Falcone, F., F. Martín, J. Bonache, R. Marqués, T. Lopetegi et al. 2004. Left handed coplanar waveguide band pass filters based on bi-layer split ring resonators. *IEEE Microwave and Wireless Components Letters*, 14: 10–12.
[20] Marqués, R., F. Martin and M. Sorolla. 2011. *Metamaterials with Negative Parameters: Theory, Design, and Microwave Applications*. John Wiley & Sons.
[21] Markets, R. a. 2021. *The Global Market for Metamaterials 2021*. Available: https://finance.yahoo.com/news/worldwide-industry-metamaterials-2030-detailed-102300743.html.
[22] Cui, T. J., D. R. Smith and R. Liu. 2010. *Metamaterials*. Springer.
[23] Tamim, A. M., M. R. I. Faruque, M. J. Alam, S. S. Islam, M. T. Islam et al. 2019. Split ring resonator loaded horizontally inverse double L-shaped metamaterial for C-, X- and Ku-band microwave applications. *Results in Physics*, 12: 2112–2122.
[24] Hossain, T. M., M. F. Jamlos, M. A. Jamlos, P. J. Soh, M. Islam et al. 2018. Modified H-shaped DNG metamaterial for multiband microwave application. *Applied Physics A*, 124: 1–7.
[25] Ahamed, E., M. R. I. Faruque, M. F. B. Mansor and M. T. Islam. 2019. Polarization-dependent tunneled metamaterial structure with enhanced fields properties for X-band application. *Results in Physics*, 15: 102530.
[26] Hoque, A., M. Tariqul Islam, A. F. Almutairi, T. Alam, M. Jit Singh et al. 2018. A polarization independent quasi-TEM metamaterial absorber for X and Ku band sensing applications. *Sensors*, 18: 4209.
[27] Saravanan, M. and S. Umarani. 2018. Gain enhancement of patch antenna integrated with metamaterial inspired superstrate. *Journal of Electrical Systems and Information Technology*, 5: 263–270.
[28] Pandey, A. K., M. Chauhan, V. K. Killamsety and B. Mukherjee. 2019. High-gain compact rectangular dielectric resonator antenna using metamaterial as superstrate. *International Journal of RF and Microwave Computer-Aided Engineering*, 29: e21968.
[29] Ojo, R., M. F. Jamlos, P. J. Soh, M. A. Jamlos, N. Bahari et al. 2020. A triangular MIMO array antenna with a double negative metamaterial superstrate to enhance bandwidth and gain. *International Journal of RF and Microwave Computer-Aided Engineering*, 30: e22320.
[30] Moniruzzaman, M., M. T. Islam, M. R. Islam, N. Misran, M. Samsuzzaman et al. 2020. Coupled ring split ring resonator (CR-SRR) based epsilon negative metamaterial for multiband wireless communications with high effective medium ratio. *Results in Physics*, 18: 103248.
[31] Morea, M., K. Zang, T. I. Kamins, M. L. Brongersma, J. S. Harris et al. 2018. Electrically tunable, CMOS-compatible metamaterial based on semiconductor nanopillars. *Acs Photonics*, 5: 4702–4709.
[32] Qiao, Z., X. Pan, F. Zhang and J. Xu. 2021. A tunable dual-band metamaterial filter based on the coupling between two crossed SRRs. *IEEE Photonics Journal*, 13: 1–7.
[33] Ma, L., D. Chen, W. Zheng, J. Li, W. Wang et al. 2021. Thermally tunable high-Q metamaterial and sensing application based on liquid metals. *Optics Express*, 29: 6069–6079.
[34] Li, Z., Y. Jiang, Y. Liang, S. Tian, M. Chen et al. 2022. C, X, and Ku three-band tunable metamaterial absorber based on varactor diodes. *Microwave and Optical Technology Letters*, 64: 1536–1543.
[35] Kim, I. K. and V. V. Varadan. 2009. Electric and magnetic resonances in symmetric pairs of split ring resonators. *Journal of Applied Physics*, 106: 074504.
[36] Moniruzzaman, M., M. T. Islam, N. Misran, M. Samsuzzaman, T. Alam et al. 2021. Inductively tuned modified split ring resonator based quad band epsilon negative (ENG) with near zero index (NZI) metamaterial for multiband antenna performance enhancement. *Scientific Reports*, 11: 1–29.
[37] Abdel-Rahman, A. B. and A. A. Ibrahim. 2016. Metamaterial Enhances Microstrip Antenna Gain. ed: August, 2016.
[38] Bouzouad, M., S. Chaker, D. Bensafielddine and E. Laamari. 2015. Gain enhancement with near-zero-index metamaterial superstrate. *Applied Physics A*, 121: 1075–1080.
[39] Enoch, S., G. Tayeb, P. Sabouroux, N. Guérin, P. Vincent et al. 2002. A metamaterial for directive emission. *Physical Review Letters*, 89: 213902.
[40] Bozdag, G. and A. Kustepeli. 2015. Subsectional tapered fed printed LPDA antenna with a feeding point patch. *IEEE Antennas and Wireless Propagation Letters*, 15: 437–440.
[41] Zou, H. and Y. Cheng. 2019. Design of a six-band terahertz metamaterial absorber for temperature sensing application. *Optical Materials*, 88: 674–679.

[42] Liu, N. and H. Giessen. 2010. Coupling effects in optical metamaterials. *Angewandte Chemie International Edition,* 49: 9838–9852.

[43] Moniruzzaman, M., M. T. Islam, M. Samsuzzaman, N. M. Sahar, S. S. Al-Bawri et al. 2022. Gap coupled symmetric split ring resonator based near zero index ENG metamaterial for gain improvement of monopole antenna. *Scientific Reports,* 12: 1–22.

[44] Bait-Suwailam, M. M. 2019. Electromagnetic field interaction with metamaterials. *In*: *Electromagnetic Fields and Waves,* ed: IntechOpen.

[45] Moniruzzaman, M., M. T. Islam, I. Hossain, M. S. Soliman, M. Samsuzzaman et al. 2021. Symmetric resonator based tunable epsilon negative near zero index metamaterial with high effective medium ratio for multiband wireless applications. *Scientific Reports,* 11: 1–21.

[46] Pozar, D. M. 2009. Microwave Engineering USA: John Wiley & Sons.

[47] Moniruzzaman, M., M. T. Islam, M. F. Mansor, M. S. Soliman, N. Misran et al. 2022. Tuning metallic stub loaded symmetrical resonator based dual band metamaterial absorber for wave shielding from Wi-Fi frequencies. *Alexandria Engineering Journal,* 2022.

[48] Li, S.-J., P.-X. Wu, H.-X. Xu, Y.-L. Zhou, X.-Y. Cao et al. 2018. Ultra-wideband and polarization-insensitive perfect absorber using multilayer metamaterials, lumped resistors, and strong coupling effects. *Nanoscale Research Letters,* 13: 1–13.

[49] Bhattacharya, A., S. Bhattacharyya, S. Ghosh, D. Chaurasiya, K. Vaibhav Srivastava et al. 2015. An ultrathin penta-band polarization-insensitive compact metamaterial absorber for airborne radar applications. *Microwave and Optical Technology Letters,* 57: 2519–2524.

[50] Kafesaki, M., I. Tsiapa, N. Katsarakis, T. Koschny, C. Soukoulis et al. 2007. Left-handed metamaterials: The fishnet structure and its variations. *Physical Review B,* 75: 235114.

[51] Nguyen, T. T. and S. Lim. 2018. Design of metamaterial absorber using eight-resistive-arm cell for simultaneous broadband and wide-incidence-angle absorption. *Scientific Reports,* 8: 1–10.

[52] Amiri, M., F. Tofigh, N. Shariati, J. Lipman, M. Abolhasan et al. 2020. Wide-angle metamaterial absorber with highly insensitive absorption for TE and TM modes. *Scientific Reports,* 10: 1–13.

Chapter 5

Metamaterial Antennas for Ultra-wideband Applications

Mohammad Tariqul Islam[1,*] and *Md. Samsuzzaman*[2]

5.1 Introduction

With the advent of the new information age, numerous advanced communication technologies have emerged in recent years and have profoundly influenced and benefited every aspect of human life. Various communication systems, such as second generation (2G), third generation (3G), and fourth generation (4G) mobile communications, Wi-Fi, global positioning system (GPS), WLAN, Bluetooth, WiMAX, and ultra-wideband (UWB) systems, have created a revolution in wireless communications. Structures of future wireless communication systems are gradually becoming a reality, with metamaterial antennas playing a crucial role in their effective and optimal performance. Metamaterial antennas are an antenna category the inspired by metamaterials that improve the antenna's gain, impedance bandwidth, and efficiency. A metamaterial is a material with an engineered structure that exhibits electromagnetic properties not typically found in nature. These metamaterials can have some unusual properties, such as simultaneous negative permittivity and permeability over a specific frequency range, resulting in a negative refractive index. Metamaterials with negative permittivity (ε) and permeability (μ) simultaneously are referred to as double negative (DNG) metamaterials, left-handed metamaterials (LHM), or negative index metamaterials.

The need for UWB technology to serve more users and transmit data at faster rates is growing quickly. The Federal Communications Commission (FCC) of the United States had authorized UWB technology for commercial usage in 2002 [1].

---

[1] Department of Electrical, Electronic and Systems Engineering, Universiti Kebangsaan Malaysia, Bangi, Selangor, Malaysia
[2] Department of Computer and Communication Engineering, Patuakhali Science and Technology University, Patuakhali, Bangladesh.
Email: sobuz@pstu.ac.bd
* Corresponding author: tariqul@ukm.edu.my

Due to its favourable qualities, including low cost, low complexity, high accuracy ranging, low spectrum power density, minimal interferences, and extremely fast data rates, a UWB system has been discovered for extensive use in short-distance wireless communication systems. Data transmission at high rates will be revolutionized by UWB technology, which will also enable the personal area networking sector to spur new advancements and improve services for customers. With its distinctive features encouraging significant advancements in wireless body area networks, wireless communications, and sensor networks, wireless personal area network biomedical instruments, positioning systems, radars, and detectors, UWB is regarded as a very promising and quickly emerging low-cost technology.

Antennas are constantly researched for wide bandwidth, small size, high gain, better efficiency, simplicity of integration and manufacturing, cheap cost, and independence from applications. The unique features of the metamaterials used in engineered materials and constructions are not found in nature. These unusual electromagnetic properties of the metamaterials are brought about by artificially created homogeneities or inclusions contained in or related to a host medium. While the tailoring and manipulation of wave properties are carried out by metamaterials, effective reductions in the weight and size of devices, components, and systems are found along with an augmentation in performances. Radiation efficiency and bandwidth in antennas both drop as the antenna size decreases. The antenna design architecture maintains this fundamental constraint indifferently. In some recent studies, the efficiency of metamaterial applications for enhancing antenna performances has been described and addressed [2, 3]. These fundamental restrictions might be disregarded because of the special properties of the metamaterials. The use of metamaterials may present a novel approach to overcome UWB antenna limitations by improving antenna attributes such as directivity, bandwidth, gain, radiation pattern, efficiency, and coupling.

Planar UWB antennas are desirable for a variety of antenna applications, despite the considerable hurdles posed by their wide working bandwidth. Antenna designers have one of their greatest challenges in developing compact planar antennas with wide impedance bandwidth, steady radiation patterns, acceptable gain, and improved time domain performance. Several types of antennas are listed for UWB applications, with planar antennas printed on PCBs gaining popularity since they can be easily implanted into wireless devices or combined with other RF circuits. Typically, a planar design can be used to minimize the volumetric size of a UWB antenna by substituting its three-dimensional radiating elements with their planar counterparts [4–6]. In order to achieve a broad operational bandwidth, the radiating element, size and form of the ground plane, and feeding structure can be optimized in the design of printed UWB antennas [7–9]. The printed antenna, which comprises of a planar radiator and ground plane, is imbalanced, however. The electric currents in these antennas are dispersed on both the radiator and the ground plane, and radiation from the ground plane cannot be avoided. Consequently, the performance of the printed UWB antenna in terms of operational bandwidth, gain, and radiation patterns is significantly impacted by the size and shape of the ground plane [10].

Utilizing new planar-patterned manufactured materials to improve these antenna characteristics is a novel approach to circumventing the constraints exhibited by certain well-established techniques for lowering antenna size [11, 12]. The possibility of artificial material design allows antenna designers unprecedented degrees of flexibility. To satisfy the ever-increasing need for compact, efficient microstrip antennas, fresh metamaterial-based techniques for patch antenna shrinking should be researched. Using design and construction procedures that are compatible with cutting-edge planar antenna technology, these novel materials can be used to produce tiny antennas with high efficiency and bandwidth. This chapter introduces and analyses a unique metamaterial structure to improve the fundamental properties of UWB antennas.

5.2 Metamaterial and UWB antennas

5.2.1 Metamaterials

In the last decade, electromagnetic engineers, microwave engineers, and physicists have shown a great deal of interest in electromagnetic metamaterials, which constitute a broad category of electromagnetic structures. Originally, the term "metamaterial" referred to any artificial (engineered) structure with effective electromagnetic properties not observed in natural materials. Metamaterials, also known as artificial electric and magnetic materials, are electromagnetic media whose physical properties are engineered by assembling microscopic and nanosomic structures in unusual combinations. Typically, these materials derive their properties from their structure rather than their composition, with the incorporation of minute inhomogeneities resulting in effective macroscopic behaviour. There are numerous types of metamaterials, which are categorized according to their fundamental properties, i.e., by the signs of their permittivity (ϵ) and permeability (μ). The classifications are shown in Figure 5.1. The double positive (DPS) metamaterials have both $\epsilon > 0$ and $\mu > 0$. The epsilon-negative (ENG) metamaterials have $\epsilon < 0$ and $\mu > 0$. The mu-negative (MNG) metamaterials have $\epsilon > 0$ and $\mu < 0$. The double negative (DNG) metamaterials have both $\epsilon < 0$ and $\mu < 0$.

In 1898, Bose attempted for the first time to investigate the concept of artificial materials with twisted structures; geometries that were essentially artificial chiral elements according to contemporary terminology [13]. Based on the signs of permittivity and permeability [14], Lamb proposed in 1905 the existence of backward waves from reflection directions. W. E. Kock created the first artificial dielectric materials with metal-lens antennas [15] and metallic delay lenses [16] at the end of the 1940s. In 1968, Victor Veselago was the first to outline the general properties of wave propagation in metamaterial structures. He predicted that their physical properties [17] would exhibit simultaneously negative permittivity (ϵ) and permeability (μ). According to his theoretical analysis, the direction of the Poynting vector in such materials is perpendicular to the direction of phase velocity. In addition, he predicted the existence of properties such as the inversion of Snell's law, the Doppler Effect, negative refraction, and the amplification of evanescent waves. 1999 saw the introduction of plasmonic-type (negative-ε/positive-μ or positive-ε/negative-μ) structures, which can be engineered to have plasmonic frequencies in the microwave range. The average cell size of both of these structures is significantly smaller than the guided wavelength, making them homogeneous structures [18].

126 *Metamaterial for Microwave Applications*

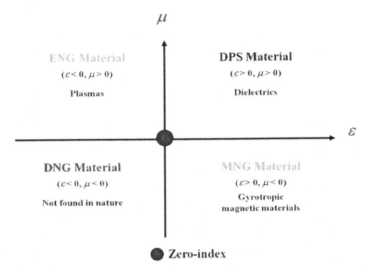

Figure 5.1. Classification of metamaterials based on the signs of their effective permittivity and permeability.

The first experimental verification of LH materials' phenomena was established by Smith in 2000 at UCSD (University of California, San Diego) [19]. By using metal split rings and thin straight metallic cables, they demonstrated that and had simultaneous negative values (ε and μ), and therefore attained a negative refractive index in the microwave area within a limited frequency range. The results of UCSD's

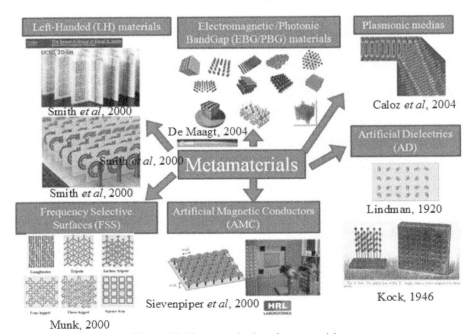

Figure 5.2. The categorization of metamaterials.

experiments were corroborated by further study in this region [20–23]. The lossy and limited bandwidth properties, on the other hand, make microwave implementation difficult. Because of this, they could only be used in a limited number of situations. An outline of the many types of extant metamaterials may be seen in Figure 5.2.

5.2.2 Ultra wideband (UWB) technology

The requirements for the UWB technology to be able to accommodate a greater number of users and provide data at faster speeds are rapidly increasing. Ultra-wideband technologies were given the green light for usage in commercial settings by the Federal Communications Commission (FCC) of the United States in the year 2002 [1]. These uses include wireless sensor networks, high-precision radars as well as imaging systems, wireless body area networks (WBANs), wireless personal area networks (WPANs), and wireless local area networks (WLANs).

UWB radio was invented by Guglielmo Marconi in the late 1800s with the Spark Gap radio, a pulse-based system. Throughout the course of many decades, this radio equipment was put to use to send Morse code via the airways. The early 1900s saw the development of narrowband (continuous wave) radio systems [24, 25]. Spark Gap radios were banned from most uses in 1924 due to their high emissions and interference with narrowband systems. Hewlett-Packard, in 1962, was the company that first developed the sampling oscilloscope [25]. At the beginning of the 1960s, there was a boom in enthusiasm in the electromagnetics time domain. This made it possible to conduct an examination of the pulse responses of microwave networks and stimulated the development of technologies for the generation of sub-nanosecond pulses. In the late 1960s, methods for measuring impulses were considered as the process of designing broadband antennas. These wideband antennas led to the creation of short-pulse radar and communication systems [26]. It wasn't until 1973 [27] that a short-pulse receiver received the first ultra-wideband communication patent. Even though in the 1980s, UWB technologies were known as baseband, carrier-free, or impulsive technologies, the United States Department of Defence did not adopt the term "ultra-wideband" until 1989. The Federal Communications Commission (FCC) made the Industrial Scientific and Medicine (ISM) frequencies available for unlicensed wideband communication usage in the middle of the 1980s. Because of the innovative way in which the spectrum is allocated, WLAN has seen phenomenal expansion in recent years. Through the allocation of this resource, research on the communication sectors is being carried out to investigate the benefits and repercussions of greater bandwidth communications. In 1989, UWB theory, methods, and various implementation approaches had been established for a wide variety of applications, such as radar, communications, positional systems, and so on. These advancements had been made in preparation for widespread use of UWB technology. In this approach, the United States government has given funding for the majority of the uses and research and development of UWB that fall under secret military projects [28, 29]. During the latter half of the 1990s, UWB technology made significant strides toward commercialization and saw significant advances in its research and development.

The year 2002 was a pivotal turning point in the course of UWB's long and illustrious history. The Federal Communications Commission issued rules that established the initial radiation constraints for Ultra-wideband transmissions and authorized the unlicensed deployment of UWB devices [1]. According to part 15 of the FCC's guidelines, a UWB system occupying a fractional bandwidth larger than 20% is defined as a wireless system with an absolute bandwidth greater than 500 MHz. As can be seen in Figure 5.3, the FCC has also granted permission for UWB radio broadcasts to take place in the unlicensed frequency range spanning from 3.1 GHz to 10.6 GHz, however they are limited to a transmission of power of −41.3 dBm/MHz.

The need for UWB technologies to handle a greater number of subscribers and to transfer data at faster speeds is increasingly growing. In the year 2002, the United States FCC gave its approval for the utilization of UWB technologies in commercial and industrial applications [1]. Imaging systems, wireless sensor networks, WBANs, WPANs, and WLANs are some of the applications that fall under this category.

Guglielmo Marconi is credited with developing the first ultra-wideband (UWB) radio, which was a pulse-based Spark Gap radio. This invention was made in the late 1800s. For many decades, this radio technology was put to use in order to send Morse code across the various frequencies. Narrowband radio techniques, which use continuous waves, were created in the early part of the 20th century [24, 25]. Spark Gap radios were banned from use in the majority of applications in 1924 because of the high emissions they produced and the interference they caused with narrow-band systems. The sampling oscilloscope was first developed by Hewlett-Packard in 1962 [25] as a result of a growing interest in time domain electromagnetics during the 1960s. Because of this, it was possible to conduct a study of the impulse response of microwave networks and it made it easier to create strategies for the creation of sub-nanosecond pulses. In the late 1960s, methods for measuring impulses were employed for the purpose of designing wideband antennas, which ultimately resulted in the creation of short-pulse radar and communication systems [26]. In 1973, the first invention for an ultra-wideband (UWB) communications device, a short-pulse receiver, was granted [27]. However, the United States Department of Defence did not endorse the word "ultra-wideband" till 1989, UWB technology was known as baseband, carrier-free, and impulsive technology in the early 1990s. The title "ultra-wideband" did not become widely used until 1989. The Federal Communications Commission (FCC) made the Industrial Scientific and Medicine (ISM) frequencies available for unauthorized wideband communication in the middle of the 1980s. The phenomenal success of WLAN may be directly attributed to the ground-breaking spectrum allocation that was implemented. By virtue of this allocation, the communication industries are investigated in order to ascertain the advantages as well as the drawbacks associated with larger bandwidth communication. UWB theory, methodologies, and multiple implementations approaches for a broad variety of applications were developed in 1989. These applications include radar, communications, and positioning systems, amongst others. The bulk of the research and applications into UWB were supported by the United States government as part of top-secret military projects [28, 29]. In the late 1990s, ultra-wideband (UWB)

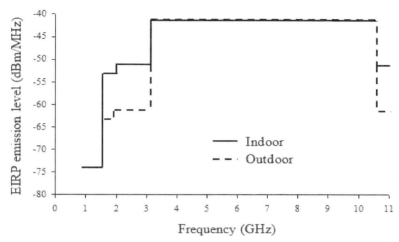

Figure 5.3. The FCC's Part 15 limits and the UWB spectral mask.

technology began to see more commercialization, and its growth also made major strides during this time.

In 2002, a significant change happened in the history of UWB. The FCC made decisions that approved the use of unlicensed UWB devices and imposed the very first radiation restrictions for UWB transmissions [1]. Any wireless system that uses more than 20 percent of a fractional bandwidth (FBW) or an absolute bandwidth of more than 500 MHz falls within the definition of UWB in Part 15 of the FCC guidelines. The FCC however has granted permission for UWB radio broadcasts in the unlicensed frequency range from 3.1 to 10.6 GHz with a restricted transmission power of –41.3 dB/MHz, as illustrated in Figure 5.3.

When compared to traditional narrow band systems, UWB systems employ extremely short impulses for transmissions, which results in an ultra-wideband spectral range with very limited power spectral density. This is in contrast to narrowband schemes, which use much longer pulses. Because of the presence of a UWB signal that occupies an extraordinarily high bandwidth, the radio frequency (RF) energy is dispersed throughout an incredibly vast spectrum. In comparison to current narrowband wireless systems, it is far more expansive and its radiated power is just a tenth of that of other narrowband systems. If the whole 7.5 GHz spectrum were to be used in the most efficient manner possible, the greatest power that could be achieved by UWB transmitters would be around 0.556 mW or less. This represents a negligible portion of the total transmitted power that is accessible in the ISM bands (which stand for industrial, scientific, and medical). This basically limits the UWB system to communications that take place inside and over short distances at large bandwidths or communications that take place over medium distances at low data rates. Applications such as wireless personal area networks (WPANs) and wireless USBs have been proposed with data transfer rates ranging from hundreds of megabits per second to several gigabits per second at distances ranging from one to four meters, and somewhere between one hundred and five hundred megabits per

130 *Metamaterial for Microwave Applications*

second over a distance of ten meters. For ranges greater than 20 meters, the data rate that can be achieved with UWB is lower than what can be achieved with present WLAN systems [30, 31].

5.3 Metamaterial based antenna design

5.3.1 Unit cell configuration

The design of the intended metamaterial antenna begins with a metamaterial unit cell that can be used for UWB applications. The resonance of the unit cell is meant to fall somewhere inside the ultra-wideband (UWB) spectrum, which spans from 3.1 to 10.6 GHz. There are well-established approaches to metamaterial design that may simultaneously give negative permittivity and permeability via the use of SRRs [18, 32]. The SRR is comprised of two loops, each of which is organized as a pair of oppositional concentric split rings [18]. The SRR is a magnetic resonating structure that, when activated, produces a magnetic field that is perpendicular to the direction in which it is applied. This results in a negative permeability. The internal ring is given a split gap, which enables capacitance to be produced and also affects the resonant qualities of the structure. The designed rectangular SRR having a CLS structure is shown in Figure 5.4(a). The enclosing of the outer ring results in a change to the design that was proposed. As a result of this alteration, the SRR structure's series capacitance is reduced, which, in turn, increases coupling between inner and outer rings. This unit cell has a thickness of 1.6 millimeters, and its permittivity is 4.6. It is fabricated on cheap FR-4 dielectric material. SRR metamaterial unit cells are changed by adding a CLS to attain resonance properties within their working UWB ranges. A CLS, also known as an I-shaped strip line, is a line that is imitative of long metallic wires and functions as electric dipoles [2]. Owing to the SRR resonance occurring by a perpendicular magnetic field and the CLS resonance occurring through the parallel electric field, the combined effect of SRR and CLS makes it possible to experience

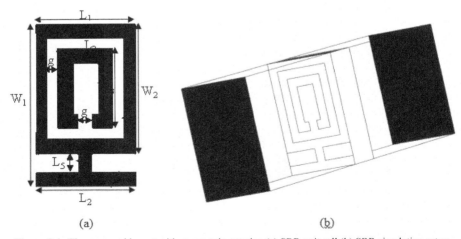

Figure 5.4. The proposed layout with parametric notation (a) SRR unit cell (b) SRR simulation setup.

Table 5.1. Design parameters of the metamaterial unit cell.

Design parameter	Dimension (mm)	Design parameter	Dimension (mm)
L_1	3.2	W_2	4.0
L_2	3.2	W_3	2.4
L_3	1.6	W_4	0.4
L_4	0.4	g	0.4
L_5	0.6	d_1	0.4
W_1	5.0	d_2	0.4

concurrent electromagnetism [33]. The combined induced current aids in lowering the overall structure's resonance by allowing the two resonances to work together [34]. Table 5.1 is a listing of the factors that were considered while designing the metamaterial unit cell.

In this work, the LHM is carried out by utilizing the finite-difference time-domain (FDTD) software that is based on computer simulation technology (CST) in order to collect the S-parameters that include the reflection coefficient (S_{11}) and the transmission coefficient (S_{21}). Figure 5.4(b) depicts the simulation configuration for the LHM unit cell. An electromagnetic (EM) wave traveling along the x-axis excites the testing component, unit cell, which is sandwiched between waveguide ports on the positive and negative axes. Figure 5.4(b) shows the simulation setup that defines a perfect electric conductor boundary condition along the y-axis walls, and a perfect magnetic conductor boundary condition along the z-axis walls. Incident waves propagate along the x-axis, their E-fields along the y-axis, and their H-fields along the z-axis. A frequency domain simulator was employed to do the simulations. A value of 50 ohms was considered for the normalized impedance. The simulation was conducted at frequencies between 2 and 16 GHz. Following the completion of the simulation, the S parameters were imported into the Math CAD program. Figure 5.5 illustrates a left-handed band with a transmission peak at 9.4 GHz.

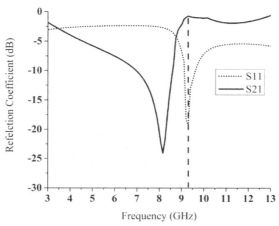

Figure 5.5. Simulated values of S-parameters for the unit cell depicted in Figure 5.3.

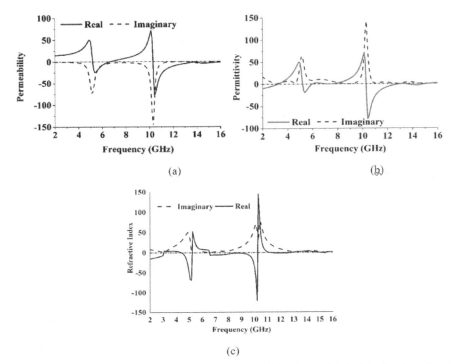

Figure 5.6. The analyzed effective properties like (a) permeability (b) permittivity (c) refractive index of the simulated unit cell.

Table 5.2. Negative index frequency region of the retrieved effective parameter.

Parameter	Negative index frequency region (GHz)
Permeability, μ	5.28–6.61, 10.31–13.92
Permittivity, ε	5.3–6, 7.27–7.37, 10.31–13.26
Refractive index, n	4.52–5.18, 6.61–10.22, 11.26–12.78

Comparing the proposed metamaterial magnetic response with conventional SRRs, the principal enhancement is the increased current, self-resonance, as well as overlapping responses. The retrieving algorithm that is described in [35, 36] is employed in order to acquire the fundamental significant parameters that are dependent on the transmission and reflection coefficient characteristics in order to validate the electromagnetic characteristics of the mentioned left-handed metamaterials. This is done so that the left-handed metamaterials can be used.

Figure 5.6 displays the recovered significant factors for the intended LHM unit cell, including its refractive index, permittivity, as well as permeability. Table 5.2 provides a summary of the frequency range with a negative refractive index. A different resonance bandwidth is observed in the negative refractive index frequency regions for the LHM unit cell in Table 5.2. This response shows that the LHM structures' parameter effectiveness has been increased in comparison to the LHMs reported in [2, 37–40], making it possible to achieve negative values throughout a large frequency range.

5.3.2 *Antenna geometry*

Using a unit cell as the radiating element is the key to the structural design of the proposed metamaterial antenna. The metamaterial antenna layout is depicted in Figure 5.7, both with a single element and with four individual elements. In terms of size, the metamaterial antenna measures 16 × 21 millimeters. The port supplies an impedance of 50 ohms to the radiator. The software application known as EM solver Computer Simulation Technology (CST) is used to model and optimize the metamaterial antenna construction. Figure 5.7 depicts the reflection coefficient using one unit cell element and four elements respectively. Both one-element and four-element antennas exhibit greater matching at upper resonance frequencies, as seen in Figure 5.8. The end goal is to acquire an Ultra-wideband frequency spectrum for a metamaterial loaded antenna that has the features of a negative index metamaterial so that it may be employed in microwave imaging areas.

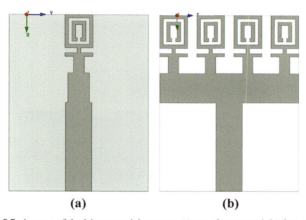

Figure 5.7. Layout of the Metamaterial antenna (a) one element and (b) four-element.

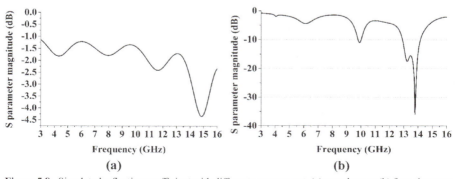

Figure 5.8. Simulated reflection coefficient with different arrangement, (a) one element (b) four elements.

Figure 5.9 provides a geometrical representation of the general layout of the proposed negative index metamaterial-based antenna. The substrate material for this antenna is affordable FR4, and it measures 1.6 millimetres in thickness and has dielectric properties (ϵ) of 4.6. Following the completion of the post-optimization process, the design parameters of the antenna are determined and are found in Table 5.3. The elements that comprise this antenna include a radiating patch made up of four-unit cells that are spaced regularly along the y-axis, a microstrip feeding line in the form of a trident, as well as a ground plane that is only partially covered. Every single unit cell is exactly the same as the others. These design strategies allow for the proposed MTM antenna to attain the increased bandwidth and radiation qualities that were mentioned before. Ground plane height, G_L, has a significant impact on the reflection coefficient, as seen in Figure 5.10(a). When G_L = 10 mm was used for the operational UWB frequency range, the best simulation results (reflection coefficient) were obtained for the presented negative index metamaterial antenna. A comparison of the reflection coefficients of partial ground planes, G_L = 10 mm, slotted partial ground planes, and etched-out antennas on top of slotted partial ground planes is

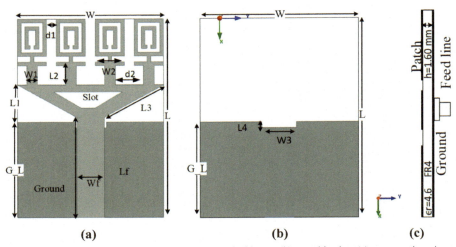

Figure 5.9. The proposed UWB antenna (a) geometrical layout (b) rare side view (c) cross section view.

Table 5.3. Antenna design specifications (according to Figure 5.8).

Parameter	Dimension (mm)	Parameter	Dimension (mm)
W	16	G_L	10
L	21	L3	7.37
d1	1.06	Wf	3
W1	1.2	Lf	10.5
L2	2	L4	0.6
d2	3.07	W3	3.5
L1	4	-	-

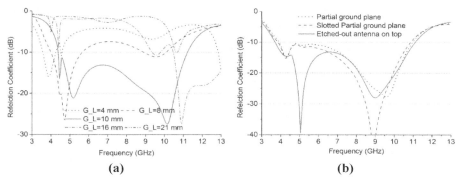

Figure 5.10. (a) The influence of ground plane height, G_L, on the reflection coefficient; (b) comparisons between the reflection coefficient of the partial ground plane, G_L = 10 mm, slotted partial ground plane, and etched-out antenna on top of the slotted partial ground plane.

shown in Figure 5.10(b). Figure 5.10(b) demonstrates that, in terms of the reflection coefficient, the suggested metamaterial antenna yields the best results when paired with a top-etched antenna.

5.3.3 Experimental validation

Using the Agilent Technologies N5230A PNA-L Network Analyzer at the Microwave Laboratory, Space Science Centre (ANGKASA), UKM Malaysia, the proposed metamaterial antenna was fabricated and experimented with to verify its ultra-wideband performance. Figure 5.11 shows the finalized antenna that was manufactured. UWB functionality has been validated, as shown in Figure 5.12(a). Following a comparison of the experimental reflection coefficients with the findings of the simulations employing CST and HFSS, a frequency range spanning 3.4–12.5 GHz was determined to have a voltage standing wave ratio (VSWR) of less than 2, and this frequency band was found to be optimal for the experiments. Manufacturing and soldering errors are probably responsible for the discrepancies that were found between both the simulated and experimental findings of the antenna. The estimated gain and estimated radiation efficiency of the proposed MTM antenna are shown in Figure 5.12(b). Figure 5.12(b) shows 88% radiation efficiency, 3.95 dBi average gain, and 5.16 dBi maximum gain.

For the proposed metamaterial antenna, observed radiation patterns are shown in Figure 5.13 at 4.0 and 6.0 GHz, as well as 9.0 and 12.0 GHZ in E-plane (xz-plane) and H-plane (yz-plane). Figure 5.13 reveals that, within the frequency range of 3.4–12.5 GHz, there are radiation patterns that are essentially omnidirectional. This is the desired outcome since it shows that the amount of cross-polarization is less than the degree of co-polarization.

5.3.4 Surface current distribution

Figure 5.14 illustrates the surface current distribution of the UWB negative index metamaterial that was proposed at four different frequencies: 4 GHz, 6 GHz, 9 GHz, and 12 GHz. Figure 5.14 shows that the x-axis is the predominant direction of flow for the currents. This flow implies that the antenna is positioned in an omnidirectional

Figure 5.11. A snapshot of the fabricated prototype including bottom and top view.

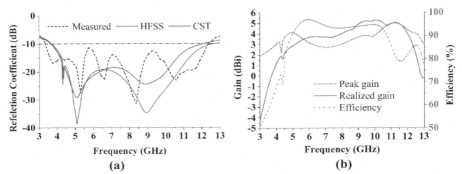

Figure 5.12. (a) Comparison between the simulated and measured reflection coefficient (b) measured gain and radiation efficiency of the proposed metamaterial UWB antenna.

manner. This figure additionally demonstrates that the circulation of currents at the metamaterial unit cell is robust due to the fact that it has a negative index frequency range.

5.3.5 Time domain performance

Signal-to-signal correlation coefficients show how much an antenna-induced pulse distortion is affecting the sent and received signals. Equation (5.1), is used to define the fidelity factor (FF) [41]:

$$F = \max_{\tau} \left| \frac{\int_{-\infty}^{+\infty} s(t) r(t-\tau)}{\sqrt{\int_{-\infty}^{+\infty} s(t)^2 \, dt . \int_{-\infty}^{+\infty} r(t)^2 \, dt}} \right| \tag{5.1}$$

where the pulses that are delivered and retrieved are denoted by $s(t)$ and $r(t)$, respectively. In order to prevent the loss of the data that has been modulated, radio impulses in ultrawide-band communications require a higher level of correlation between both the signals that are sent out and those that are received. In contrast, the fidelity factor (FF) is optional for the vast majority of the other types of communications systems. In addition to that, the time domain features of this designed antenna had been investigated and analyzed. Both a face-to-face as well as a side-by-side arrangement were considered as the two possible configurations.

Metamaterial Antennas for Ultra-wideband Applications 137

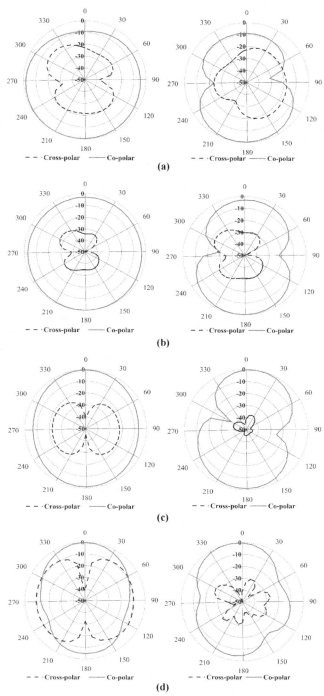

Figure 5.13. Measured radiation pattern of the proposed antenna at (a) 4 GHz (b) 6 GHz (c) 9 GHz and (d) 12 GHz.

138 *Metamaterial for Microwave Applications*

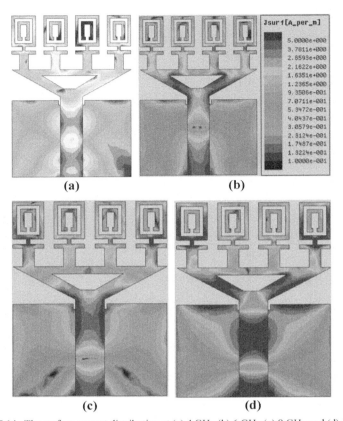

Figure 5.14. The surface current distribution at (a) 4 GHz (b) 6 GHz (c) 9 GHz and (d) 12 GHz.

Sending and receiving antennas were placed at a distance of 300 mm from each other, and the received pulse had been measured on the receiver side. The received and broadcast pulses are shown in their respective forms in Figure 5.15. Both the pulse that was received and the pulse that was broadcast were then normalized based on their highest values. This graph illustrates how there hardly any little pulse distortions in comparison to the highest value, which is 1. The face-to-face

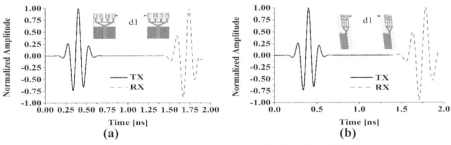

Figure 5.15. Transmitted and received pulses (a) side by side and (b) face-to-face.

configuration has a fidelity value of 0.87, while the side-by-side configuration has a fidelity factor of 0.79. As a consequence of this, the presented antenna is capable of operating with a narrow pulse that has little distortion.

5.4 Summary

Metamaterials and the applicability of this field of study are seen as a highly exciting and promising research area that is rapidly grabbing prominence. UWBs enable high-speed wireless devices with the portability and flexibility of wireless technology. Designing metamaterial antennae for ultra-wideband purposes is challenging since they have extra particular requirements over narrow band antennae. Additionally, metamaterial UWB antennas that are small and compact in profile are greatly in demand for a wide range of applications in portable devices, allowing for easy integration. So this study relies on designing and analyzing a UWB antenna based on planar metamaterials. EM solvers have been used to model the planar antenna based on metamaterials' principles. In addition, a PCB prototyping machine was used to fabricate the antenna so that it could be experimentally verified. This chapter investigates not just the frequency domain characteristics of the antenna, but also its time domain characteristics. Since the input and output signals are so closely related, the antenna's transfer function ensures that the received signal is as close to the input signal as possible, resulting in excellent fidelity. These qualities and their diminutive size qualify them as possibilities for prospective UWB application areas.

References

[1] Commission, F. C. 2002. *Federal Communications Commission Revision of Part 15 of the Commission's Rules Regarding Ultra-wideband Transmission System from 3.1 to 10.6 GHz*, 98–153. Washington, DC.
[2] Alhawari, A. R. H., A. Ismail, M. A. Mahdi and R. S. A. R. Abdullah. 2011. Miniaturized ultra-wideband antenna using microstrip negative index metamaterial. *Electromagnetics,* 31(6): 404–418.
[3] Nordin, M. A. W., M. T. Islam and N. Misran. 2013. Design of a compact ultrawideband metamaterial antenna based on the modified split-ring resonator and capacitively loaded strips unit cell. *Progress in Electromagnetics Research,* 136: 157–173.
[4] Antonino-Daviu, E., M. Cabedo-Fabres, M. Ferrando-Bataller and A. Valero-Nogueira. 2003. Wideband double-fed planar monopole antennas. *Electronics Letters,* 39(23): 1635–1636.
[5] Wu, X. H. and Z. N. Chen. 2005. Comparison of planar dipoles in UWB applications. *IEEE Transactions on Antennas and Propagation,* 53(6): 1973–1983.
[6] Ammann, M. and Z. N. Chen. 2003. A wide-band shorted planar monopole with bevel. *IEEE Transactions on Antennas and Propagation,* 51(40): 901–903.
[7] Choi, S. H., J. K. Park, S. K. Kim and J. Y. Park. 2004. A new ultra-wideband antenna for UWB applications. *Microwave and Optical Technology Letters,* 40(5): 399–401.
[8] Chung, K., H. Park and J. Choi. 2005. Wideband microstrip-fed monopole antenna with a narrow slit. *Microwave and Optical Technology Letters,* 47(4): 400–402.
[9] Liang, J., C. C. Chiau, X. Chen and C. G. Parini. 2005. Study of a printed circular disc monopole antenna for UWB systems. *IEEE Transactions on Antennas and Propagation,* 53(11): 3500–3504.
[10] Chen, Z. N., N. Yang, Y.-X. Guo and M. Y. W. Chia. 2005. An investigation into measurement of handset antennas. *IEEE Transactions on Instrumentation and Measurement,* 54(3): 1100–1110.
[11] Li, L.-W., Y.-N. Li, T. S. Yeo, J. R. Mosig, O. J. Martin et al. 2010. A broadband and high-gain metamaterial microstrip antenna. *Appl. Phys. Lett.,* 96(16): 164101.

[12] Han, X., H. Jin-Song, Z. Qing-Yi and J. Da-Lin. 2012. Compact ultra-wideband microstrip antenna with metamaterials. *Chinese Physics Letters*, 29(11): 114102.
[13] Bose, J. C. 1898. On the rotation of plane of polarisation of electric waves by a twisted structure. *Proceedings of the Royal Society of London*, 63(389-400): 146–152.
[14] Lamb, H. 1904. On group-velocity. *Proceedings of the London Mathematical Society*, 2(1): 473–479.
[15] Kock, W. E. 1946. Metal-lens antennas. *Proceedings of the IRE*, 34(11): 828–836.
[16] Kock, W. E. 1948. Metallic delay lenses. *Bell System Technical Journal*, 27(1): 58–82.
[17] Veselago, V. G. 1968. The electrodynamics of substances with simultaneously negative values of permittivity and permeability. *Physics-Uspekhi*, 10(4): 509–514.
[18] Pendry, J., A. Holden, D. Robbins and W. Stewart. 1999. Magnetism from conductors and enhanced nonlinear phenomena. *IEEE Transactions on Microwave Theory Techniques*, 47: 2075–2084.
[19] Shelby, R. A., D. R. Smith and S. Schultz. 2001. Experimental verification of a negative index of refraction. *Science*, 292(5514): 77–79.
[20] Houck, A. A., J. B. Brock and I. L. Chuang. 2003. Experimental observations of a left-handed material that obeys Snell's law. *Physical Review Letters*, 90(13): 137401.
[21] Simovski, C. R., P. A. Belov and S. He. 2003. Backward wave region and negative material parameters of a structure formed by lattices of wires and split-ring resonators. *IEEE Transactions on Antennas and Propagation*, 51(10): 2582–2591.
[22] Ozbay, E., K. Aydin, E. Cubukcu and M. Bayindir. 2003. Transmission and reflection properties of composite double negative metamaterials in free space. *IEEE Transactions on Antennas and Propagation*, 51(10): 2592–2595.
[23] Ziolkowski, R. W. and A. D. Kipple. 2003. Application of double negative materials to increase the power radiated by electrically small antennas. *IEEE Transactions on Antennas and Propagation*, 51(10): 2626–2640.
[24] Belrose, J. S. 1910. The sounds of a spark transmitter: telegraphy and telephony. *Adventures in CyberSound*.
[25] Fontana, D. R. J. 1966. *A Brief History of UWB Communications*. Multispectrum Solutions, ed: Inc.
[26] Ligthart, L. P. 2006. Antennas and propagation measurement techniques for UWB radio. *Wireless Personal Communications*, 37(3-4): 329–360.
[27] Ross, G. 1973. Transmission and reception system for generating and receiving base-band pulse duration pulse signals without distortion for short base-band communication system. ed: US Patent number 3728632.
[28] Barrett, T. W. 2000. History of ultrawideband (UWB) radar & communications: pioneers and innovators. pp. 1–42. *In: Proc. Progress in Electromagnetics Symposium*.
[29] Bennett, C. L. and G. F. Ross. 1978. Time-domain electromagnetics and its applications. *Proceedings of the IEEE*, 66(3): 299–318.
[30] Safarian, A. and P. Heydari. 2007. *Silicon-based RF Front-ends for Ultra Wideband Radios*. Springer Science & Business Media.
[31] Oppermann, I. 2004. The Role of UWB in 4G. *Wireless Personal Communications*, 29(1): 121–133.
[32] Smith, D. R., W. J. Padilla, D. Vier, S. C. Nemat-Nasser, S. Schultz et al. 2000. Composite medium with simultaneously negative permeability and permittivity. *Physical Review Letters*, 84(18): 4184.
[33] Tang, W. X., Q. Cheng and T. J. Cui. 2009. Electric and magnetic responses from metamaterial unit cells at Terahertz. *Terahertz Sci. Technol*, 2(1): 23–30.
[34] Li, L.-W., H.-Y. Yao, Q. Wu and Z.-N. Chen. 2006. Broad-bandwidth and low-loss metamaterials: theory, design and realization. *Journal of Zhejiang University Science A*, 7(1): 5–23.
[35] Chen, X., T. M. Grzegorczyk, B.-I. Wu, J. Pacheco Jr, J. A. Kong et al. 2004. Robust method to retrieve the constitutive effective parameters of metamaterials. *Physical Review E*, 70(1): 016608.
[36] Smith, D., D. Vier, T. Koschny and C. Soukoulis. 2005. Electromagnetic parameter retrieval from inhomogeneous metamaterials. *Physical Review E*, 71(3): 036617.

[37] Alici, K. B. and E. Ozbay. 2008. A planar metamaterial: Polarization independent fishnet structure. *Photonics and Nanostructures-Fundamentals and Applications*, 6(1): 102–107.
[38] Ekmekci, E. and G. Turhan-Sayan. 2009. Comparative investigation of resonance characteristics and electrical size of the double-sided SRR, BC-SRR and conventional SRR type metamaterials for varying substrate parameters. *Progress in Electromagnetics Research B*, 12: 35–62.
[39] Eleftheriades, G. V., A. K. Iyer and P. C. Kremer. 2002. Planar negative refractive index media using periodically LC loaded transmission lines. *IEEE Transactions on Microwave Theory and Techniques*, 50(12): 2702–2712.
[40] Huang, C., Z. Zhao, Q. Feng, J. Cui, X. Luo et al. 2010. Metamaterial composed of wire pairs exhibiting dual band negative refraction. *Applied Physics B*, 98(2-3): 365–370.
[41] Ojaroudi, N., M. Ojaroudi and Y. Ebazadeh. 2014. UWB/omni-directional microstrip monopole antenna for microwave imaging applications. *Progress in Electromagnetics Research C*, 47: 139–146.

Chapter 6

Flexible Metamaterials for Microwave Application

Mohammad Tariqul Islam[1,]* and *Md Atiqur Rahman*[2]

6.1 Introduction

The development of **flexible metamaterial** and **flexible antenna** is expanding with the improvement of several types of material substrates that can be easily incorporated with uninformed surfaces without losing their functional features. The progress of new materials has guided the development of several **microwave dielectric** substrates with remarkable features regarding efficiency, weight, flexibility and biodegradability. The selection of the dielectric substrate plays an important role on the metamaterial and antenna dimensions and performances such as impedance bandwidth, radiation efficiency and gain. The dielectric permittivity, loss tangent and thickness of substrate material has its significance to attain the specifications of different applications. Furthermore, advances in embedded software and digital signal processing have enabled modern personal devices to acquire many functions, such as voice and data transfer as well as positioning and entertainment. Along with this trend, advances have also occurred in the area of flexible material selection for use in metamaterials and antennas as substrates for wearable computers, military, medical telemetry, firefighters as well as emergency rescue teams' communication system applications [1–3].

At first, the nature of electromagnetic waves was quantitatively explained by world-famous scientist James Clerk Maxwell in the year 1873. However, the usefulness of radio waves in communications was separately demonstrated after twenty-two years by other two-world famous scientists Sir Jagadish Chandra

---

[1] Department of Electrical, Electronic and Systems Engineering, Universiti Kebangsaan Malaysia, Bangi, Selangor, Malaysia.
[2] Department of Electrical and Electronic Engineering, Dhaka University of Engineering & Technology, Gazipur, Bangladesh.
Email: atiq_eee@duet.ac.bd
* Corresponding author: tariqul@ukm.edu.my

Bose from India and Guglielmo Marconi from Italy in the year 1895. A few years later, prominent scientist Heinrich Hertz performed an experiment to confirm the presence of electromagnetic waves in free space. At present, in the 21st century, the development of the mobile phone and other devices marked a new era in electronics and wireless communication started to become a part of everyday life. Such progress was conditioned by advances in the technology of electronic component fabrication, which enabled the size of such components to be reduced, making devices convenient for the user. Since, antennas and in some cases, metamaterials are necessary parts of any wireless device, thus the techniques for reducing their size have consequently emerged in order to fulfil the demands of modern communication systems. Unlike electronic components, **antenna miniaturization** and flexibility with the volume, bandwidth, and efficiency are important.

As early as in 1993, the vision of so-called wearable computing was proposed at the Massachusetts Institute of Technology [4] as the next step in the development of personal electronic devices, while the first concrete steps towards its realization were taken in the late 1990s and the early years of the 21st century. Such a concept consisted of unifying the functions of various personal devices of that time (mobile phone, personal digital assistant PDA, pager, laptop, medical sensors, and more.). The philosophy of wearable computers was inspired by the concept of the cyborg (a man-computer symbiosis) from the popular literature in the 1960s [5]. Another idea present from the beginnings of wearable computers is to use everyday clothes and to integrate as much electronics as possible into them, which leads to the concept of so-called smart clothing. Such a concept has given rise to research looking for new textile materials that would support integration of various electronic components, while maintaining the primary function of clothing.

Alongside the trends in communication systems for the general public and research in **wearable computers**, communication between soldiers on the field is one of the primary needs for military teams, since having knowledge about the action on the ground and the position of other soldiers and military units, while at the same time being connected to the headquarters, clearly improves the effectiveness of the team. In addition, some future scenarios include the implementation of **flexible sensors** for monitoring the physiological condition of soldiers in any physical position such as, bending, rolling, so the action can, if needed, quickly be modified from the operating centre with regard to the situation on the field [6]. Such a communication network needs to be independent, since military teams often operate in environments where no other communication systems are present (such as forests, caverns, mountains, deserts and canyons).

The idea of remote patient monitoring has a history starting from the beginning of the 20th century when electrocardiography (EKG) was first introduced. In addition, pacemakers and cochlea implants are other examples of well-established and widely used implants in the body which have improved the quality of life for many. Moreover, the use of flexible metamaterial and antenna are increasing day by day in electronic products. Therefore, the need for a reliable flexible antenna structure that can establish links between implanted devices and external monitoring devices for constant patient monitoring. Moreover, the use of flexible metamaterial

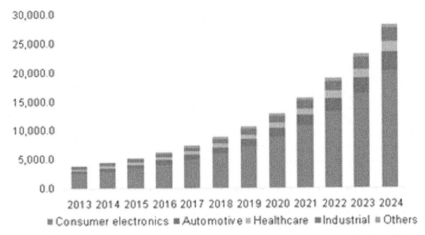

Figure 6.1. Forecasting of Asia Pacific flexible electronics market by application, 2013–2024 (USD Million).

and antenna are increasing day by day in electronics products to meet the future demand of consumer electronics' components in the automotive, medical, and healthcare, energy and power segments and other related fields. Thus, the demand for **flexible electronic gadgets** is increasing interestingly and according to Grand View Research the expected value of the global flexible electronic market will be USD 87.21 billion by 2024 as shown in Figure 6.1.

6.2 Importance of flexible substrate materials for microwave application

Despite having numerous benefits, conventional microstrip antennas have a few drawbacks, including a limited bandwidth, poor gain, low efficiency, a disproportionately large surface area, and inability to fit suitably with irregular surfaces. Increasing the substrate thickness is one way to increase bandwidth, however it has a negative impact on **radiation efficiency**. The conventional metamaterials also fabricated on rigid substrates cannot be fitted properly with modern microwave technologies' based wearable devices. Thus, huge attention has been paid on the development of flexible substrate materials to meet the future demands of flexible electronics' market components [7, 8]. During the preparation of new flexible microwave substrate materials, the researchers should keep in mind ways of overcoming the above mentioned drawbacks of conventional antennas and metamaterials.

Several flexible antennas have been reported by researchers such as paper based [9], textile, fluidic, **Kapton Polyimide** [10] and natural rubber [11] antennas. Flexible textile antennas are suitable for wearable and conformal applications but have some limitations like tendencies for fluid absorption and wrinkling. Paper based substrates are not robust enough and exhibit commencement of discontinuity in bent positions preventing their adaptation with equipment and integrated elements.

The flexible substrate of organic material based-on Kapton Polyimide has relatively poor electrical and physical properties. It also unable to offer low profile device dimensions, mechanical robustness and low loss tangent over a wide frequency range. In case of natural rubber based flexible antenna the bandwidth and return loss improve with an addition of rubber contents but the antenna efficiency and gain decrease.

Several flexible metamaterials also have been reported by the researchers such as flexible metamaterial-based sensors [12], flexible metamaterial-based absorbers [13] and flexible metamaterial-based antennae for energy harvesting [14]. Various types of metamaterials with different characteristics have been proposed, e.g., **negative refractive index**, and **magnetic conductivity**. These unusual properties play an important role in military, aerospace, solar power management, medical devices, and many other applications by providing better performance, more functions, and more flexibility. Even after decades of continuous effort in the research for metamaterials, more challenges are encountered on the road to real applications.

In the emerging field of wireless technology, a goal is to make flexible materials having special properties with respect to those of bulk or single-particle species. As a consequence in recent years, there has been an increasing interest in the synthesis and characterization of crystalline spinel oxide materials [15, 16]. This is preliminary because nanoparticles have a large surface-to-volume ratio leading to a dominance of surface behaviors over those in the interior of the particles [17]. Flexible materials offer a low dielectric constant and tangent loss, natural degradability, and easy fabrication. However, there is very little research that has been conducted on ferrite nanoparticle-based flexible materials as a substrate for microstrip patch antennae and metamaterial applications.

Most of the **conventional substrate materials** used for designing antennas and metamaterials have no flexibility, and their electromagnetic properties are also non-tunable. Hence, the conventional substrate materials are not suitable to fit with wearable microwave devices and the degree of freedom in designing metamaterials and antennae with them is very low. In the past, attempts were made to dope crystalline ferrite spinel with divalent and trivalent metal ions, and the effects of such substitutions on the structural, electric, and dielectric properties were reported [18]. Further, better optimization of the electrical and dielectric properties namely, dielectric constant and loss tangent values is important for the samples.

Considering all these problems, there is a strong demand to synthesize and characterize ferrite-based nanoparticles to develop new flexible microwave substrate materials with tunable **electromagnetic properties** for designing flexible metamaterials and antennae for microwave applications. The technique to design, analysis and develop new flexible microwave substrate materials may enlighten the scientific community with future directions to meet the growing demand in future **flexible electronics' markets**.

Typically, the microstrip antennas and metamaterials are fabricated on hard materials along with high fabrication costs and structural complexity, and their electromagnetic properties cannot be selected arbitrarily. Hence, the conventional substrate materials are not suitable to fit with wearable microwave devices and the degree of freedom in designing metamaterials and antennae with them is very low. Thus, flexible materials having adequate electromagnetic properties necessary for microwave applications. Several flexible materials, including polydimethylsiloxane, liquid crystal polymers, organic materials, liquid metal, fabric, and paper, have recently attracted a lot of interest in the context of current wireless communication systems. Since flexible material may be rolled up around any arbitrary form to get the desirable outcomes; flexible material technology is important, particularly in the **biomedical** and communication sectors.

For applications in wireless communications, many studies have been reported on the simulated performances of flexible metamaterials and antennae. Metamaterials are transformed materials and structures with unique characteristics not found in regular materials. These unusual properties play an important role in military, aerospace, solar power management, medical devices, and many other applications by providing better performance, more functions, and more flexibility. However, the microstrip patch antenna inherently offers narrow bandwidths and produces catastrophic radiation characteristics while operating at higher frequencies. With the technological drive towards portability and multiple services out of a single module, the design of the patch antenna should be light weight, small in size and should possess decent performance characteristics such as high gain, wider bandwidth, and satisfactory **radiation patterns.**

Recently, research and development in the realm of implantable devices is receiving a lot of attention. In the 21st century, people are getting more accustomed to a fast-paced lifestyle. Therefore, a variety of applications, such as **flexible displays**, health monitoring systems, and flexible sensors, have witnessed considerable advancement with flexible technology. Flexible antennas can be developed into thin, light weight, low profile structures with the ability to be installed to conformal surfaces more readily than conventional antennas made of rigid materials. Therefore, this research focuses on the preparation of flexible material used for compact metamaterial and antenna structures, which may achieve the necessary increased bandwidth, high gain, improved efficiency, and acceptable radiation properties.

In this chapter, sol-gel synthesized ferrite nanoparticle based flexible microwave substrates are prepared. The necessary design, analysis and fabrication techniques for flexible antennas with flexible metamaterials for some partial microwave applications are presented which may help future researchers in this field. In addition, due to the remarkable progress in wireless technology, at present, affordable wireless communication services have become a reality within a compact flexible electronic device, and day by day the demand for compactness and efficient performances is expected. As a result, flexible metamaterials and antennas are still a buzz topic for continuing research.

6.3 Development of flexible substrate material

With the advent of electromagnetic technologies and their massive applications, the **oxide-based flexible microwave composites** have significant potential as they offer excellent physical and chemical properties. The flexible composite materials are in high demand for microwave communication systems because they have superior properties such as ease of synthesis, and low fabrication costs. They contain several organic and inorganic materials, polydimethylsiloxane, liquid metals, and liquid crystal polymers. Metal composite-based materials are given more consideration than other composites; they contain Ca, Zn, Mg, Fe, or Ag particles randomly dispersed in porous ferrite hosts and synthesized to achieve negative electromagnetic properties in the range of radiofrequency applications. They have broadly been used in the application of **energy harvesting**, electromagnetic absorbers and/or antenna design, filter design and meta-composites with negative permittivity. In recent decades, there has been a surge in interest for flexible materials that offer lightweight, high attenuation intensity, flexible metamaterials, and thin flexible antennas having wide bandwidths [19].

In the fields of physics, chemistry, and material science, the function that metal oxides play is particularly essential. Among them metal oxides having general formula AB_2O_4, crystallize into the cubic close packed (CCP) structure of oxygen ions in which the divalent ions occupy tetrahedral A-sites and trivalent ions occupy octahedral B-sites, respectively. They have a crystal structure related to that of the ferrite mineral AFe_2O_4 (where, A = Ca, Mg, Co, Mn, Zn), which is a compound of a group known as spinel [20]. Certain groups of spinel compounds show interesting magnetic, mechanical, catalytic, and electrical properties. These properties have led to the technological importance of **spinel oxide materials**.

The basis for numerous technological applications of spinel oxides is related to their **structural flexibility**, because various properties of these materials can be modified suitably by changing the chemical composition of the compound through cation redistribution or by substitution with a suitable dopant [21]. The chemical relationship between the compounds becomes clearer, while considering the valences of the atomic species involved in these binary oxide ceramics. For example, the mineral spinel is $A^{2+}Fe^{3+}O_4^{2-}$ and it turns out that a wide range of divalent cations may substitute A^{2+} at the tetrahedral A-sites. These substitutions can be used to derive a series of ferrites, which constitute a remarkable group of spinel compounds exhibiting fascinating physical behaviour and thus provide possibilities for a wide range of practical applications [22]. Moreover, due to their immobilized characteristics, ferrite-based spinels possess high electrical resistivity, low dielectric constant, and loss tangent.

6.3.1 Synthesis of $Mg_xZn_{(1-x)}Fe_2O_4$ nanoparticles

The $Mg_xZn_{(1-x)}Fe_2O_4$ **nanoparticles** were synthesized using the sol-gel chemical process. All raw ingredients, including magnesium nitrate, zinc nitrate, iron nitrate, and others, were bought in hydrate form from SIGMA-ALDRICH, USA. A precision balance was used to weigh the required amounts of nitrates, which were then dissolved in 50 ml of distilled water according to the x values (x = 20, 40, 60, and 80) in a beaker. The beaker was put on top of a magnetic hot plate stirrer (Cimarec^{+TM} Hotplate, Thermo Fisher Scientific) and stirred continuously at 90 degrees Celsius to make a homogeneous solution. As a **chelating agent**, citric acid was also added to the solution. By evaporating some water, the solution became a reddish gel after four hours, and the nitrates bonded together. The solution was then dried in an oven for

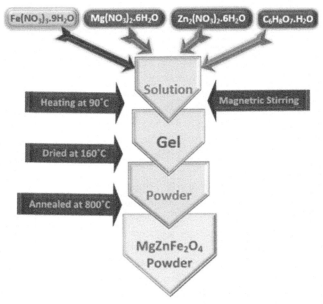

Figure 6.2. Sol-gel synthesis flow diagram of $Mg_xZn_{(1-x)}Fe_2O_4$ nanoparticles.

one hour at 160 degrees Celsius. The dried chemicals were then hand crushed with a marble mortar pestle. The pulverized powders were then transferred to crucible pots and calcined for one hour at 800 degrees Celsius in MTI Corporation's 1200 C Muffle Furnace. As a result, $Mg_xZn_{(1-x)}Fe_2O_4$ nanoparticles were prepared as shown in Figure 6.2.

6.4 Characterization of $Mg_xZn_{(1-x)}Fe_2O_4$ nanoparticles

The structural and morphological properties of the proposed ferrite-based nanoparticles are characterized through XRD, FESEM, and TEM analysis. The tunable electromagnetic properties are also investigated by **Dielectric Assessment Kit (DAK)** and **Vibrating Sample Magnetometer (VSM).**

6.4.1 Structural analysis

The XRD patterns of the samples were recorded with a Siemens D8 Advance X-ray diffraction meter as shown in Figure 6.3. An X-ray diffraction meter usually consist of three basic elements: an X-ray tube, a sample holder, and a X-ray detector. The X-ray tube emitted an X-ray onto the sample placed on a sample holder at an angle in the range 0°~90° and the reflected X-ray was detected through the X-ray detector. Then using the Bragg's Law [23], the authors could predict when diffraction would actually take place. The Bragg's condition for the constructive interference of the scattered X-ray from the successive atomic planes formed by the crystal lattice of the material is given by,

$$2d \sin \theta = n\lambda \tag{6.1}$$

where 'λ' is the wavelength, 'd' is distance between the interatomic planes, 'θ' is the glancing angle and 'n' is an integer representing the order of diffraction. The angle of incidence 'θ' is equal to the angle of reflection. The crystallite size 'D' was calculated from the dominant peak of the XRD pattern according to the Scherrer equation [24].

$$D = \frac{k\lambda}{B \cos \theta} \tag{6.2}$$

where, 'λ' is the X-ray wavelength (1.54059 A°), 'B' is the full width at the half height of the peak in radians and 'θ' is the diffraction angle of the peak. The average crystallite size at each annealing temperature of ferrite nanoparticles was calculated within the range of 0°~90° (2θ) from the broadening of the XRD lines.

Firstly, to validate the phase development of $Mg_xZn_{(1-x)}Fe_2O_4$, a Siemens D500 **X-ray diffractometer** with a Cu K anode (40 kV, 20 mA) at a **2 Theta angle range** of 20° to 80° was employed, and the recorded XRD plots for the synthesised samples is shown in Figure 6.4.

With all the samples, i.e., Mg_{20}, Mg_{40}, Mg_{60}, and Mg_{80}, the spinel structure of $Mg_xZn_{(1-x)}Fe_2O_4$ is recognized and indexed with a principal peak of (311). The overall **diffraction peaks** are found at 2 Theta = 32° (220), 35° (331), 37° (311), 44° (400), 55° (442), 58° (551), 65° (440), and 77° (533). From the line width of the (311) reflection, the Scherrer's equation [36] is used to derive the average **crystallite size** of $Mg_xZn_{(1-x)}Fe_2O_4$ nanoparticles.

Where, λ is the X-ray wavelength (1.54060 nm), is the full width at half maximum values in radians, and θ is the **diffraction angle** corresponding to the most intense reflection plane of (311) in radians. The nano spinel crystallite sizes for all the samples are calculated from the above equation and the values are 27.45 nm for Mg_{20}, 25.75 nm for Mg_{40}, 23.84 nm for Mg_{60}, and 22.17 nm for Mg_{80}, which are extremely near to the tabular values published by S. B. Somvanshi et al.

Figure 6.3. Photograph of the Bruker D8 Advance XRD machine.

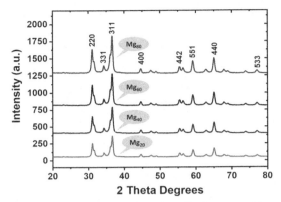

Figure 6.4. XRD patterns of $Mg_xZn_{(1-x)}Fe_2O_4$ nanoparticles.

[35]. The following equation was used to compute the lattice constraint (a) using Miller indices ($h\ k\ l$) and interplanar spacing (d) values corresponding to the leading peak (311):

$$a = d\sqrt{h^2 + k^2 + l^2} \tag{6.3}$$

The assessed values of the lattice constraint are $a = 8.433$ for Mg_{20}, $a = 8.425$ for Mg_{40}, $a = 8.415$ for Mg_{60}, and $a = 8.405$ for Mg_{80}, which are extremely comparable to those reported in [35]. Table 6.1 contains the computed values of lattice constant (a) and average crystallite size (D), and the compositional change of the lattice constant and average crystallite size in relation to Mg^{2+} substitution. Figure 6.5(a) shows that when the proportion of Mg^{2+} content x increases, the values of the lattice constant (a) and crystalline size decrease. This event occurred as a result of Vegard's law [36]. With the assistance of the lattice constant (a), the volume (V) of the unit cell of the synthesised samples may be calculated as follows:

$$V = a^3 \tag{6.4}$$

Since the values of the unit cell volume are exactly proportional to the values of the lattice parameter, they follow the same pattern as the values of the lattice constant. As a result, calculating the X-ray density factor is required to describe materials. As a result, the following equation can be used to calculate the X-ray density (d_X):

$$d_x = \frac{Z \times M}{V \times N_A} \tag{6.5}$$

where, Z is the cubic **lattice coordination number**, M is the molecular weight of individual concentrations, V is the unit cell volume, and N_A is the Avogadro's number ($N_A = 6.022 \times 10^{-23}$). Figure 6.5(b) depicts the fluctuation in **X-ray density** (d_X) and bulk density (d_B) with varied compositions. With a rise in Mg^{2+} ions, the values of the X-ray density (d_X) and the bulk density (d_B) are found to drop. It might be related to the manufactured samples' **molecular weight** reduction. Finally, using the

Table 6.1. Values of Lattice parameter (*a*), and Average crystallite size (*D*) for prepared $Mg_xZn_{(1-x)}Fe_2O_4$ samples.

Mg Concentration (%)	a (Å)	V (Å³)	D (nm)	d_x (gm/cm³)	d_B (gm/cm³)	P (%)
20	8.433	599.72	27.45	5.306	3.627	31.64
40	8.425	598.01	25.75	5.103	3.577	29.91
60	8.415	595.88	23.84	4.925	3.513	28.66
80	8.405	593.76	22.17	4.747	3.412	28.13

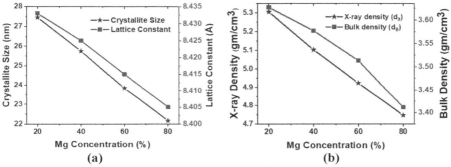

Figure 6.5. Variation of (a) crystallite size and lattice constant and (b) X-ray density and bulk density of the prepared $MgZnFe_2O_4$ samples.

equation below, the percentage porosity (*P*) is determined using the values of the X-ray density (d_X) and the bulk density (d_B):

$$P = 1 - \frac{d_B}{d_X} \quad (6.6)$$

Table 6.1 also includes the computed X-ray density (d_X) and bulk density (d_B), as well as the percentage 5 Variation (a) crystallite size and lattice constant and (b) X-ray density and bulk density of the prepared $MgZnFe_2O_4$ samples with different Mg concentrations.

6.4.2 Morphological analysis

The **structural morphology** of the synthesized powder was then observed using FESEM (Carl Zeiss Supra 55 VP). Figure 6.6 shows the image and operational structure of the scattered **field emission scanning electron microscope (FESEM)** mentioned in this chapter. Among the morphological characteristics that can be determined are the composition, shape and size of the grains through observed image of the FESEM using electron energy radiation to study objects at a fine scale.

The shape of the produced nanoparticle is investigated using the FESEM (Field Emission Scanning Electron Microscopy) method. Mg_{20}, Mg_{40}, Mg_{60}, and Mg_{80} FESEM images and **particle size histograms** are shown in Figure 6.7 and Figure 6.8. The $Mg_xZn_{(1-x)}Fe_2O_4$ species were ground to nanocrystals in a high-energy ball mill EMAX (Retsch, Germany), which enables quicker grinding with a maximum rotational speed of 2000 rpm and produces unique sized particles.

152 *Metamaterial for Microwave Applications*

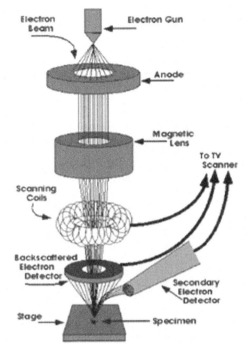

Figure 6.6. Working principle of FESEM.

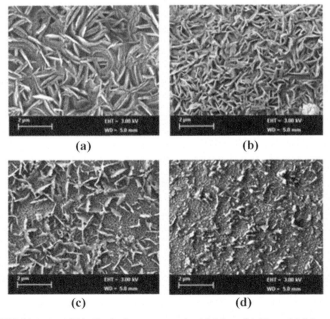

Figure 6.7. FSEM images of Mg-Zn ferrite nano powder (a) Mg_{20}, (b) Mg_{40}, (c) Mg_{60}, and (d) Mg_{80} respectively.

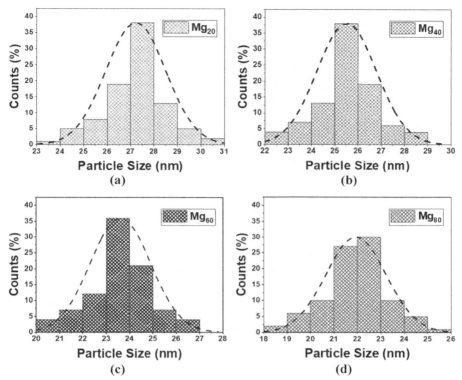

Figure 6.8. Particle size histogram of Mg-Zn ferrite nano powder (a) Mg_{20}, (b) Mg_{40}, (c) Mg_{60}, and (d) Mg_{80} respectively.

The mean grain size of the produced nanoparticles is around 27.30 nm for Mg_{20}, 25.60 nm for Mg_{40}, 23.70 nm for Mg_{60}, and 22.02 nm for Mg_{80}, which are all in excellent agreement with the XRD results. It is also seen that with an increasing value of Mg content the **grain size** and porosity decrease and offer tuneable behaviours.

6.4.3 Dielectric properties analysis

The prepared flexible substrates were analysed using **Vector Network Analyzer (VNA)** along with the DAK 3.5 measurement kit as shown in Figure 6.9. The water is used to calibrate for nullifying the **measurement error**. The Sweep command is sent to the VNA through the DAK 3.5 software and observations of dielectric constant, loss tangent and electrical conductivity are taken by it over the frequency range of 200 MHz to 20 GHz. The measurement method is fast and non-destructive to the material under test. The total system has,

a. A DAKS-3.5 Dielectric Probe.
b. Portable 1-port VNA.
c. DAK-3.5/1.2 Short.
d. DAK-3.5/1.2 MSS.

Figure 6.9. DAK 3.5 Dielectric measurement kit.

e. Thermometer.
f. DAK verification kit.
g. Probe stand.

The flexible substrates are placed under the **dielectric probe** and the required data related to dielectric parameters is collected by the VNA. By post-processing the data using DAK 3.5 software, the parameters are calculated.

The prepared, flexible substrates' dielectric constants and loss tangents are measured at a frequency range of 4–10 GHz with the DAK 3.5 (200 MHz to 20 GHz) dielectric assessment kit, manufactured by Schmid & Partner AG, Switzerland, and the plots are shown in Figure 6.10(a–b), respectively. The values of the dielectric constants and loss tangents change when the applied frequency on the specimen under test increases, which is compatible with **Koop's phenomenological theory** and the **Maxwell-Wagner model** of interfacial polarization [37, 38]. The computed **relative permittivity** values are 6.01 for Mg_{20}, 5.10 for Mg_{40}, 4.19 for Mg_{60}, and 3.28 for Mg_{80}, respectively, while the loss tangents values are 0.002 for Mg_{20}, 0.004 for Mg_{40}, 0.006 for Mg_{60}, and 0.008 for Mg_{80}.

With increasing Mg concentration, the porosity and value of the dielectric constants decline from 6.01 (Mg_{20}) to 3.28 (Mg_{80}), while the values of **loss tangents** rise. This difference is due to grain size changes in the synthesized nanoparticles, which can also be seen in XRD and FESEM studies. Thus, the produced $Mg_xZn_{(1-x)}Fe_2O_4$ composites' dielectric characteristics may be modified by adjusting the molar ratios, which is highly useful when predetermined or arbitrary values of various dielectric properties are needed. Finally, since the produced, flexible substrates based on $Mg_xZn_{(1-x)}Fe_2O_4$ have low relative permittivity (< 15) and extremely low loss tangent, they are suitable for application as microwave dielectric materials.

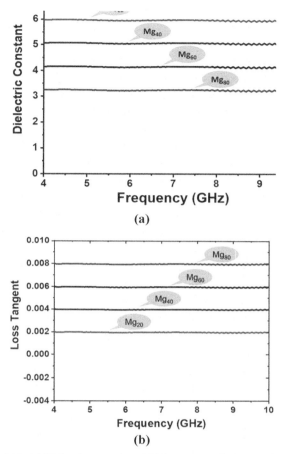

Figure 6.10. (a) Dielectric constant, and (b) Loss tangent (T_δ) versus frequency.

6.4.4 Optical and photoluminescence analysis

The UV–Vis **spectrophotometer** was used to record the absorption spectra of UV–Vis rays as a function of wavelength (300 nm–800 nm) in order to determine the optical bandgap energy (E_g) parameter of the prepared $Mg_xZn_{(1-x)}Fe_2O_4$ samples. The most common "Tauc plot" [$(h\nu)^2$ v/s (E_g)] was drawn using UV–Vis absorbance spectral data to calculate the optical bandgap energy (E_g) values of the prepared $Mg_xZn_{(1-x)}Fe_2O_4$ samples. The "Tauc plot" was drawn using the following formula, and the results are shown in Figure 6.11(a).

$$\alpha = \frac{A(h\nu - E_g)^{\frac{n}{2}}}{h\nu} \tag{6.7}$$

The tangent drawn at the X-axis of the "Tauc plots" was used to establish the **optical bandgap energy (E_g) parameters**, as illustrated in Figure 6.11(a). The values of 'E_g' were discovered to be between 2.20 and 2.36 eV. The manufactured

156 *Metamaterial for Microwave Applications*

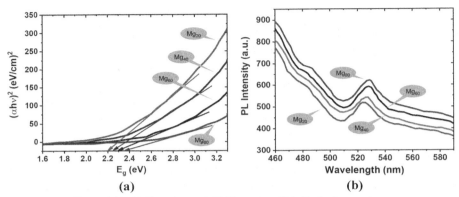

Figure 6.11. (a) Tauc plots and (b) PL spectra of Mg-Zn ferrite samples.

samples' **photoluminescence (PL) spectra** were recorded to investigate their photoluminescent capabilities and to look for inherent flaws as well as other polluted entities. The PL spectra collected at 300 K are shown in Figure 6.11(b). All the measurements were taken using a wavelength of 400 nm to excite all the samples. A distinctive near band-edge emission (NBE) was seen in all the samples in the wavelength range of 524 nm–530 nm.

6.4.5 Magnetic properties analysis

To verify the proposed nanoparticles' microwave applicability, it is also very important to know their magnetic characteristics, such as the magnetization (M)-magnetic field (H) hysteresis loop, permeability, and magnetic loss tangent. The values of M (emu/g) and H (Oe) were determined using Superconducting quantum interference vibrating sample manometer (SQUID-VSM) as shown in Figure 6.12 [25]. Moreover, the values of the permeability and magnetic loss tangent of the

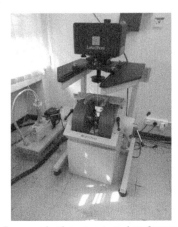

Figure 6.12. Photograph of Superconducting quantum interference vibrating sample manometer (SQUID-VSM), UKM.

Flexible Metamaterials for Microwave Application 157

prepared samples were estimated by post-processing the scattering parameters S_{11} and S_{21} using the Nicolson-Ross-Weir method [26, 27]. The scattering parameters S_{11} and S_{21} are collected from the vector network analyzer by placing the prepared substrate materials between two waveguides.

The fluctuation in magnetization M (emu/g) of $Mg_xZn_{(1-x)}Fe_2O_4$ ferrite nanoparticles with an applied magnetic field H (kOe) are recorded through superconducting quantum interference vibrating sample manometer (SQUID-VSM) at room temperature and the **M-H hysteresis loops** are illustrated in Figure 6.13(a). All the ferrite nanoparticle samples had narrow loops with practically no hysteresis, showing that the ferrite nanoparticles were a **soft magnetic material**. Figure 6.13(a) also demonstrates that when the Mg content increases, the maximum magnetization

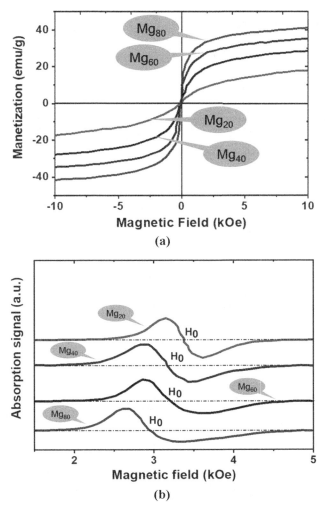

Figure 6.13. (a) Hysteresis loops of magnetization against magnetic field (M-H) and (b) ferromagnetic resonance (FMR) spectra of Mg-Zn ferrite samples.

of the $Mg_xZn_{(1-x)}Fe_2O_4$ nanoparticles also increases. Even when the applied magnetic field was 10 kOe, the magnetic hysteresis loops of the Mg–Zn ferrite nanoparticles did not attain the complete saturation stage. Spinel ferrite nanoparticles often have this property. It's due to the existence of a spin-disordered layer on nanoparticle surfaces, which necessitates a strong magnetic field to saturate, as well as influence the ultrafine ferrite particles' size. Figure 6.13(b) shows the **ferromagnetic resonance (FMR)** spectra of four Mg-Zn ferrite nanoparticle samples. The resonance field was discovered, and the values were varied in-between 2.65 and 3.20 kOe. It is seen that the **magnetic properties** also vary with the variation of material concentration. Hence, tunable magnetic properties can be archived by varying the **compositional ratio** of the materials. Thus, the overall magnetic analysis suggested that the prepared Mg–Zn ferrite nanoparticles developed can be a suitable candidate for microwave applications. This material may be capable of reducing the human body's influence on wearable devices.

6.5 Flexible metamaterial design technique

6.5.1 Metamaterials on $Mg_xZn_{(1-x)}Fe_2O_4$ nanoparticles-based flexible substrate

Two Mg-Zn ferrite nanoparticle-based flexible substrates, namely Mg_{40} and Mg_{60}, are developed by mixing **Polyvinyl Alcohol (PVA)** glue, -and a unit cell of metamaterials is built on top of them. Figure 6.14(a), shows the geometrical aspects of the designed unit cell. This unit cell comprises of two **SRR (split-ring resonators)** on the front side of the flexible substrate, which are primarily responsible for determining resonances, similar to the LC resonator structure [28]. Subwavelength criteria were used to choose the physical dimensions of the metamaterial unit cell, which range from $\lambda/10$ to $\lambda/4$ [29]. The inductance (L) for the LC resonator corresponding to the metamaterial unit cell is determined by the dimensions of the metallic strips, while the capacitance (C) is determined by the diameters of the split gaps. Each of the split gaps is considered as a capacitor and the metallic strips are considered as an inductor to develop the unit cell model in Figure 6.14(a) and its equivalent circuit is shown in Figure 6.14(b). For example, the capacitor C1 is assumed to correspond to the gap g1, whereas the inductor L1 is assumed to correspond to the outer resonator's L-shaped bottom metallic strip. Similarly, the capacitor C5 is assumed to correspond to the gap g2, while the inductor L3 is assumed to correspond to the outer resonator's L-shaped top metallic strip. The inductor L2 is represented by the stair-shaped connecting metallic strips between the outer and inner resonators, while the gaps g3 and g4 between them are represented by the capacitors C3 and C4. The dielectric gap S2 between the margins of the outer resonators and the substrate ends is represented by the capacitor C2. The Table 6.2 lists the dimensions of the various components of the metamaterial unit cell. The metamaterial unit cells are made up of conductive components with splits/gaps and dielectric materials in the gaps between the conductive elements. When an electromagnetic field is applied to the conductive components, they operate as inductors (L), and the splits act as capacitors (C). These

Flexible Metamaterials for Microwave Application 159

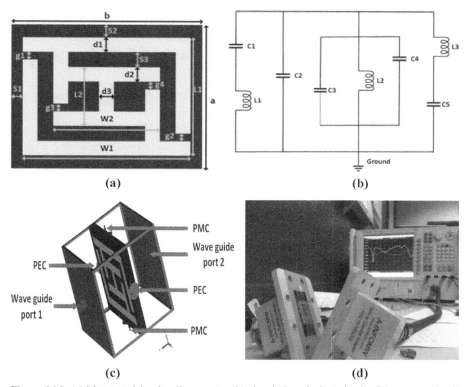

Figure 6.14. (a) Metamaterial unit cell geometry, (b) electrical equivalent circuit of the proposed unit cell, (c) simulation view with boundary conditions, and (d) measurement setup with the network analyzer.

Table 6.2. Geometrical dimensions of the proposed unit cell.

Dimension	Length (mm)	Dimension	Length (mm)
a	10	L1	8
b	13	L2	4
W1	11	s1, s2, s3, d1, d2, d3	1
W2	7	g1, g2, g3, g4	0.5

inductors and capacitors are LC resonators, and their resonance frequencies (f) may be calculated using the following equation [30]:

$$f = \frac{1}{2\pi\sqrt{LC}} \quad (6.8)$$

The following equation from **quasi-state theory** is used to calculate the value of the effective capacitance produced between the metal distances or splits:

$$C(F) = \varepsilon_0 \varepsilon_r \frac{A}{d} \quad (6.9)$$

where, ε_r is the **relative permittivity**, A is the area formed by the conductive gap, and d is the distance between the conductive gaps, and $\varepsilon_0 = 8.854 \times 10^{-12}$ F/m is the permittivity in free space. The transmission line concepts may be used to calculate the effective values of inductances as follows [31]:

$$L(nH) = 2\times 10^{-4} l \left[\ln\left(\frac{l}{w+t}\right) + 1.193 + 0.02235\left(\frac{w+t}{l}\right)\right] K_g \quad (6.10)$$

$$K_g = 0.57 - 0.145 \ln \frac{w'}{h'} \quad (6.11)$$

where Kg is the correction factor, w' is the substrate width, and h' is the substrate thickness. The length and width of the microstrip line are indicated by l and w, respectively, while the thickness of the microstrip line is denoted by t.

Figure 6.14(c) depicts the CST simulation perspective of wave propagation. The **S-parameters** were measured using a vector network analyzer (VNA) brought from Agilent, Santa Clara, CA, USA to extract the effective parameters of metamaterial unit cells, as shown in Figure 6.14(d). The X-walls are configured as ideal electric boundaries in the CST simulation setup, whereas the Y-walls are configured as perfect magnetic boundaries. In both the positive and negative Z-directions, an electromagnetic wave is applied. The TEM (transverse electric and magnetic) mode of operation was used to explore the polarization characteristics.

Two waveguide ports of the desired frequency range were coupled to ports 1 and 2 of the vector network analyzer in the experimental setup, and the metamaterial unit cells were put in-between these waveguides, as illustrated in Figure 6.14(d). The S-parameters were then measured, and the effective parameters were calculated using the Nicolson-Ross-Weir (NRW) technique [32].

Therefore, extractions of effective permittivity, permeability and refractive index parameters are performed according to the NRW method given as follows [33]. The equations for reflection coefficient (S_{11}) and transmission coefficient (S_{22}) are:

$$S_{11} = \left\{ \frac{R_1 \left(1 - e^{-2j\theta}\right)}{1 - R_1^2 e^{-2j\theta}} \right\} \quad (6.12)$$

$$S_{21} = \left\{ \frac{e^{-2j\theta}\left(1 - R_1^2\right)}{1 - R_1^2 e^{-2j\theta}} \right\} \quad (6.13)$$

where, 'R_1' is the **normal impedance**, 'j' is the imaginary operator, and 'θ' is the **polar angle**. If 'V_1 & V_2' are arbitrarily chosen vectors for simplification, they can be expressed as:

$$V_1 = S_{21} + S_{11} \quad (6.14)$$

$$V_2 = S_{21} - S_{11} \quad (6.15)$$

The equations for the effective permittivity (ε_r) and the effective permeability (μ_r) can be written as:

$$\varepsilon_r \approx \frac{2}{jk_0 d} \times \left\{\frac{1-V_1}{1+V_1}\right\} \quad (6.16)$$

$$\text{or } \varepsilon_r \approx \frac{c}{j\pi fd} \times \left\{ \frac{1 - S_{21} - S_{11}}{1 + S_{21} + S_{11}} \right\} \quad (6.17)$$

$$\mu_r \approx \frac{2}{jk_0 d} \times \left\{ \frac{1 - V_2}{1 + V_2} \right\} \quad (6.18)$$

$$\text{or } \mu_r \approx \frac{c}{j\pi fd} \times \left\{ \frac{1 - S_{21} + S_{11}}{1 + S_{21} - S_{11}} \right\} \quad (6.19)$$

where 'd' is the thickness of the dielectric substrate material, 'k_0' is the **wave vector**, and 'c' is the velocity of light. Thus, refractive index n_r can be written as:

$$\eta \approx \sqrt{\mu_r \varepsilon_r} \quad (6.20)$$

$$\text{or } \eta_r \approx \frac{c}{j\pi fd} \times \sqrt{\left\{ \frac{(S_{21} - 1)^2 - S_{11}^2}{(S_{21} + 1) - S_{11}^2} \right\}} \quad (6.21)$$

An electrical circuit that is **analogous** to the proposed metamaterial unit cell is shown in Figure 6.14(b). The inductors L1–L3 are used to counteract conductive parts, whereas the capacitors C1–C5 are used to counteract conductive gaps. In addition, the **Advance Design System (ADS)** simulator was used to determine the values of the inductances and capacitances indicated in Figure 6.15(a–b) for Mg_{40} and Mg_{60}, respectively. From these electrical equivalent circuits, the S-parameters corresponding to the proposed metamaterial unit cells were also further analysed.

Figure 6.16 represents the significant steps of **flexible substrates** and then metamaterial unit cell preparation from the obtained Mg-Zn ferrite nanoparticles. The flexible substrates are prepared by adding 10 ml PVA (polyvinyl alcohol) glue per gram of Mg-Zn nanoparticles and applying necessary heat to the sticky mixer. Lastly, the unit cell of the metamaterials was printed using **copper sputtering** on these Mg_{40} and Mg_{60} flexible substrates.

6.5.2 Metamaterial measurement method

At first, it is necessary to measure the S-parameters (both S_{11} and S_{21}) of the proposed metamaterials to determine the permittivity, **permeability**, absorption, and **refractive index**. A commercially available Vector Network Analyzer (VNA) (model: Agilent PNA Network Analyzer N5227A) is used to measure the S-parameters. Two coaxial cables are connected with the VNA at one end, and a waveguide-to-coaxial adapter is connected at the other end of each cable. Before starting any measurement, an electronic calibration module (model: Agilent N4694-60001) is first connected at the ends of both cables for the calibration required for the desired operation. The number of sweeping frequencies is set on the VNA as per the number of frequency samples found from the **CST simulator**. When the setup is ready after calibration, the waveguide ports are attached with the free end of the cables. The photograph of the calibration kit and a sample measurement setup is shown in Figure 6.17.

The metamaterial unit cell is placed in-between the waveguides and an EM wave is applied to measure the S-parameters. The values of the measured S-parameters are

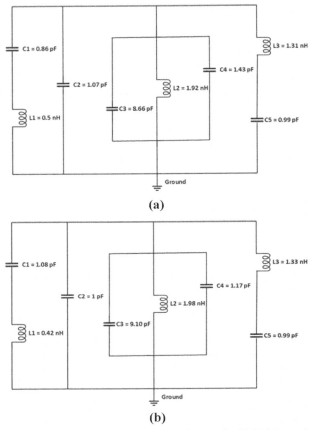

Figure 6.15. Electrical equivalent circuit obtained from ADS with (a) Mg_{40} and (b) Mg_{60}.

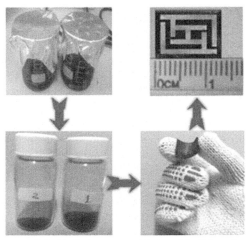

Figure 6.16. Flow diagram of Mg-Zn ferrite nanoparticle-based flexible metamaterial unit cell design.

Figure 6.17. Metamaterial unit cell measurement setup.

collected from the VNA. After that, this data is post-processed by the **MATLAB software** by loading a program related to the **NRW (Nicolson-Ross-Weir) method** discussed earlier to extract the values of the absorptions, permittivity, permeability and refractive index of the metamaterial unit cell. The MATLAB software was further used to export the data in the excel files with an appropriate code to calculate the absorptions, permittivity, permeability and refractive index to other methods for cross-checking.

6.5.3 Performance of flexible metamaterial with $Mg_xZn_{(1-x)}Fe_2O_4$ nanoparticles with Mg_{40}

The reflection parameter (S_{11}) and transmission parameter (S_{21}) for the metamaterial with Mg_{40} determined from the CST simulation is shown in Figure 6.18(a). The Nicolson-Ross-Wire approach is used to extract the effective parameters from the simulated data, and the associated graphs are drawn in Figure 6.18(b). Later, the transmission parameter (S_{21}) was measured using VNA and justified by ADS analysis. Figure 6.18(c) shows the simulated, measured, and ADS-obtained transmission (S_{21}) coefficients. During the simulation, four resonances were discovered at distinct frequencies and bandwidths: 2.92 GHz (S-band), 4.42 GHz (C-band), 7.75 GHz (C-band), and 9.84 GHz (X-band). On the other hand the measured resonances occurred at 2.90 GHz (S-band), 4.34 GHz (C-band), 7.50 GHz (C-band), and 9.57 GHz (X-band) were found to be much closer. During the measurement at 11.72 GHz (X-band), an extra resonance was discovered (X-band).

Table 6.3 lists the negative values of the effective electromagnetic parameters. Negative values of effective permittivity were discovered at numerous frequency ranges of 2.85~3.85 GHz, 4.45~7.55 GHz, 7.70~8.45 GHz, and 10.15~12 GHz, while negative values of **effective permeability** were found at 2.35~3.15 GHz, 3.65~4.75 GHz, 5.80~7.80 GHz, and 7.95~12 GHz, as shown in Table 6.3. As both the **effective permittivity** and effective permeability had negative values at 2.85~3.35 GHz, 4.45~5.15 GHz, 6.65~6.80 GHz, and 8.50~12 GHz, they resulted in a negative refractive index. Thus, the metamaterial unit cells on flexible Mg_{40} were explored for **double-negative (DNG) characteristics** throughout a broad frequency range of 2.85~3.15 GHz, 4.45~4.75 GHz, 7.70~7.80 GHz, and 10.15~12 GHz.

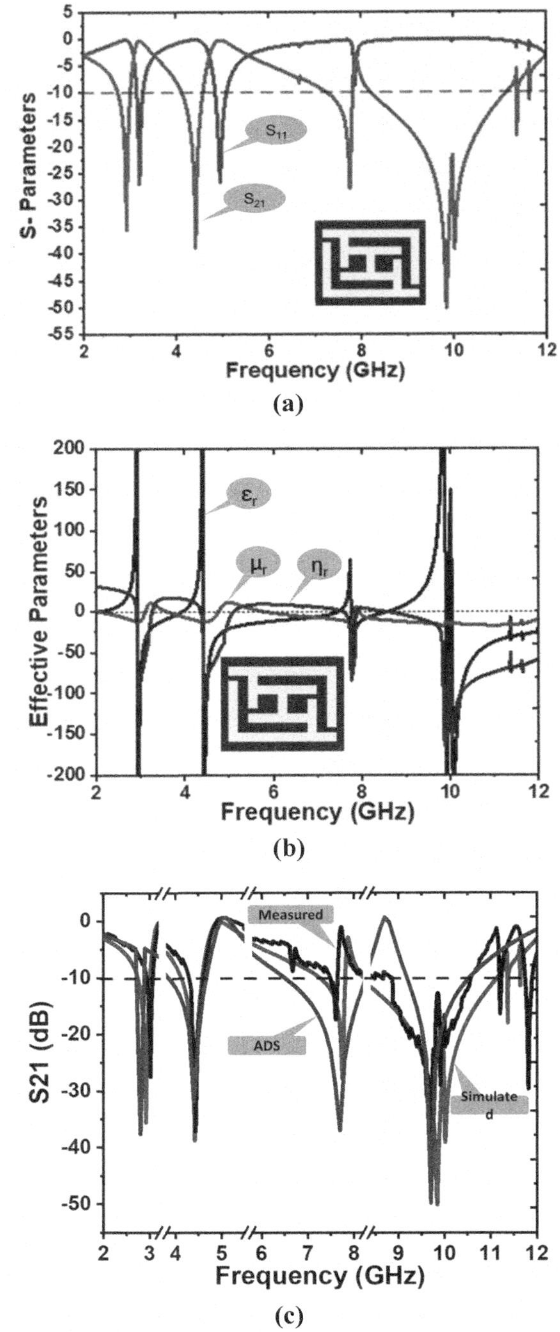

Figure 6.18. The amplitude of the planned metamaterial unit cell with Mg$_{40}$ (a) simulated S-parameters, (b) extracted effective parameters, and (c) simulated, measured, and ADS derived S21.

Flexible Metamaterials for Microwave Application   165

Table 6.3. Negative electromagnetic region of several effective parameters with Mg$_{40}$.

Effective Electromagnetic Parameters	Negative Bandwidth (GHz)	DNG Region (GHz)
Permittivity (ε_r)	2.85 ~ 3.85, 4.45 ~ 7.55, 7.70 ~ 8.45, 10.15 ~ 12	2.85 ~ 3.15, 4.45 ~ 4.75, 7.70 ~ 7.80, 10.15 ~ 12
Permeability (μ_r)	2.35 ~ 3.15, 3.65 ~ 4.75, 5.80 ~ 7.80, 7.95 ~ 12	
Refractive Index (n_r)	2.85 ~ 3.35, 4.45~5.15, 6.65~6.80, 8.50 ~ 12	

6.5.4 Electromagnetic properties analysis of the flexible metamaterial with Mg$_{40}$

The **electric field (E), magnetic field (H), and surface current distributions** for the metamaterial with Mg$_{40}$ at various resonances are shown in Figure 6.19(a–c), respectively, to examine electromagnetic characteristics. Waveguide port excitations cause the generation of electric and magnetic fields on the unit cell's inductors and capacitors. Electromagnetic wave propagation is caused by the interaction of these electric and magnetic forces. The fluctuations of the electric (E) and magnetic (H) fields can be represented using Maxwell's curl equations in terms of electric field density (D), magnetic field density (B), and surface current density (J):

$$\nabla \times H = J + \frac{\partial D}{\partial t} \tag{6.22}$$

$$\nabla \times E = -\frac{\partial B}{\partial t} \tag{6.23}$$

$$\nabla = \left[\frac{\partial}{\partial x}, \frac{\partial}{\partial y}, \frac{\partial}{\partial z}\right] \tag{6.24}$$

$$B = \mu_0 \mu H \tag{6.25}$$

$$D = \varepsilon_0 \varepsilon E \tag{6.26}$$

The relative permeability and permittivity of the medium, are **frequency-dependent variables.** As a result, the **electric field density (D)** and **magnetic field density (B)** change with the frequency, and the electric field intensity (E) and magnetic field intensity (H) change as well, according to the equations above. Figure 6.19(a–c) shows how the E-field, H-field, and surface current distributions change with frequency.

It's worth noting that the electric field is maximum when the magnetic field is non-existent or negative, implying that they are inversely related. The magnetic field strength is precisely proportional to the surface current density, as shown in Figure 6.19(a–c). As a result, when the density of the surface current is maximum, the magnetic field strength is observed to be maximum also. The magnetic field also travels in the same direction as the current. The resonances at lower frequencies are

166 *Metamaterial for Microwave Applications*

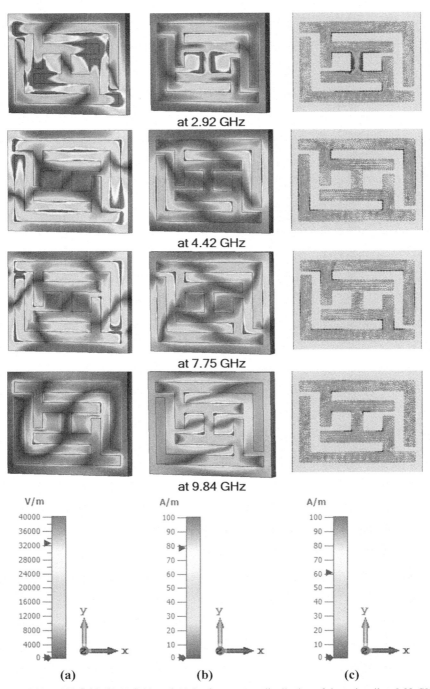

Figure 6.19. (a) E-field, (b) H-field, and (c) Surface current distribution of the unit cell at 2.92 GHz, 4.42 GHz, 7.75 GHz, and 9.84 GHz with a 40 percent concentration of Magnesium (Mg$_{40}$).

caused by the elements of the metamaterial unit cell with a bigger size, while the resonances at higher frequencies are caused by the elements of the metamaterial unit cell with a smaller size. The electric fields are concentrated around the components that cause the resonances at the corresponding frequencies. With an increase in applied frequencies, the strength of the electric field will shift from the bigger components of the unit cell to the smaller elements of the unit cell. The magnetic field strength and surface current density are maximum in the reversal region, where the electric field intensity is weakest, and the magnetic flux density of the conductors decreases further. The Biot–Savart law can be used to describe the relationship between the magnetic flux density (B) and the surface current (I) at a distance r from the conductor:

$$B = \frac{\mu I}{2\pi r} \qquad (6.27)$$

6.5.5 *Electromagnetic field interaction of the metamaterial properties with Mg_{40}*

The proposed DNG unit cell is put between two DPS media to examine the electromagnetic field interaction between a **double-positive (DPS) medium** and a double-negative (DNG) medium, as illustrated in Figure 6.20. Air with a positive permittivity and permeability is considered as the DPS medium. The electromagnetic wave is applied to the whole structure in the z-direction. The E-field interaction at various **resonances** is shown in Figure 6.21, and the equivalent H-field interaction is shown in Figure 6.22 across the DPS and DNG media. The E and H fields are largely concentrated at the DNG medium, with modest spreading in the DPS medium, as shown in these figures.

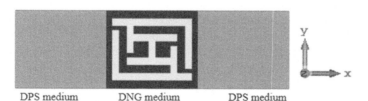

Figure 6.20. The electromagnetic wave propagation between the DNG and DPS mediums was investigated using this simulation model.

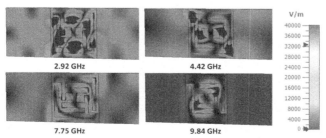

Figure 6.21. E-field, and (b) H-field interaction between the DPS and DNG mediums.

168 *Metamaterial for Microwave Applications*

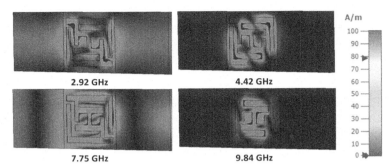

Figure 6.22. H-field interaction between the DPS and DNG mediums.

6.5.6 Performance analysis of flexible metamaterial properties with Mg_{60}

The **reflection parameter** (S_{11}) and **transmission parameter** (S_{21}) for the metamaterial with Mg_{60} determined from the CST simulation are shown in Figure 6.23(a). The Nicolson-Ross-Weir approach is used to extract the effective parameters from simulated data, and the associated graphs are displayed in Figure 6.23(b). Later, the transmission parameter (S_{21}) was measured using VNA and ADS analysis was used to justify it. Figure 6.23(c) shows the simulated, measured, and ADS-obtained transmission S_{21} coefficients. At various frequencies and bandwidths, four resonances have been detected. During simulation, these frequencies were 3.16 GHz (S-band), 4.71 GHz (C-band), 8.21 GHz (X-band), and 10.52 GHz (X-band). The measured resonances occurred at 3.17 GHz (S-band), 4.63 GHz (C-band), 8.11 GHz (X-band), and 10.21 GHz (X-band) and were found to be much closer. Table 6.4 also lists the negative values of the effective electromagnetic parameters. It is seen that the negative effective permittivity values can be found at several frequency ranges, including 3.15~4.25 GHz, 4.75~7.85 GHz, 8.30~8.90 GHz, and 10.55~12 GHz, whereas negative effective permeability values can be found at 2.50~3.45 GHz, 4.10~5.05 GHz, 6.25~8.40 GHz, and 8.45~12 GHz. At 3.15~3.50 GHz, 4.75~5.20 GHz, 8.25~8.40 GHz, and 10.45~12 GHz, both effective permittivity and permeability have negative values that resulted in negative values of the refractive index. Thus, the DNG properties of **metamaterial unit cells** on flexible Mg_{60} are explored across a large frequency range of 3.15~3.45 GHz, 4.75~5.05 GHz, 8.30~8.40 GHz, and 10.55~12 GHz.

Table 6.4. Negative electromagnetic region of several effective parameters with Mg_{60}.

Effective Electromagnetic Parameters	Negative Bandwidth (GHz)	DNG Region (GHz)
Permittivity (εr)	3.15 ~ 4.25, 4.75 ~ 7.85, 8.30 ~ 8.90, 10.55 ~ 12	3.15 ~ 3.45, 4.75 ~ 5.05, 8.30 ~ 8.40, 10.55 ~ 12
Permeability (μr)	2.50 ~ 3.45, 4.10 ~ 5.05, 6.25 ~ 8.40, 8.45 ~ 12	
Refractive Index (nr)	3.15 ~ 3.50, 4.75 ~ 5.20, 8.25 ~8.40, 10.45 ~ 12	

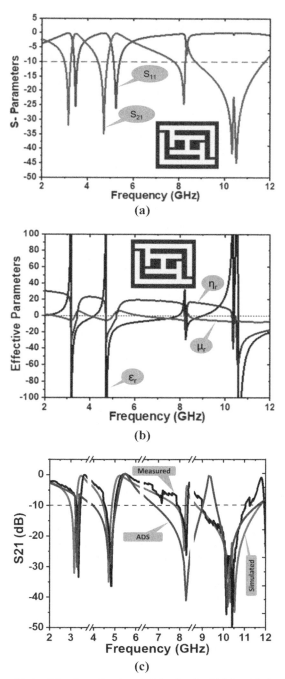

Figure 6.23. The amplitude of the planned metamaterial unit cell with Mg$_{60}$ (a) simulated S-parameters, (b) extracted **effective parameters**, and (c) simulated, measured, and ADS derived S21.

170   Metamaterial for Microwave Applications

6.5.7 Comparison of $Mg_xZn_{(1-x)}Fe_2O_4$ nanoparticles-based proposed flexible metamaterials with Mg_{40} and Mg_{60}

An overview of fabricated flexible metamaterial unit cells employing Mg_{40} and Mg_{60} substrate is given in Table 6.5, which includes the following information: (i) overall dimension, (ii) **electrical dimension**, (iii) **EMR (effective medium ratio)**, (iv) substrate material composition, (v) operational frequency band, (vi) resonance frequencies, (vii) kind of metamaterials developed, and (viii) microwave band of applications.

Table 6.5. Summarized features of the metamaterials with Mg_{40} and Mg_{60}.

Parameters	Flexible Metamaterial with Mg_{40}	Flexible Metamaterial with Mg_{60}
Physical Dimensions	13×10 mm^2	13×10 mm^2
Electrical Dimensions	$0.127 \lambda \times 0.097 \lambda$	$0.137 \lambda \times 0.10 \lambda$
EMR	7.90	7.30
Materials Compositions	Magnesium zinc ferrite ($Mg_{0.4}Zn_{0.6}Fe_2O_4$)	Magnesium zinc ferrite ($Mg_{0.6}Zn_{0.4}Fe_2O_4$)
Operating Frequency	2.75 ~ 11.16 GHz	2.98 ~ 11.75 GHz
Resonance Frequency	2.92 GHz, 4.42 GHz, 7.75 GHz, and 9.84 GHz	3.16 GHz, 4.71 GHz, 8.21 GHz, and 10.52 GHz
Type of Metamaterial	Double Negative	Double Negative
Applications	S-, C-, and X-Band	S-, C-, and X-Band

6.5.8 Comparison of $Mg_xZn_{(1-x)}Fe_2O_4$ nanoparticles-based proposed flexible metamaterials with existing metamaterials

Table 6.6 shows a brief comparison of the proposed flexible metamaterials with the existing metamaterials. Although a flexible metamaterial is proposed with $NiAl_2O_4$-based materials its size is bigger (25×20 mm^2) with a narrow bandwidth and a low EMR value of 3.75 [39]. Another flexible metamaterial based on $NiAl_2O_4$ nanoparticles with a smaller size of 12.5×10 mm^2 has also been suggested. Even yet, its EMR value (2.88) is relatively low, and it only provides a dual-band operation [40]. Moreover a metamaterial has also been developed with a smaller surface area (10×8 mm^2), however it only has a single operating band with a low EMR value of 3.40, has no flexibility, and no polarisation independence [41]. Besides, a meta atom is reported with a modest dimension (9×9 mm^2) on Rogers RT 5880, however its EMR value (5) is likewise insufficient and has no flexibility [42]. A left-handed metamaterial has also been fabricated on a FR4 substrate. Its size (5×5 mm^2) is small, but the EMR value (4.95) is low, and it only provides dual-band operations [43]. Overall, our proposed metamaterial is innovative in that it proposes polarization-independent metamaterials with multiband double-negative properties, having great flexibility, and low weight. They also have tunable microwave dielectric characteristics, with small electrical dimensions, and a lower EMR (effective medium ratio). Furthermore, they are capable of improving metamaterial performance for the S-, C-, and X-bands of microwave regimes.

Table 6.6. Comparison of some current literature with proposed metamaterials.

Ref. No.	No. of Resonance	EMR	Material Type	Band	Observations
Rahman et al. 2018	7	3.75	Flexible	S-, C-, X	• Dimension: large • Bandwidth: narrow • EMR value: low
Faruque et al. 2019	2	2.88	Flexible	X-, Ku	• Number of bands: two • EMR value: very low
Ahamed et al. 2019	1	3.40	Rigid	X	• Polarization: dependent • Flexibility: no • EMR value: low
Hasan et al. 2017	3	5.00	Rigid	C-, X-, Ku	• Flexibility: no • EMR value: poor
Hossain et al. 2014	2	5.55	Rigid	S-, C	• Number of bands: two • Flexibility: no • EMR value: very poor
Liu et al. 2016	2	4.95	Rigid	C-, X	• Number of bands: two • Flexibility: no • EMR value: low
Proposed	4	7.30 (Mg_{40}) 7.90 (Mg_{60})	Flexible	S-, C-, X	• Flexibility: high • Weight: low • Electrical dimension: low • EMR value: high • Polarization: independent

6.6 Summary

Development of flexible communication systems is most significant for the future and the biggest challenge for meeting the upcoming demand for flexible electronic antennas. An unavoidable part of such systems is flexible antennas and metamaterials, which are required to ensure transfer of signals, depending on the application. This chapter deals with the preparation of ferrite nanoparticle-based flexible microwave substrate materials for developing flexible metamaterials and antennas. The $Mg_xZn_{(1-x)}Fe_2O_4$, $Co_xZn_{(0.90-x)}Al_{0.10}Fe_2O_4$ is synthesized using the sol-gel chemical process with different concentrations and characterized through **X-ray diffraction (XRD)** and scanning electron microscopy (SEM) analysis. The tunable electromagnetic properties of the ferrite-based nanoparticles are investigated with a Dielectric Assessment Kit (DAK) and **Vibrating Sample Magnetometer (VSM)**. A detailed explanation of these preparation procedures, dielectric and magnetic characteristics as well as the way to control or moderate these properties on demand and electromagnetic performances are given. After fruitful investigations, a set of metamaterial and antenna prototypes have been fabricated, and their performances in microwave applications have been tested. The flexible substrate is prepared by adding $Mg_xZn_{(1-x)}Fe_2O_4$ nanoparticles with PVA glue and these offer tunable dielectric permittivity values that vary from 3.28 to 6.10 and loss tangents from 0.002 to 0.008. A double negative (DNG) metamaterial is developed on $Mg_xZn_{(1-x)}Fe_2O_4$ nanoparticle-based flexible substrate and its electromagnetic properties are investigated. This metamaterial offers better performances over the conventional FR4 and Rogers RO4533 substrates and covers the S-, C-, and X-bands of microwave regimes. The overall results ensure that the ferrite nanoparticle-based flexible microwave substrates offer tunable electromagnetic properties and are suitable for developing flexible metamaterials for flexible microwave technology. The dielectric and magnetic properties of the synthesized materials are varied with the variation of the compositional ratio of the materials as well as the synthesis parameters. So, the necessary electromagnetic parameters can be achieved by maintaining synthesis conditions and compositional ratios. The ferrite-based materials can also reduce human body influences on wearable devices in the case of wireless body area network (WBAN) applications. Thus, ferrite-based flexible materials can be potential candidates for meeting the demands for flexible electronic communication systems instead of conventional materials.

References

[1] Kirtania, S. G., A. W. Elger, M. R. Hasan, A. Wisniewska, K. Sekhar et al. 2020. Flexible antennas: a review. *Micromachines,* 11: 847.
[2] Kim, S.-J., M. Seong, H.-W. Yun, J. Ahn, H. Lee et al. 2018. Chemically engineered Au–Ag plasmonic nanostructures to realize large area and flexible metamaterials. *ACS Applied Materials & Interfaces,* 10: 25652–25659.
[3] Thangaselvi, E. and K. Meena alias Jeyanthi. 2019. Implementation of flexible denim nickel copper rip stop textile antenna for medical application. *Cluster Computing,* 22: 635–645.
[4] Amft, O. and P. Lukowicz. 2009. From backpacks to smartphones: Past, present, and future of wearable computers. *IEEE Pervasive Computing,* 8: 8–13.
[5] Starner, T. 2001. The challenges of wearable computing: Part 1. *Ieee Micro,* 21: 44–52.
[6] Kaija, T., J. Lilja and P. Salonen. 2010. Exposing textile antennas for harsh environment. pp. 737–742. In: *2010-Milcom 2010 Military Communications Conference.*
[7] Zhang, H., Y. Lan, S. Qiu, S. Min, H. Jang et al. 2021. Flexible and stretchable microwave electronics: past, present, and future perspective. *Advanced Materials Technologies,* 6: 2000759.
[8] Alqadami, A. S., N. Nguyen-Trong, B. Mohammed, A. E. Stancombe, M. T. Heitzmann et al. 2019. Compact unidirectional conformal antenna based on flexible high-permittivity custom-made substrate for wearable wideband electromagnetic head imaging system. *IEEE Transactions on Antennas and Propagation,* 68: 183–194.
[9] Aziz, A. A. A., A. T. Abdel-Motagaly, A. A. Ibrahim, W. M. El Rouby, M. A. Abdalla et al. 2019. A printed expanded graphite paper based dual band antenna for conformal wireless applications. *AEU-International Journal of Electronics and Communications,* 110: 152869.
[10] Wang, Z., L. Qin, Q. Chen, W. Yang, H. Qu et al. 2019. Flexible UWB antenna fabricated on polyimide substrate by surface modification and *in situ* self-metallization technique. *Microelectronic Engineering,* 206: 12–16.
[11] Ruslan, A. A., S. Y. Mohamad, N. F. A. Malek, S. H. Yusoff, S. N. Ibrahim et al. 2020. Design of flexible microstrip patch antenna using rubber substrate for brain tumor detection. pp. 1–5. In: *2020 IEEE Student Conference on Research and Development (SCOReD).*
[12] Zhang, X., K. Chen, Y.-S. Lin and B.-R. Yang. 2021. Flexible metamaterial nanograting for biosensing application. *Optical Materials,* 122: 111679.
[13] Ha, D. T., B. S. Tung, B. X. Khuyen, T. S. Pham, N. T. Tung et al. 2021. Dual-band, polarization-insensitive, ultrathin and flexible metamaterial absorber based on high-order magnetic resonance. In: *Photonics,* p. 574.
[14] Elwi, T. A. 2020. A further realization of a flexible metamaterial-based antenna on nickel oxide polymerized palm fiber substrates for RF energy harvesting. *Wireless Personal Communications,* 115: 1623–1634.

[15] Thakur, P., D. Chahar, S. Taneja, N. Bhalla, A. Thakur et al. 2020. A review on MnZn ferrites: Synthesis, characterization and applications. *Ceramics International*, 46: 15740–15763.
[16] Kumar, S., M. Kumar and A. Singh. 2022. Synthesis and characterization of iron oxide nanoparticles (Fe_2O_3, Fe_3O_4): a brief review. *Contemporary Physics*, pp. 1–21.
[17] Siegel, R. W. 1994. What do we really know about the atomic-scale structures of nanophase materials? *Journal of Physics and Chemistry of Solids*, 55: 1097–1106.
[18] Almessiere, M., Y. Slimani, A. Trukhanov, A. Sadaqat, A. D. Korkmaz et al. 2021. Review on functional bi-component nanocomposites based on hard/soft ferrites: structural, magnetic, electrical and microwave absorption properties. *Nano-Structures & Nano-Objects*, 26: 100728.
[19] Xu, F., D. Zhang, Y. Liao, F. Xie and H. Zhang. 2019. Dispersion of LiZnTiBi ferrite particles into PMDS film for miniaturized flexible antenna application. *Ceramics International*, 45: 8914–8918.
[20] Manohar, A., V. Vijayakanth, S. P. Vattikuti and K. H. Kim. 2022. A mini-review on AFe_2O_4 (A= Zn, Mg, Mn, Co, Cu, and Ni) nanoparticles: Photocatalytic, magnetic hyperthermia and cytotoxicity study. *Materials Chemistry and Physics*, p. 126117.
[21] Kapoor, P. N., A. K. Bhagi, R. S. Mulukutla and K. J. Klabunde. 2004. Mixed metal oxide nanoparticles. *Dekker Encycl. Nanosci. Nanotechnol.*, pp. 2007–2015.
[22] Katoch, G., G. Rana, M. Singh, A. García-Peñas, S. Bhardwaj et al. 2021. Recent advances in processing, characterizations and biomedical applications of spinel ferrite nanoparticles. *Ferrite: Nanostructures with Tunable Properties and Diverse Applications*, 112: 62–120.
[23] Elton, L. and D. F. Jackson. 1966. X-ray diffraction and the Bragg law. *American Journal of Physics*, 34: 1036–1038.
[24] Tatarchuk, T., M. Myslin, I. Mironyuk, M. Bououdina, A. T. Pędziwiatr, R. Gargula et al. 2020. Synthesis, morphology, crystallite size and adsorption properties of nanostructured Mg–Zn ferrites with enhanced porous structure. *Journal of Alloys and Compounds*, 819: 152945.
[25] Shemerliuk, Y., Y. Zhou, Z. Yang, G. Cao, A. U. Wolter, B. Büchner et al. 2021. Tuning magnetic and transport properties in quasi-2D (Mn1–xNix) 2P2S6 single crystals. *Electronic Materials*, 2: 284–298.
[26] Saini, A., A. Thakur and P. Thakur. 2016. Matching permeability and permittivity of $Ni_{0.5}Zn_{0.3}Co_{0.2}In_{0.1}Fe_{1.9}O_4$ ferrite for substrate of large bandwidth miniaturized antenna. *Journal of Materials Science: Materials in Electronics*, 27: 2816–2823.
[27] Narang, S. B. and K. Pubby. 2016. Single-layer & double-layer microwave absorbers based on Co–Ti substituted barium hexaferrites for application in X and Ku-band. *Journal of Materials Research*, 31: 3682–3693.
[28] Schurig, D., J. Mock and D. Smith. 2006. Electric-field-coupled resonators for negative permittivity metamaterials. *Applied Physics Letters*, 88: 041109.
[29] Yoo, M., H. K. Kim and S. Lim. 2015. Angular-and polarization-insensitive metamaterial absorber using subwavelength unit cell in multilayer technology. *IEEE Antennas and Wireless Propagation Letters*, 15: 414–417.
[30] Gay-Balmaz, P. and O. J. J. J. o. a. p. Martin. 2002. Electromagnetic resonances in individual and coupled split-ring resonators. *Journal of Applied Physics*, 92: 2929–2936.
[31] Hoque, A., M. Tariqul Islam, A. F. Almutairi, T. Alam, M. Jit Singh et al. 2018. A polarization independent quasi-TEM metamaterial absorber for X and ku band sensing applications. *Sensors*, 18: 4209.
[32] Luukkonen, O., S. I. Maslovski and S. A. Tretyakov. 2011. A stepwise Nicolson–Ross–Weir-based material parameter extraction method. *IEEE Antennas and Wireless Propagation Letters*, 10: 1295–1298.
[33] Chen, X., T. M. Grzegorczyk, B.-I. Wu, J. Pacheco Jr, J. A. Kong et al. 2004. Robust method to retrieve the constitutive effective parameters of metamaterials. *Physical Review E*, 70: 016608.
[34] Rahman, S., K. Nadeem, M. Anis-ur-Rehman, M. Mumtaz, S. Naeem et al. 2013. Structural and magnetic properties of ZnMg-ferrite nanoparticles prepared using the co-precipitation method. *Ceramics International*, 39: 5235–5239.
[35] Somvanshi, S. B., M. V. Khedkar, P. B. Kharat and K. Jadhav. 2020. Influential diamagnetic magnesium (Mg^{2+}) ion substitution in nano-spinel zinc ferrite ($ZnFe_2O_4$): thermal, structural, spectral, optical and physisorption analysis. *Ceramics International*, 46: 8640–8650.
[36] Ashcroft, N. and A. Denton. 1991. Vegard's law. *Phys. Rev. A*, 43: 3161–3164.

[37] Koops, C. 1951. On the dispersion of resistivity and dielectric constant of some semiconductors at audiofrequencies. *Physical Review*, 83: 121.
[38] Wagner, K. W. 1913. Zur theorie der unvollkommenen dielektrika. *Annalen der Physik*, 345: 817–855.
[39] Rahman, M. A., E. Ahamed, M. R. I. Faruque and M. T. Islam. 2018. Preparation of $NiAl_2O_4$-based flexible substrates for metamaterials with negative dielectric properties. *Scientific Reports*, 8: 1–13.
[40] Faruque, M. R. I., E. Ahamed, M. A. Rahman and M. T. Islam. 2019. Flexible nickel aluminate ($NiAl_2O_4$) based dual-band double negative metamaterial for microwave applications. *Results in Physics*, 14: 102524.
[41] Ahamed, E., M. R. I. Faruque, M. F. B. Mansor and M. T. Islam. 2019. Polarization-dependent tunneled metamaterial structure with enhanced fields properties for X-band application. *Results in Physics*, 15: 102530.
[42] Hasan, M. M., M. R. I. Faruque and M. T. Islam. 2017. Multiband left handed biaxial meta atom at microwave frequency. *Materials Research Express*, 4: 035015.
[43] Liu, S.-H., L.-X. Guo and J.-C. Li. 2016. Left-handed metamaterials based on only modified circular electric resonators. *Journal of Modern Optics*, 63: 2220–2225.

Chapter 7
Microwave Head Imaging and 3D Metamaterial-inspired Antenna

Mohammad Tariqul Islam* and *Mohammad Shahidul Islam*

7.1 Introduction

In developed nations, strokes are the second most common cause of death and disability. It is essential to remember that every year, over 16 million individuals suffer from **strokes** globally, of whom 6 million pass away and another 6 million get crippled [1–3]. Each year, around 796,000 people in the USA have a stroke, of which 650,000 are categorised as the first and 187,000 as recurrent strokes [4]. Because of an underlying interruption in blood flow to the brain, a stroke is characterized by an immediate neurological impairment [4, 5]. The ischemic strokes (IS) account for around 80% of all strokes, while haemorrhagic strokes accounts for about 20% [6]. The two medical diagnostic imaging techniques that are currently accessible to detect a stroke are computed tomography (CT) and magnetic resonance imaging (MRI). Apart from select centres providing CT imaging with mobile ambulances, existing technologies, despite being very sensitive, widely inaccessible or uneconomical for patients in rural hospitals and can only be used once the patient has been admitted to the hospital. However, roughly three-quarters of the world's population lacks access to trustworthy and inexpensive medical imaging technologies, according to the World Health Organization (WHO) [7]. Additionally, these current imaging systems are rather large [8]. For a neurological disability to be reduced and possibly reversed, reperfusion needs to start as soon as possible. CT scans have issues with radiation exposure [9], and cancer hazards [10]. MRIs cost more and are less accessible

---

Department of Electrical, Electronic and Systems Engineering, Universiti Kebangsaan Malaysia, Bangi, Selangor, Malaysia.
Email: shaahidul.mohammad@gmail.com
* Corresponding author: tariqul@ukm.edu.my

than CT scans [11, 12]. Although ultrasound imaging can be used to measure the structure and blood flow of intracranial and extracranial blood vessels, it is difficult and time-consuming [13]. Although not frequently employed, PET imaging has been used to evaluate the ischemic penumbra [11, 12]. Therefore, there is an urgent need for a mobile, non-ionizing, affordable, and highly precise imaging system for the identification of strokes.

Due to its low cost, non-ionizing, non-invasive, and portable characteristics, **microwave head imaging** is now an emerging technology for medical diagnostic systems that has generated substantial interest among researchers across the world [13–18]. This method uses variations in the electrical characteristics of the human cranium to identify early malignant tissues, haemorrhages, and other interior abnormalities [19, 20]. Both radar scanning and tomography are methods for reconstructing images, and in both instances, the antenna array is one of the key elements because the system's performance depends on how well it performs. The main concerns in **antenna** geometry for microwave head imaging are compactness and directionality because compactness enables more elements to fit within the available volume with portable and deployable features. Directionality ensures that the received scattered signal strength increases without causing backward signal interference [19, 21–23]. Wideband properties within the lower frequencies are also necessary to achieve appropriate image resolution for effective data capture. Several antennas have been constructed for microwave imaging applications using the **metamaterial** (MTM) structure to obtain compactness and wideband features. MTMs are becoming increasingly popular due to their bandwidth enhancement, gain enhancement, and isolation capabilities [24–35]. By influencing electromagnetic waves, they provide either negative permittivity or negative permeability, or even both, which makes them appropriate for use in any antenna by enhancing the isolation between antenna parts. The microwave head imaging system has used various wideband antennas, including MTM-loaded antennas, patch antennas, flexible antennas, and **3D antennas**. However, the size, radiation properties, number of array components, confirmation with a head phantom, and specific absorption rate of each individual antenna are all constrained (SAR). A triangle-shaped patch antenna is shown in [36] for brain imaging with microwaves. The imaging system extracts the scattering parameter using two antennas. However, the reduced amount of data points collected from the two-antenna array element limits the image reconstruction skills. Additionally, the performance that does not replicate the realistic head model is validated using the homogeneous phantom. We present a semi-flexible monopole submerged brick antenna in [37] for a microwave brain imaging system. The antenna's operational frequency ranges from 800 MHz to 1.2 GHz. The head model and two prototypes are used to examine the scattering performance of the antenna; however, there is no image reconstruction analysis, which restricts the system's potential. In addition, the antenna's dimensions are not small, and resonance frequencies below 1 GHz reduce image resolution by producing hazy images even though they increase penetration. A bendable antenna was shown in [38], proposed as an on-body imaging device. The investigation shows that the presented antenna's maximum SAR value is 0.80 W/kg, which is extremely high when compared to other recent head imaging antennas [22, 39]. Additionally, no study has been done on microwave image reconstructions. In [40], despite the suggested flexible antenna's strong image reconstruction capabilities, its homogeneous properties make it unsuitable for realistic head phantom verification. The article limits the analysis of harmful effects caused by antenna radiation and does not include the SAR analysis. Another clinical prototype is being created by EMTensor GmbH [13, 41]. The 160 antenna arrays of the prototype are based on high relative permittivity technology and operate between 0.9 GHz and 1.8 GHz. The use of a big antenna array ensures improved image resolution. However, because there are so many antenna arrays involved, it has a low bandwidth and requires complicated processing. Additionally, the overall performance may decline over time due to the coupling medium's lack of dielectric characteristics. A laboratory prototype using frequency-constrained sparse electromagnetic tomography (mfEMT) has been presented for detecting acute strokes in the human brain by applying frequency constrained sparse Bayesian learning algorithm [42]. To produce **microwave signals**, induction coils are utilized instead of antenna arrays. The changes in the coil's high mutual inductive reactance, however, could cause damage to the brain tissues. The excessive noise created as a sinusoidal signal travelling through the coil may also lower the resolution of the formed image.

The prototype of the antennas shown in [43–45] was built using several multilayer substrate layers. Although the use of many dielectric layers in an antenna design substantially impacts performance, the superstrate combination must be carefully chosen since dielectric loss can negatively affect the precision of the antenna gain and operating bandwidth. The SAR analysis of the consequences of the antenna prototypes' detrimental radiations also have not been discussed. Although the SAR analysis is presented in [22], the superstrate layer's actions cause the antenna bandwidth to drop. In a microwave imaging system, the scattering parameters are measured using an unhomogeneous phantom instead of defending the reconstructed images. In [44], Despite the antenna having a high realized gain and a small footprint, the research can only pinpoint the dispersion factors. The analysis shows no reason for the experimental validation of the microwave imaging system. An MTM-based flexible wearable wideband antenna is presented at [46] for brain stroke diagnosis. The MTM unit cell expands the antenna's bandwidth and makes unidirectional radiation possible. However, it omits the radiation gain and efficiency analysis when the antennas are positioned using the head model. Another metasurface enhanced antenna is presented in [47] for microwave brain imaging systems. The imaging performance is evaluated using a tissue-imitating gel phantom, and the antenna operates between 0.5 and 2.0 GHz. The investigation comes to the conclusion that adding metasurface elements can significantly reduce return loss and increase the efficiency of image reconstruction. However, problems are caused by the tissue-imitating gel phantom, which is inadequate in its replication of genuine head tissue. Additionally, the article does not investigate the metasurface antenna's radiation consequences. A metamaterial-loaded monopole antenna is also presented in [48] for a feasibility study of the head imaging system. This article presents the metamaterial analysis and its effects on the penetration signals. The imaging target is examined using a liquid phantom, which does not accurately replicate the milieu of real tissue. In addition, RF coils in the head imaging system are a more established technology than head imaging systems based on microwaves. The fundamental idea behind RF coils is that they act as "antennae" to transmit a RF signal to a target and receive a response. The RF coil has many uses, but there are still a few technological restrictions. Typically, the overall acquisition time exceeds eight (8) hours, making it difficult to maintain uniformity for many live patients [49]. The microwave imaging system is a real-time imaging system with portable features, where individuals may be conveniently positioned for diagnosis, as opposed to the RF coil system. In addition to Maxwell's law, the potential RF coil effect was divided into thermal and non-thermal forms. In the RF range, non-thermal effects such as non-ionizing radiation harm and unrelated tissue heating are observed. In contrast, thermal effects may affect physiological changes like thermoregulation, discomfort, or burns because nuclei only absorb less than 2% of the transmitted RF power.

This chapter provides a MTM-loaded small 3D antenna for portable **microwave head imaging** systems. It has an 80 percent fractional bandwidth and operates from 1.95 GHz to 4.5 GHz. The antenna's maximum efficiency has been improved from 66 percent to 89.6 percent, and its maximum realized gain has increased from 4.55 dBi to 6.01 dBi, respectively, using a finite array of MTM unit cell structures. In addition, mathematical modelling is investigated to determine the antenna power distribution with the metamaterial unit cell's optimization. A parametric analysis is carried out to emphasize the design factors and investigate the influences on the sensitivity of the skull tissue. The E-field and H-field distributions are studied with the effective medium theory to support the antenna proximity performance. The antenna's radiation properties, effectiveness, and gain are analysed when it is placed close to the head model.

Additionally, the SAR distribution is examined to demonstrate antenna compatibility. Prototypes of the antenna and tissue simulating phantom are made and measured to verify the total imaging system performance. An improved Iterative Correction of the Coherence Factor Delay-Multiply-and-Sum (IC-CF-DMAS) beam forming technique is used to create the bleeding pictures from the scattering parameters. The haemorrhages' reconstructed images show the system's advantages as a **portable platform** and its promise for microwave head imaging devices.

7.2 CCSRR based metamaterial structure design

7.2.1 Design and analysis of CCSRR unit cell structure

The **concentric crossed line split ring resonator (CCSRR)** unit cell structure is presented in Figure 7.1, which is a modification of the conventional concentric split-ring resonators (SRR). To design and analyze the properties of the proposed CCSRR unit cell structure, frequency domain-based electromagnetic simulation software CST has been used. Initially, the conventional SRR was drawn where two concentric metal rings are separated by a gap and split into opposite sides. The proposed CCSRR structure consists of a perfect electric conductor (PEC) printed on the Rogers RT5880 substrate material, creates electromagnetic radiation, and matches

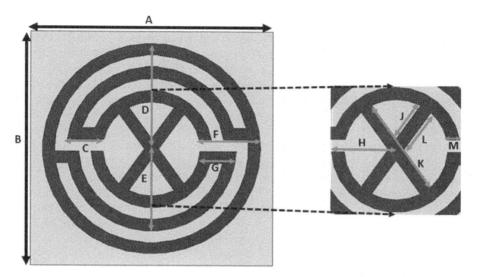

Figure 7.1. CCSRR unit cell structure.

Table 7.1. Specifications of CCSRR unit cell structure.

Parameters	Mm	Parameters	mm
A	10	G	1.5
B	10	H	2.5
C	1.5	J	1.65
D	4.5	K	5.95
E	5.5	L	1.8
F	2.5	M	0.5

the impedance. The dielectric constant of the substrate material is 2.2, where the loss tangent is 0.009. The thickness of the substrate and PEC is 1.575 mm and 0.035 mm, respectively, which are responsible for creating the capacitance and inductance by fixing the resonance frequency. The design specification of the proposed CCSRR unit cell structure is illustrated in Table 7.1.

7.2.2 Effective medium parameters of CCSRR unit cell

The **electromagnetic field** interactions with the metallic inclusions, their placements, and distributions play a significant role in the characterization of a metamaterial structure. The orientation in the specific directions of the electric and magnetic fields enhances the effective parameters like permittivity and permeability. In addition, the tangent inclusions of the electric field and the normal direction of the magnetic field to the surface structure also enhance the influential parameters. Thus, the electromagnetic field distribution must be specified accurately during the characterisation procedure. To get the ideal electromagnetic field, the boundary conditions are set in the two axial directions, X-direction and Y-direction, respectively.

These characteristics direct the CCSRR unit cell structure's responses. The upper and lower walls of the CCSRR are perpendicular to the E-vector and the back and front walls of the CCSRR are perpendicular to H-vector;they have been applied with the open boundaries to operate in the free space. The negative Z-axis and positive Z-axis are used as the waveguide port's position, and the CCSRR unit cell structure is placed between them, which is later energised towards the Z-axis by the electromagnetic wave. Figure 7.2 represents the boundary conditions of the proposed CCSRR unit cell structure with incidence, reflection (R1, R2), and transmission (T). There are two slabs to be considered for **reflection coefficients**; one is the air to metamaterial slab, and the other is the metamaterial slab to air. The **transmission coefficients** are the same in both directions, and a plane wave propagates in the air as the incident wave from either side. This process helps to observe the field features of the resonance frequencies by joining each wall's magnetic, electric, and open space conditions. The simulation procedure of the proposed CCSRR structure has been performed with a Frequency-domain solver where the tetrahedral meshing is applied within the frequency range of 1 GHz to 4 GHz.

To extract the effective medium parameters from the simulation data, the Nicolson-Ross-Weir (NRW) method is used, which is one of the popular electromagnetic characterization techniques [50–52]. The following equations are used to extract the effective parameters. The representations: the reflection and transmission coefficients S_{11}, and S_{21}, the permittivity and refractive index ε_r, and η_r respectively, the wavenumber k_0 and the thickness d of the substrate.

$$\varepsilon_r = \frac{2}{jk_0 d} \frac{1 - S_{21} + S_{11}}{1 + S_{21} - S_{11}} \quad (7.1)$$

$$\eta_r = \frac{2}{jk_0 d} \sqrt{\frac{(S_{21} - 1)^2 + S_{11}^2}{(S_{21} + 1)^2 - S_{11}^2}} \quad (7.2)$$

The retrieved normalized **effective permittivity** and refractive index of the CCSRR unit cell structure are shown in Figure 7.3. The negative real permittivity frequency ranges from 2.82 GHz to 4 GHz. The negative real **refractive index**

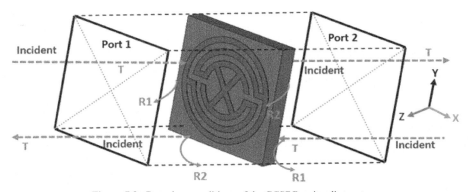

Figure 7.2. Boundary conditions of the CCSRR unit cell structure.

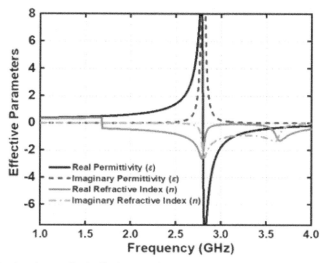

Figure 7.3. Retrieved normalized effective permittivity and refractive index of the CCSRR unit cell structure.

Table 7.2. Normalized values and frequency range of the imaginary permittivity and refractive index.

Parameter	Frequency Range	Peak Value
Positive Permittivity	2.0 GHz to 5.15 GHz	+7.96
Negative Refractive Index	2.72 GHz to 5.67 GHz	−2.75

frequency ranges from 1.65 GHz to 3 GHz and 5.42 GHz to 4 GHz, respectively. The imaginary values of the permittivity and refractive index are presented in Table 7.2

7.2.3 Equivalent circuit model of CCSRR unit cell structure

The transmission line principle is used for the single patch element that reflects the RLC circuit in series form. The following equation can present the relationship between the resonance frequency and LC circuit: L has lumped inductance, and C is lumped capacitance.

$$f = \frac{1}{2\pi\sqrt{LC}} \qquad (7.3)$$

When the **electromagnetic wave** propagates through any structure, electric and magnetic resonances are created by combining splits with the electric field and metal loops with a magnetic field. The capacitance and equivalent inductance can also be presented through the following equations where ε_0 represents free space permittivity and ε represents relative permittivity, A represents the area of the splits, d represents the split length, l is the length of the microstrip line, w is the width of the microstrip line, t is the thickness of microstrip line, correction factor, Kg is 0.57–0.145$ln(w'/h')$, w' represents the width of the substrate and h' represents the thickness of the substrate, respectively. The equivalent capacitance and inductance follow the concept of the transmission line principle.

$$C = \varepsilon_0 \varepsilon \frac{A}{d} (F) \qquad (7.4)$$

$$L(nH) = 2 \times 10^{-4} l[ln\frac{l}{(w+t)} + 1.193 + 0.02235\frac{w+t}{l}]K_g \qquad (7.5)$$

Initial L and C magnitudes of the designed circuit were calculated based on Equation 7.4 and Equation 7.5 [31, 33, 53]. After that, the approximated unit cell circuit with calculated values is designed with the electromagnetic simulation software ADS for extracting the scattering parameter values. Finally, the optimized L and C values were achieved as shown in Figure 7.4. The equivalent reflection coefficient is presented in Figure 7.5, showing resonance frequency within 2.67 GHz to 5.12 GHz with good agreement, although there are some deviations due to the approximation precision laggings.

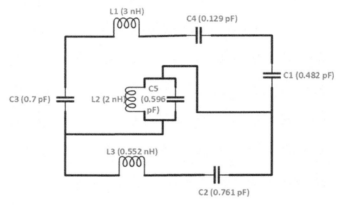

Figure 7.4. Equivalent circuit model of CCSRR unit cell structure.

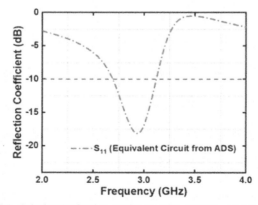

Figure 7.5. Calculated reflection coefficient of the CCSRR unit cell structure.

7.2.4 Parametric study of CCSRR unit cell structure

The **electric field, magnetic field, and surface current** circulations of the CCSRR unit cell structure at 2.6 GHz are depicted in Figure 7.6. It is observed that the current concentration is maximum at the outer central circle of the CCSRR structure at 2.6 GHz, and it exactly shows the opposite density for the magnetic field circulation within the identical frequency regime, which meets the Maxwell equation criterion. The 1 × 3 array of the CCSRR unit cell structure is depicted in Figure 7.7. The negative Z-axis and positive Z-axis are used as the waveguide ports position to analyze the effective parameter performances of the array structure. The effective parameters of permittivity and refractive index are retrieved using the same NRW method, which is depicted in Figure 7.8. It is observable that the results are almost identical to the single CCSRR unit cell structure, which implies the effectiveness of the CCSRR array structure.

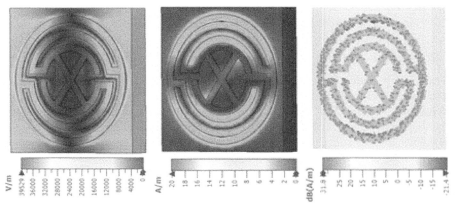

Figure 7.6. Electric field, magnetic field, and surface current distribution of CCSRR **unit cell** structure at 2.6 GHz.

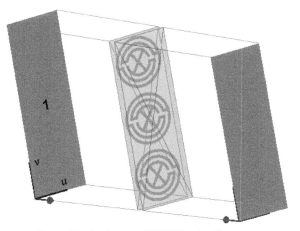

Figure 7.7. 1 × 3 array of CCSRR unit cell structure.

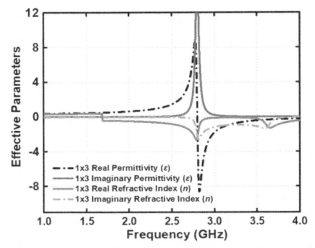

Figure 7.8. Retrieved effective parameters of the 1 × 3 array of CCSRR unit cell structure.

7.2.5 Design and analysis of CCSRR loaded 3D antenna

The CCSRR loaded 3D antenna and its analysis are explained in detail in this section. The CCSRR loaded 3D antenna schematics with perspective, and the top view are depicted in Figure 7.9. The design mechanism of the antenna starts with printing on the low loss Rogers RT5880 substrate slab where the permittivity is 2.2, loss tangent is 0.0009, and thickness is 1.57 mm. The **antenna** consists of a slotted dipole element with a finite array of MTM unit cell structures and a folded parasitic structure with two vertical and one bottom copper walls. The vertical and bottom parasitic walls use copper plates of 0.2 mm thickness. The bottom layer parasitic wall acts as a parasitic reflector that has the same dimensions, L × W.

The feed and ground layer parasitic elements have a coupling effect with the antenna feed and ground layers, which drives current onto the parasitic walls, resulting in lower operating frequency. In addition, the slotted dipole element has symmetrical top and ground elements with a 1 × 3 finite array MTM unit cell structure where a 50Ω **microstrip feeding** line is used. The antenna specifications are shown in Table 7.3.

Table 7.3. Specifications of CCSRR loaded 3D antenna.

Parameter	mm	Parameter	Mm	Parameter	mm	Parameter	mm
L	70	L_4	5.5	L_9	3	f_5	11
W	30	L_5	3	f_1	16	f_6	6
L_1	3	L_6	22.5	f_2	20	f_7	4
L_2	6	L_7	3	f_3	5	f_8	16
L_3	17	L_8	4	f_4	4	m_1	10
m_2	1						

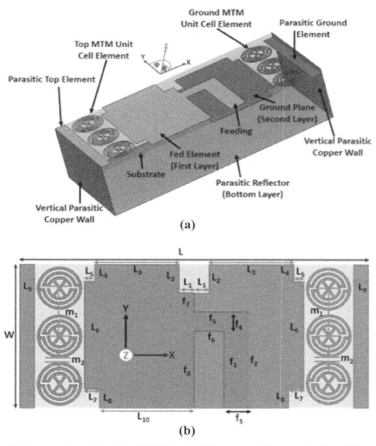

Figure 7.9. Schematic profile of the CCSRR loaded 3D antenna (a) Perspective view (b) Top view.

7.2.6 Mathematical modeling of the CCSRR loaded 3D antenna

The proposed 3D antenna mathematical modelling starts developing the **equivalent circuit** based on the transmission line principle [54]. It is noteworthy that all the components have been approximated by following standard lumped element parameter equations mentioned in [55, 56]. Figure 7.10(a) shows the equivalent circuit model where L1, C2, L2, and C3 represent the top and ground layers with the substrate and the air gap of the 3D antenna structure equivalence to C1 and C4 capacitors, respectively. The copper plate parasitic elements along with horizontal and vertical sections, are characterized by three consecutive LC series circuits from L3 to L5 and C5 to C7, respectively. For mathematical modelling, let us consider the simplified equivalent circuit following the lumped element Transmission line principle [56–58], as shown in Figure 7.10(b).

In addition, the impact of the EM field contribution from the parasitic element is minimal rather than the patch element; hence, the simplified model ignores it. In Figure 7.10(b), let the initial **voltage** be $V_i(x,t)$, initial **current** be $i(x,t)$, **capacitance**

186 *Metamaterial for Microwave Applications*

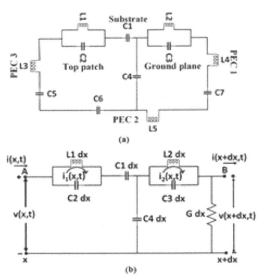

Figure 7.10. Analytical representation of 3D antenna (a) Equivalent circuit of the CCSRR loaded antenna (b) Simplified model for mathematical modelling.

including ground plane be C1 *dx* to C4 *dx*, inductance be L1 *dx* and L2 *dx* and the conductance of the 3D antenna be G *dx*. Now, the voltage drop across the inductor and the capacitor is generally expressed as,

$$v_L = -L\frac{di}{dt} \text{ and } v_c = \frac{1}{c}\int i\, dt \qquad (7.6)$$

Differential length (*dx*) is assumed across the x-axis. According to feeding or excitation voltage to the antenna $V(x,t)$ at point A and consumed voltage in the entire antenna $V_{consume}(x,t)$ would be equal to the differential voltage at point B, which represents $V(x+dx,t)$. So,

$$V(x, t) + V_{consume}(x, t) = V(x + dx, t) \qquad (7.7)$$

Otherwise,

$$V(x+dx,t) - V(x,t) = -[[-L_1 dx]\frac{di}{dt} + \frac{1}{c}\int i_1\, dt + [-L_2 dx]\frac{di_2}{dt}] \qquad (7.8)$$

Taking, dx → 0, in Equation 7.8 and doing a partial differentiation of the equation with respect to x we get,

$$\frac{\partial v}{\partial x} = L_1\frac{\partial i_1}{\partial t} - \frac{i_1}{L_1 C_2} - \frac{i}{C_1} - \frac{i}{C_4} - L_2\frac{\partial i_2}{\partial t} - \frac{i_1}{L_2 C_3} \qquad (7.9)$$

For 3D antenna currents along with the simplified circuit, terminal B current must be equal to current at A minus the current passing through the entire patch and it flows through the capacitors and conductance G.

$$\frac{\partial i}{\partial x} = -G_v - i_{1L1} - i_{1C2} - i_{2L2} - i_{2C3} \tag{7.10}$$

As we know, the current through the lumped capacitor and the inductor is

$$\left.\begin{array}{l} i_{1C2} = C_2 \dfrac{\partial v_{C2}}{\partial t} \\[4pt] i_{1C3} = C_3 \dfrac{\partial v_{C3}}{\partial t} \\[4pt] V_{C2} = -L_1 \dfrac{\partial i_{1L1}}{\partial t} \\[4pt] V_{C3} = -L_1 \dfrac{\partial i_{2L2}}{\partial t} \end{array}\right\} \tag{7.11}$$

From Equation 7.10,

$$v = -\frac{1}{G}\frac{\partial i}{\partial x} - \frac{1}{G}i_{1L1} - \frac{1}{G}i_{1C2} - \frac{1}{G}i_{2L2} - \frac{1}{G}i_{2C3} \tag{7.12}$$

Then again, Equation 7.9 differentiated with respect to x gives,

$$\frac{\partial^2 v}{\partial x^2} = L_1 \frac{\partial^2 i_1}{\partial x \partial t} - \frac{1}{L_1 C_2}\frac{\partial i_1}{\partial x} - \frac{1}{L_1 C_4}\frac{\partial i}{\partial x} - L_2 \frac{\partial^2 i_2}{\partial x \partial t} - \frac{1}{L_2 C_3}\frac{\partial i_2}{\partial x} \tag{7.13}$$

and the partial differential of Equation 7.12, gives,

$$\frac{\partial v}{\partial t} = -\frac{1}{G}\frac{\partial^2 i}{\partial x \partial t} - \frac{1}{G}\frac{\partial i_{1L1}}{\partial t} - \frac{1}{G}\frac{\partial i_{1C2}}{\partial t} - \frac{1}{G}\frac{\partial^2 i_{2L2}}{\partial t} - \frac{1}{G}\frac{\partial i_{2C3}}{\partial t} \tag{7.14}$$

Now using Equations 7.11 and 7.13, we can conclude that,

$$\left(\frac{1}{GL_1} + \frac{1}{GL_1}\right)\frac{\partial^2 v}{\partial x^2} = -\frac{C_2}{G}\frac{\partial^2 v}{\partial t^2} - \frac{C_3}{G}\frac{\partial^2 v}{\partial t^2} - \frac{\partial v}{\partial t} + \frac{v}{L_1 L_2 G} \tag{7.15}$$

Equation 7.15 is a one-dimensional partial differential equation for the proposed **3D antenna**. Equation 7.15 addresses the computational parameter of the 3D antenna to estimate the power or voltage requirement more precisely for sophisticated applications like EM imaging. However, the metamaterial unit cell incorporated in the antenna has been chosen to expedite the mutual coupling between the intra-unit cell and antenna patches at the top layer. The unit cell has a complete circle, two mirror-reflexed gaps, a circular shape, patch circumference, and a lumped element equivalent circuit representing a LC tank circuit. A typical LC tank circuit resonance, as in Equation 7.3, is calculated following the microstrip circuit design principle. According to the **microstrip transmission line** [57–59], the simulated inductance is 2.39 nH, and capacitance is 1.385 pF.

The corresponding transmission coefficient f_r evaluated as 2.78 GHz with −52 dB magnitude. So, a single metamaterial unit cell with such magnitude is likely to enhance the overall antenna response. Moreover, the E field response in the 3D antenna reveals that excitations through the SMA port gradually pass through the

188 *Metamaterial for Microwave Applications*

entire patch and get significant couplings between the patch and parasitic elements. Hence, identifying the three-unit cell in proximity optimizes the antenna performance and attains maximum coupling.

7.2.7 Parametric study of CCSRR loaded 3D antenna

The analysis of the evolution of the CCSRR loaded 3D antenna is depicted in Figure 7.11, where it represents the **reflection coefficient, efficiency, and realized gain** of the five different structures associated with the measured results.

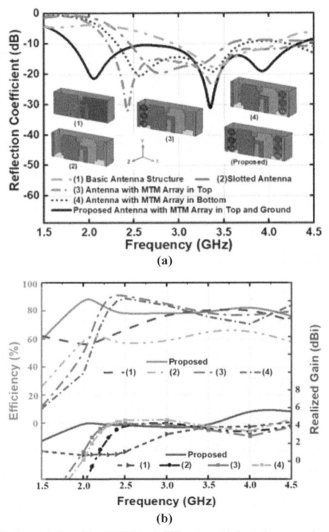

Figure 7.11. Design evolution of the CCSRR loaded 3D antenna (a) Scattering parameters (b) Efficiency and realized gain.

Figure 7.11(a) shows the analysis of antenna structure changes with the reflection coefficient. Antenna-1 presents the fundamental dipole element that generates resonance frequency at 5.5 GHz, whereas Antenna-2 shows resonance at 2.7 GHz with a slotted fed and ground. The CCSRR metamaterial unit cell structure array is placed between the fed and parasitic element (Antenna-3), which increases the coupling effect that lowers the resonance frequency and creates a wideband frequency range covering 2.25 GHz to 4.5 GHz. The CCSRR metamaterial unit cell structure array is placed between the ground and parasitic element (Antenna-4), which also shows the lowering of resonance frequency compared to Antenna-2. The proposed structure contains an array of metamaterial unit cell structures in both the fed and ground sides that show a strong coupling effect by increasing the frequency range and lowering the resonance frequency. From Antenna-1 to the proposed antenna, all have parasitic walls for conduction connected with them. The conducting parasitic walls directly impact wideband impedance matching and gain enhancement with the directional radiation pattern. The analysis in Figure 7.11(b) shows the evolution of efficiency and realized gain. The simulated antenna shows 89.6% and 78.5% efficiency with 4.34 dBi and 5.36 dBi realized gains at 2.1 GHz and 5.34 GHz, respectively. It is noticeable that the array of CCSRR metamaterial unit cell structures increases the maximum efficiency from 58% to 89.6% and maximum realized gain from 0.8 dBi to 4.2 dBi, respectively, at the lower 2.2 GHz frequency.

Furthermore, it is noticeable that the folded parasitic element with dipole metamaterial unit cell structure increases the electrical current paths. The gaps between the patch and **metamaterial structure** are induced to match the input impedance. The dipole structure is responsible for the high resonant frequencies, whereas the folded structure is responsible for antenna radiation characteristics towards unidirectionality. Also, the array of finite metamaterial unit cell structures impacts the overall antenna bandwidth, efficiency, and realized gain, which also ensures a significant level of compactness.

7.2.8 Antenna fabrication and measurement

After completing the design and analysis of the metamaterial-loaded **3D antenna**, it is required to fabricate and measure the designed prototype. The geometric parameters of the metamaterial-loaded antennas are collected from the CST simulation software for further processing with fabrication. Furthermore, the reflection coefficient, realized gain, efficiency, and radiation pattern are measured after completing the fabrication process.

A power network analyzer (PNA) is one of the standard equipments used to measure antenna performance. This PNA is a form of radiofrequency (RF) network analyzer widely used for RF designs and applications. It enables the user to measure the RF performance of the microwave devices to be characterized as scattering parameters (S_{11} and S_{21}). The Agilent P-series PNA (Agilent N5227A) is used to collect the measured results from the two equipped measurement ports that are shown in Figure 7.12(a). The frequency range of the PNA is 10 MHz to 67 GHz.

Moreover, this PNA is connected to the personal computer using a USB-General purpose interface bus (GPIB) port on the back panel that allows the personal computer

190 *Metamaterial for Microwave Applications*

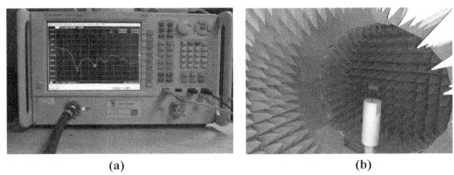

Figure 7.12. Measurement setup (a) Power network analyzer (b) Satimo near field.

to control the PNA with Standard commands for programmable instruments (SCPI). In addition, calibration over the desired frequency range is required to get accurate results. **Scattering parameters** like reflection coefficient, transmission coefficient, voltage standing wave ratio, and phase are calculated as per the requirements and stored in graphic and text format for further analysis.

The performance in terms of realized gain, efficiency, and radiation patterns is measured in the Satimo UKM StarLab, which is depicted in Figure 7.12(b). According to the Satimo UKM StarLab.

The XZ-plane is considered to be a magnetic field, and the YZ-plane is considered to be an electric field. It is a standalone system to test the antenna and wireless devices integrated into the subsystems.

7.2.9 Head phantom fabrication and measurements

This section presents the preparation and fabrication procedure of the tissue-mimicking **head phantom**. The Hugo model is considered as the layout of the human head model, as depicted in Figure 7.13. The model consists of CSF, dura, grey matter, white matter, and blood (represented as stroke). These phantom elements are fabricated by using different chemical mixtures that imitate the electrical properties of real head tissues (**CSF, dura, gray matter, white matter, and blood/stroke**) over the frequency band of 1 to 4 GHz. Individual phantom elements are then placed step by step in a three-dimensional skull. The calibration and measurement procedures are also explained in this section.

7.2.10 Preparation and fabrication of tissue mimicking head phantom

Different compositions of water, corn flour, gelatin, agar, Sodium Azide (NaN_3), Propylene Glycol, and Sodium Chloride (NaCl) are used to fabricate the head phantom components, dura, CSF, gray matter, white matter, and blood or stroke. The composition is summarized in Table 7.4. The fabrication process starts with adding propylene glycol into the water in a small beaker for all the **phantom** components. The quantity is used as per Table 7.4, and all the processes are performed at room temperature. Then corn flour is added to the mixture by stirring gradually in small

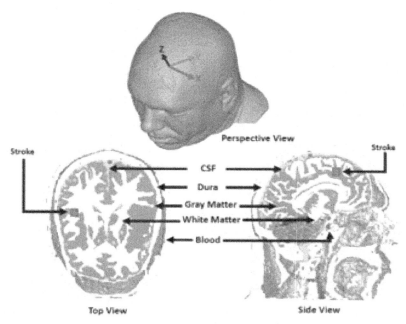

Figure 7.13. Layout of human head (Hugo Model).

Table 7.4. Composition of the tissue-mimicking phantom elements.

Components	Dura (500 gm)	CSF (500 gm)	Gray Matter (500 gm)	White Matter (500 gm)	Blood/Stroke (100 gm)
Water (gm/ml)	361.90	418.75	405.25	355.35	81.97
Corn Flour (gm)	120.65	10.15	82.95	134.30	2.73
Gelatin (gm)	0.00	0.00	0.00	7.05	0.00
Agar (gm)	4.58	56.20	5.2	0.00	12.75
Sodium Azide (gm)	1.8	1.85	1.75	1.75	0.36
Propylene Glycol (gm)	9.65	7.45	4.6	5.55	0.91
Sodium Chloride (gm)	1.20	5.60	2.30	0.00	1.28

portions to make it a thick gelatinous syrup (component 1). Water is a primary permittivity source due to its high dielectric properties over the wideband frequency range. Propylene glycol is added as a humectant and stabilizing agent, where it also helps to preserve the tissue by lowering the freezing temperature. Besides, the corn flour increases the mixture's thickness so that a strong bond is created among the different layers of the tissue. In another beaker, agar is added to the water and heated gradually up to 90–95°C. NaCl is then added to the mixtures when the temperature rises to that point, and the heating process is continued for another 3–4 minutes approximately until the agar melts (component 2). This procedure with agar is applicable for fabricating Dura, CSF, gray matter and blood.

192 *Metamaterial for Microwave Applications*

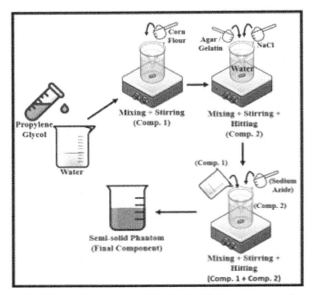

Figure 7.14. Schematic representation of the semi-solid phantom preparation process.

The gelatin is used alternatively to fabricate the white matter with the same procedure except for NaCl. NaCl controls the conductivity of the mixtures. The agar and gelatin control the **relative permittivity** and help form the final tissue's semi-solid structure. The viscous syrup from component 1 is added to the mixtures of component 2, where the burner and stirring process are kept running until the whole mixture becomes nearly semi-solid. Finally, the heating is stopped, and the mixture is allowed to cool down to 40–50°C. At this point, NaN3 is added as a preservative to the mixture, and the mixture forms a semi-solid structure. The overall schematics of fabricating the phantom materials are depicted in Figure 7.14. The materials used in fabricating the phantom components possess high mechanical properties and make it easy to produce the **head phantom** by considering different components in different layers. The fabricated components are presented in Figure 7.15, where they are measured in the next step to validate the electrical properties.

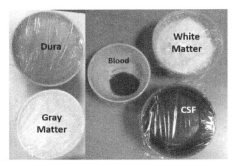

Figure 7.15. Fabricated elements of the head phantom.

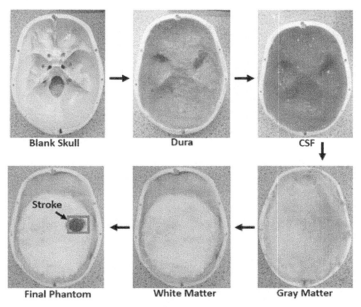

Figure 7.16. Overall placement process of the phantom layers.

The step-by-step adding procedure of the phantom components into a 3D skull after fabricating the phantom elements is depicted in Figure 7.16. Dura is the first component to be filled into, which later continues with **CSF, gray matter, white matter, and blood**. The rectangular mark represents the area of the stroke inside the head phantom.

7.2.11 Electrical properties measurement technique

The open-ended coaxial probe technique measures the **dielectric properties** of the fabricated tissue-mimicking head phantom. This technique is simple, non-destructive, and applicable for both *in vivo* and *ex vivo* measurements over a broad frequency range. However, the limitations towards accurate measurements are observed because of the complex heterogeneous structures and uneven surfaces in the homogeneous structures. Calibration procedures and measurement devices like vector network analyzers (VNA) are the key factors that limit the measuring contents.

The standard calibration procedure is applied with the three most common standards consisting of the open circuit, short circuit, and a broadband load while the probe (KEYSIGHT 85070E) is connected directly with the VNA (PNA-L N5232A; 300 kHz to 20 GHz). The aim is to correct the post-calibration measurement and make it reliable by analyzing the relationship between the measured complex **reflection coefficient** and the expected one. Besides, environmental factors like temperature, pressure, and humidity and system components like the cleanliness of the probe tip should be considered for a reliable measurement. Figure 7.17 represents the coaxial probe's cross-section schematics with its electric field orientation. The probe consists of a truncated section of the transmission line where the EM waves propagate through

194 *Metamaterial for Microwave Applications*

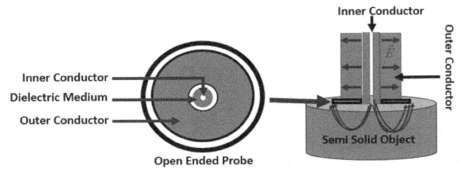

Figure 7.17. Cross-section schematics of the probe with electric field orientation.

the coaxial line. The impedance mismatch between the probe and targeted tissue sample generates reflected signals later converted into complex permittivity values.

The **vector network analyzer** converts the reflected signal into complex permittivity. Figure 7.17 represents the starting phase of calibration with 25 cm³ sterile water using the VNA and coaxial probe. Next, all the samples of the head phantom are sliced separately to ensure enough contact between the sample and the coaxial probe while performing the measurements. The visuals of the inner and outer parts of the sample are analyzed to assess the consistency as the outer flat part which is polished to ensure no gap between the coaxial probe and sample component. The **coaxial probe** is placed multiple times randomly on the sample surface to collect the data for better accuracy. The final decision is then made by calculating the mean value from the multiple data taken. It is observed that the measurement is as accurate as possible as the average percentage error is less than 2%.

7.3 EM head imaging system

7.3.1 Imaging setup with nine antennas

The proposed EM head imaging system to check the feasibility of brain stroke detection is demonstrated in Figure 7.18. The operating frequency range is 30 MHz to 8 GHz. A **metamaterial-loaded antenna** array of nine elements is installed on a rotatable circular ABS plastic holder platform, which is the imaging system's key element. Besides, SP8T RF switch, stepper motor, motor driver, a microcontroller with a portable mounting stand, PNA, and data controlling and image processing unit with a personal computer are also installed in the imaging system.

From the installed nine antenna array elements, one antenna acts as a transmitter, and the other eight antennas act as receivers. The head phantom is placed in the middle with a portable stand where the mechanical rotation platform rotates in polar coordinates from 0 to 2π around it using the stepper motor. The antennas are connected to a SP8T non-reflective positive control switching network using low-loss coaxial cables. The data (S_{21}, S_{31}, S_{41}, S_{51}, S_{61}, S_{71}, and S_{81}) are collected at each 7.2° and 50 equal points, covering 360°. Port one of the PNA generates EM signals in the frequency domain and transmits them to the head phantom. Another port of the PNA

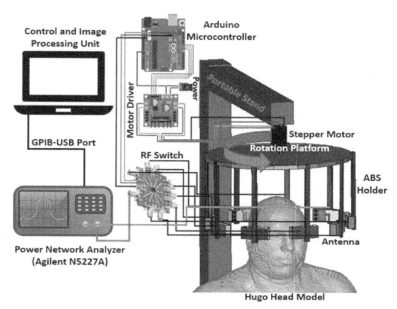

Figure 7.18. The design architecture and connected components of the proposed experimental EM head imaging system.

receives the backscattered signals via the RF switch. The **scattering parameters** are then sent to the image processing unit. The personal computer controls all the data acquisition processes through a USB-GPIB port. Finally, the collected scattered data is processed by using the **frequency domain algorithm** to reconstruct the detected and located brain stroke images.

7.3.2 Antenna phase center optimization

The distance and height from the center of the bottom surface of the imaging domain are optimized in this section for the nine circular antennae setup. Two sets of data are recorded from the empty setup for the calibration process, which is calculated as a measure of the generated noise. The generated noise is later processed through the frequency domain imaging algorithm to reconstruct the intensity map. Any recorded scattering intensity must be considered to remove the false positives due to the noise. The radius of the observation array is then optimized along with the radial length of the antenna. During the optimization, the circular disk's radius is considered the upper limit of the radial dimension, and the difference between disc radius and antenna length is considered a lower limit of the radial dimension.

Moreover, the height dimension of the optimization is considered from the feed point of the antennas. It is considered within the continuous closed bound $[-h/2, h/2]$, where h is the width of the antenna. A multi-start particle swarm optimization (MSPSO) is run for 120 iterations with a population size of 4, resulting in 480 fitness function evaluations (FFEs). Since MSPSOs have natural exploratory behaviour with the random re-initializations, higher convergence rates are implemented using

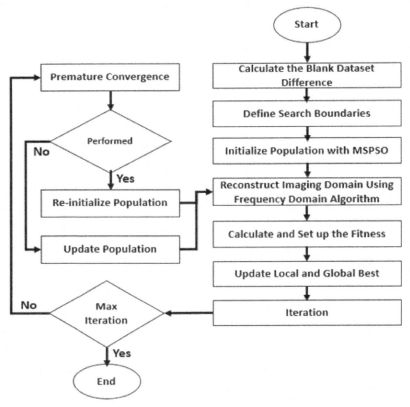

Figure 7.19. Overall flowchart of antenna phase center optimization.

low inertia constants, w. c1 and c2 set to 1.2 so that a 'damped oscillation' occurs around the local and global best, ensuring enough exploitation. The criterion for early convergence detection is the standard deviation of the current fitness of the population members falling below ε, where ε is sufficiently small constant to indicate the degree of accuracy desired in the solution. In this study, ε = 10 to 4 is used. The optimal focal point or phase centre is determined for the proposed antennas. A detailed algorithm flowchart for determining the **optimum focal point** is presented in Figure 7.19.

7.4 Image reconstruction technique

7.4.1 IC-CF-DMAS image reconstruction algorithm

The collected scattering data must be post-processed to reconstruct images in the EM head imaging system. Several image **reconstruction algorithms** have previously been used for EM imaging systems, which are discussed in the literature. This section introduces the iteratively corrected coherence factor delay multiply and sum (IC-CF-DMAS) to reconstruct the scattering object images. The IC-CF-DMAS is the modification of the previously developed iteratively corrected coherence factor delay and sum (IC-CF-DAS) algorithm [60], where the modification is applied to enhance the computing performance for more accurate results.

In this study, the scattered sample parameters, $S(f, rx, \varphi)$ will be split into two matrices based on odd and even φ values, or $S_{odd}(f, rx, \varphi_{odd})$ and $S_{even}(f, rx, \varphi_{even})$, individually, where φ odd = 1,3,5,...Nφ–1, and φeven = 2,4,6,...Nφ. Thus, S_{odd} can be considered as the initial illumination, and S_{even} the 'offset' illumination. Furthermore, the rotation subtraction is implemented by simply calculating the difference between the two matrices according to Equation 7.16.

$$S(f, rx, \varphi_{odd}) = S_{odd}(f, rx, \varphi_{odd}) - S_{even}(f, rx, \varphi_{even}) \qquad (7.16)$$

The signals will be translated into time-domain mode by applying the Inverse Fourier Transform for generating $\Gamma(t, rx, \varphi_{odd})$. Then, the time domain signals in the $\Gamma(t, rx, \varphi_{odd})$ domain will be processed using the proposed IC-CF-DMAS algorithm to reconstruct the images.

The Cartesian 3D coordinates of each point of the **imaging domain** contained in the matrices, C, and I display the complete set of points. The transmitter and receiver antenna coordinates are A_{Tx} and A_{Rx}, respectively. The imaging domain is static; hence, the position of the rotating antenna array needs to shift from the reconstruction point. The $P_{Tx\varphi odd-C}$ and $P_{C-Rx\varphi odd}$ are calculated from C, $A_{Tx\varphi odd}$, and $A_{Rx\varphi odd}$, counting the space between each point and transmitting and receiving antennas. The proper delay is generated by dividing the total distance l, by the background medium air, and the dielectric constant ε_r.

$$\tau(i, tx, rx, \varphi_{odd}) = \frac{\sqrt{\varepsilon_b}(P_{Tx\varphi_{odd}-C}(i, tx, l) + P_{C-Rx\varphi_{odd}}(i, rx, l))}{C} \qquad (7.17)$$

Here, c is the speed of light. The delay is calculated from the smallest anticipated distance and the reflected signal from C(i). The delays are further added to the signals for delivering the proper delayed signal. After multiplying the paired delayed signal, they are summed to determine the **scattering intensity** at the allotted point in the region of interest, as shown in the following equation.

$$\Upsilon_{DMAS}(i) = \int_{-\infty}^{\infty} \sum_{\varphi_{odd}=1}^{N/2} \sum_{tx=1}^{Tx} \sum_{rx=1}^{Rx} \sum_{\varphi'_{odd}=\varphi_{odd}}^{N/2} \sum_{tx'=tx}^{Tx} \sum_{rx'=rx+1}^{Rx} \begin{bmatrix} \Gamma(t-\dfrac{\tau(i,tx,rx,\varphi_{odd})}{\Delta t}, tx, rx, \varphi_{odd}) \\ \times \Gamma(t-\dfrac{\tau(i,tx',rx',\varphi'_{odd})}{\Delta t}, tx', rx', \varphi'_{odd}) \end{bmatrix} dt \quad (7.18)$$

The Coherence Factor Delay-Multiply-and-Sum makes use of a weighted sum of the channels where the coherence factor (CF) will become modelled as an incentive for more coherent channels at every single stage in the **imaging domain** with more significant weights. The equation below computes it:

$$CF(i) = \frac{\gamma_{DMAS}(i)}{\int_{-\infty}^{\infty} \sum_{\varphi_{odd}=1} \sum_{rx=1} \left[\Gamma(t - \dfrac{\tau(i, rx, \varphi_{odd})}{\Delta t}, rx, \varphi_{odd}) \right] dt} \qquad (7.19)$$

From then on, the scattering intensity map will be computed with the next formula:

$$\gamma_{CF-DMAS}(i) = CF(i) \cdot \gamma_{DMAS}(i) \tag{7.20}$$

The inverse distance weighting will be utilized for the reflection of the 3D Green function for electromagnetic waves.

$$\gamma'(i) = \int_1^n \frac{\gamma_{CF-DMAS}^{n-1}(i)}{1 + P_C - C(i,j)} dj \tag{7.21}$$

After that, the modified delay will be estimated by the subsequent formula:

$$\tau(i, tx, rx, \varphi_{odd}) = \tau(i, tx, rx, \varphi_{odd}) + \frac{\gamma'(i)}{C} \tag{7.22}$$

The **Coherence Factor** will be computed, and the scattering intensity map will be assessed as below:

$$\gamma_{CF-DMAS}^n(i) = CF(i)\gamma_{DMAS}^n(i) \tag{7.23}$$

In accordance with the modified delays, the scattering strength map will be reconstructed. Lastly, the closure requirements test for convergence. Equations 7.21–Equation 7.23 are assessed iteratively for n = 1, 2.... 7.

$$E_\gamma = \sum_{vl} \left| \gamma_{CF-DMAS}^n - \gamma_{CF-DMAS}^{n-1} \right| \tag{7.24}$$

The iterative process will be prematurely terminated when E_Υ decreases to the preferred precision standard as convergence would have been realized. In this study, $E\Upsilon < 10^{-5}$ will be applied to produce high-quality and noiseless images.

7.4.2 Matching medium consideration

The matching medium for the proposed imaging system will be optimized by considering the noise and distance of the phantom from the radiating element. So, the background medium is assumed to be an Additive White Gaussian Noise (AWGN) channel. This assumption supports the Power Spectral Density (PSD) and IC-CF-DMAS algorithm's proper delay. Technically, the proposed system's hypothetical noise insertion possibility or noise source is categorized into three types: **Electrical noise, Vibrational noise, and Shot noise**. The noise PSD is a square root of the noise power, and the stated noise in this scenario is either a left-sided or right-sided PSD. However, AWGN considers both-sided spectral power and a general mathematical model is used to model the timing error as an ideal situation. A general AWGN is presented as in Equation 7.25.

$$P_W = N_0 \frac{F_S}{2} \tag{7.25}$$

Here, P_w is the noise power, N_0 depends on the noise power over the bandwidth and F_s is the sampled limited band signal frequency.

In other words, the equation is modified as stated in Equation 7.26, where the matching medium dielectric constant is multiplied by the power of the AWGN channel power.

$$\tau(i,tx,rx,\varphi_{odd}) = \frac{\sqrt{E_{AWGN}} P_W (P_{Tx\varphi_{odd}-C}(i,tx,l) + P_{C-Rx\varphi_{odd}}(i,rx,l))}{C} \quad (7.26)$$

7.4.3 Internet of things framework for em head imaging system

The **IoT framework** for the portable **EM head imaging system** is depicted in Figure 7.20. This framework presents the approaches toward the monitoring of stroke-affected patients after the detection of brain stroke. The system relates to the PNA using the RF cables, microcontroller, and RF switch. The GPIB port is applied to connect with the PC for post-processing and reconstructing the images. The presence of the IoT framework also implies the proposed system's capability to operate remotely with a computer. Two methods are proposed in this section to transfer and monitor the reconstructed images of **brain injuries**. The first method comes with the involvement of cloud computing. The image processing unit collects the generated signal from the PNA to reconstruct the images using the IC-CF-DMAS algorithm. The reconstructed images are later transferred to the cloud platform from the image processing unit that can be later accessed by authorized personnel through the internet.

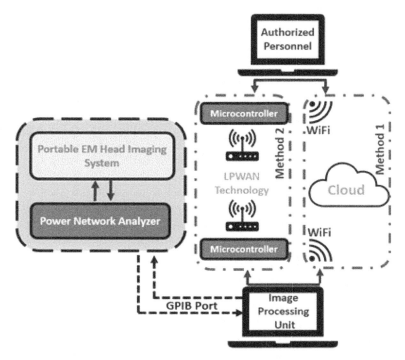

Figure 7.20. IoT framework for portable EM head imaging system.

The second method uses the LPWAN technologies, another suitable communication technology in the IoT field. This approach is most suitable for remote areas without an internet connection. The representative low-power long-distance communication technologies that could be used are Sigfox, LoRa, NB-IoT, and LTE Cat.M1. This approach transfers the data initially generated in the PNA to the image processing unit. This unit reconstructs the images from the generated signals using the IC-CF-DMAS algorithm. The **microcontroller** then collects the reconstructed images. An external modem will be used to connect and apply the stated LPWAN communication. On the other side, the receiver that contains the same elements receives the images and lets authorized personnel to continue processing them for further monitoring.

7.5 CCSRR loaded 3D antenna with head model

Initially, the analysis starts with checking the performance of the scattering parameters of the CCSRR loaded 3D antenna when it is placed near the head model with and without a stroke element inside it. A stroke element is inserted into the head model, where an average dielectric permittivity of 52 and conductivity of 2.48 are considered. Figure 7.21 represents the scattering parameter performance of the CCSRR loaded 3D antenna with and without the stroke element. It is noticeable from the results that the CCSRR loaded 3D antenna maintains a good return loss with and without the placement of the stroke element.

The **scattering parameters** in terms of different antenna locations and various sizes and locations of strokes are depicted in Figures 7.21(b) and 7.21(c). Strokes with six different sizes and locations are analyzed with the CCSRR-loaded 3D antennas which are placed in four different positions. This analysis investigates the depth sensitivity of the antenna in terms of any changes inside the **brain tissue**. The radius of the circular-shaped strokes ranges from 9 mm to 12 mm. The analysis shows that the responses of the scattering parameters are different compared to the scattering parameters in free space. The significant changes in the scattering parameters with the placement of stroke elements indicate the possibility of useful identification of strokes with various sizes and locations from the reflected scattered signals. The overall scanning setup for the identification of brain stroke with a perspective viewpoint is shown in Figure 7.21(a).

Nine CCSRR loaded 3D antennas are set adjacent to each other that surround the Hugo head model to verify the antenna performances in terms of scattering parameters and mutual coupling effects, which is shown in Figure 7.22(b). One antenna acts as a transmitter, whereas the other eight antennas act as receivers. Despite the placement with proximity, the receiving antennas show reflection coefficients below –20 dB, which indicates that the CCSRR loaded 3D antennas demonstrate a stable and comparable performance of scattering parameters and mutual coupling effects within the frequency regime of 1 GHz to 4 GHz.

The E and H fields are other vital considerations to be carried out to check the wave penetration inside the human head. Figure 7.23(a) represents the E-field distribution inside the realistic head model at 2.20 GHz, 5.20 GHz, and 5.85 GHz, respectively. The CCSRR loaded 3D antenna shows unidirectional wave propagation

Microwave Head Imaging and 3D Metamaterial-inspired Antenna 201

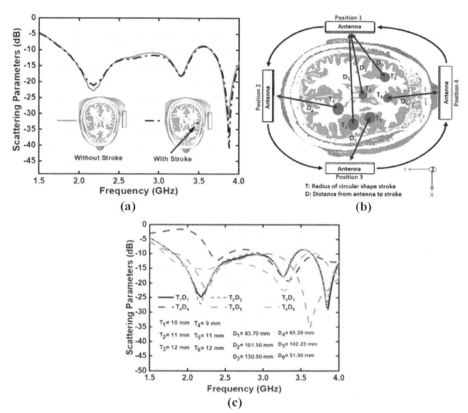

Figure 7.21. CCSRR loaded antenna with head model (a) Scattering parameters with and without stroke, (b) placement of the antenna on four different positions with different size and location strokes (c) Scattering parameters with the different positions, size, and locations.

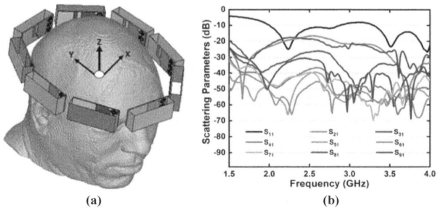

Figure 7.22. (a) Perspective view of nine antenna setup with CCSRR loaded 3D antenna (b) Antenna scattering parameters with mutual coupling effect.

202 *Metamaterial for Microwave Applications*

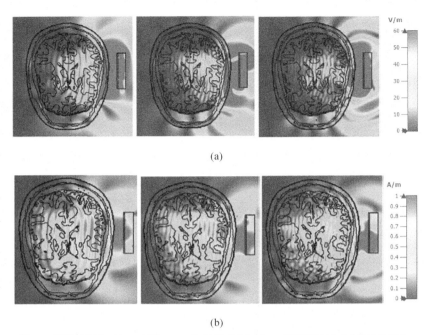

Figure 7.23. (a) Electric and (b) magnetic field distribution of CCSRR loaded 3D antenna.

towards the realistic head model, which shows the EM wave penetration inside the lossy head tissue. There would be a propagation difference of the E-field with and without stroke due to the changes in the dielectric properties and tissue surroundings. The H-field distribution shows similar wave propagation characteristics towards the head model, which is shown in Figure 7.23(b). The observed fact is that the E and H fields decrease inside the lossy head tissue. As the development of the EM head imaging system consists of nine antenna array elements that cover the whole head, the wave propagation depth is sufficient for successful image reconstruction with all the conveying tissues' information.

The simulated antenna radiation patterns at 2.2 GHz, 5.2 GHz, and 5.85 GHz, respectively, when the antenna is placed near the head model, are depicted in Figure 7.24. The analysis is performed by applying the CCSRR unit cell structure with and without integration of the E and H planes to the antenna.

The CCSRR loaded 3D antenna shows the maximum directionality with a reduction of back lobe radiations by filtering the metamaterial electromagnetic radiations. It is observed from the analysis that the CCSRR unit cell structures to the antenna increase the performance of antenna directionality when placed in close proximity to the head. The performances in terms of radiation efficiency and gain with similar placement are presented in Table 7.5. It is noticeable that the CCSRR unit cell structure within the antenna increases the radiation efficiency and gain when placed near the head model. The radiation efficiency and realized gain increases continuously over the upper frequencies, which also implies the effectiveness of the CCSRR unit cell integration with the 3D antenna.

Microwave Head Imaging and 3D Metamaterial-inspired Antenna   203

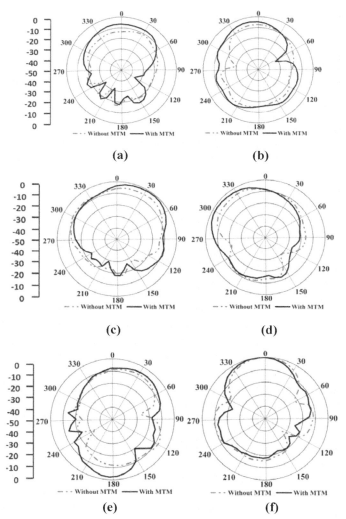

Figure 7.24. Antenna radiation with and without CCSRR structure in close proximity to the head phantom. (a) and (b) At 2.2 GHz, (c) and (d) At 5.2 GHz, (e) and (f) 5.85 GHz. (a), (c) and (e) E-plane. (b), (d) and (f) H-plane.

Table 7.5. Realized gain and radiation efficiency of the CCSRR loaded 3D antenna with and without CCSRR structure in close proximity to the head model.

Frequency (GHz)	Radiation Efficiency (%)		Realized Gain (dBi)	
	Without CCSRR	With CCSRR	Without CCSRR	With CCSRR
2.20	11.5	18	−5.03	−2.16
5.20	30	42.8	0.42	2.40
5.85	36.7	55.8	2.50	4.97

7.5.1 Specific absorption rate (SAR) analysis

The specific absorption rate (SAR) distribution is an important consideration to protect the human body from harmful radiation effects. As per the standard regulations [61] the maximum SAR must not exceed 2 W/kg for 10 g and 1.6 W/kg for 1 g of tissue. As the CCSRR loaded 3D antenna are placed near the head model, the SAR must be calculated and analyzed by following Equation 7.27. The input power for the antenna is 1 mW, and only one antenna acts as a transmitter at a particular time. Figure 7.25 represents the metamaterial loaded 3D antenna and its setup with the head model to calculate the SAR distribution. Four different positions are considered for the CCSRR loaded 3D antenna,

$$SAR = \frac{|E|^2 \sigma}{M} \quad (7.27)$$

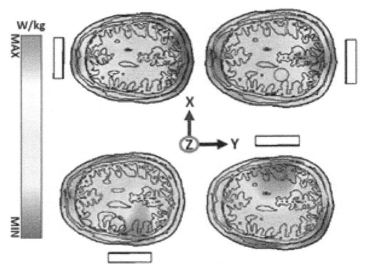

Figure 7.25. The specific absorption rate (SAR) inside the head phantom with metamaterial loaded antenna operation in different positions within XY-plane.

7.5.2 SAR analysis of CCSRR loaded 3D antenna

The analysis to calculate the SAR distributions with CCSRR loaded 3D antenna for 1 g, and 10 g of tissue in four different positions with different resonance frequencies as per the scattering parameters is shown in Table 7.6. Notably, 10 g tissue absorbs less electromagnetic energy compare to 1 g of tissue. The maximum SAR positions are identified at the exterior head tissue layer. The maximum values are 0.117, 0.152, 0.154, and 0.087 of 1 g tissue, and 0.050, 0.081, 0.071, and 0.056 of 10 g tissue, respectively, for the four different antenna positions which satisfy the IEEE public radiation exposure limit of 1.6 W/kg.

Table 7.6. SAR value distribution when CCSRR loaded 3D antenna is in XY-plane near the head model.

Antenna Position	Frequency	1 g	10 g
	2.20	0.085	0.045
	5.20	0.117	0.050
	5.85	0.070	0.030
	2.20	0.135	0.073
	5.20	0.118	0.062
	5.85	0.152	0.081
	2.20	0.134	0.063
	5.20	0.154	0.071
	5.85	0.152	0.061
	2.20	0.073	0.045
	5.20	0.076	0.056
	5.85	0.087	0.054

7.5.3 Measurements of CCSRR loaded 3D antenna fabricated prototype

After getting the proper CCSRR loaded 3D antenna structure, it is fabricated and measured to investigate the performance in free space. Figure 7.26 depicts the fabricated prototype of the CCSRR loaded 3D antenna with a top and perspective view. The fabricated CCSRR loaded 3D antenna is then set up with the PNA for the measurement of scattering parameters, which is shown in Figure 7.27(a). The RF flexible coaxial cable is used to connect the CCSRR loaded 3D antenna with the PNA. The measurement is performed within 1.5 GHz to 4.5 GHz. The measured reflection coefficient compared with the simulated reflection coefficient is presented in Figure 7.27(b). The simulation result shows that the antenna operates over the band of 1.8 GHz to 4.34 GHz, where the fractional bandwidth is approximately 83%

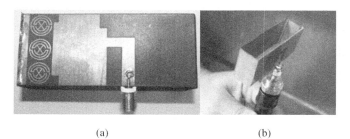

(a)   (b)

Figure 7.26. Fabricated prototype of the CCSRR loaded 3D antenna (a) Top view (b) Perspective view.

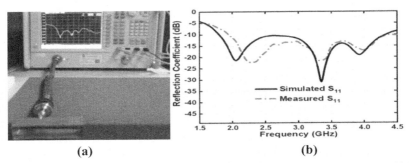

Figure 7.27. Reflection coefficient measurement of CCSRR loaded 3D antenna in free space (a) PNA setup (b) Simulated and measured reflection coefficient.

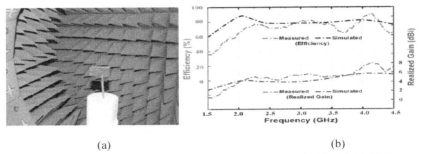

Figure 7.28. Efficiency realized gain and radiation pattern measurement of CCSRR loaded 3D antenna in Satimo StarLab (a) Antenna setup (b) Simulated and measured efficiency and realized gain in free space.

with respect to the center frequency of 5.07 GHz. The measurement result indicates the approximate 80% of the fractional bandwidth with respect to the 5.22 GHz center frequency within the 1.95–4.5 GHz frequency region, which implies a good agreement between the simulated and measured results.

The CCSRR loaded 3D antenna is then set up in a Satimo near-field measurement system to measure the realized gain, efficiency, and radiation pattern, shown in Figures 7.28–7.29. The analysis in Figure 7.28(b) shows a good agreement between the simulated and measured results on efficiency and realized gain. The simulated antenna shows 89.6% and 78.5% efficiency with 4.34 dBi and 5.36 dBi realized gain at 2.1 GHz and 5.34 GHz, respectively, whereas the measurement shows 71.7% and 77.9% efficiency with 4.52 dBi and 5.18 dBi realized gain at the same frequencies. The maximum efficiency and gain achieved from the measured results are 87.3% and 7.22 dBi, respectively, within the 1 GHz to 4 GHz frequency regime.

Figure 7.29(a–c) depicts the far-field radiation characteristics of the antenna with E (XZ) and H (YZ) plane patterns at 2.2 GHz, 5.2 GHz, and 5.34 GHz, respectively. The 3D far-field simulation view is also presented along with the plotted results. The analysis shows that the simulated and measured radiation patterns have a good match as the CCSRR loaded 3D antenna has a stable directional radiation pattern with a

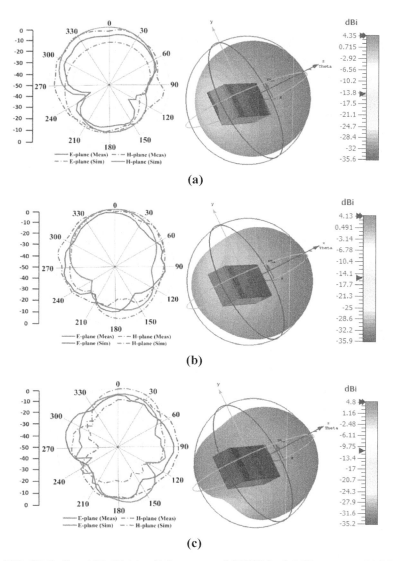

Figure 7.29. 2D (Left) and 3D (Right) radiation pattern of CCSRR loaded 3D antenna at (a) 2.2 GHz, (b) 5.2 GHz, and (c) 5.85 GHz.

boresight direction with an average gain of 3 dBi along the Z-axis. It is noticeable that the H-plane has wider beamwidths compare to the E-plane radiation pattern, and the cross-polarization level is more than 10 dB in both the planes along the Z-directions, which indicates high polarization purity. In addition, the antenna attains an average of 10.5 dB of the front-to-back ratio along the Z-axis.

7.6 Electrical properties of tissue mimicking head phantom

Dielectric properties of the fabricated head phantom components are measured and compared with the reference relative permittivity and conductivity, depicted in Figure 7.30. It is noticeable from the first measurement that the relative dielectric permittivity of the CSF, dura, gray matter, white matter, and blood ranges from 68–63, 48–42, 50–45, 40–34, and 57–49, respectively. The conductivity of the mentioned consecutive phantom components ranges from (in S/m) 1.9–4.5, 0.9–5.6, 0.8–5.3, 0.5–1.8, and 1.5–5.2, respectively. The open-ended coaxial probe is placed at two more random positions to evaluate the effectiveness of the electrical properties with the components. Table 7.7 represents the lists of data and its mean value which are taken from multiple positions of the phantom components at the center frequency of 1 GHz to 4 GHz.

The relative permittivity reference values range from 68–60, 47–43, 50–45, 40–35, and 58–50 for the CSF, dura, gray matter, white matter, and blood, respectively. The conductivity values (in S/m) are 2–4.8, 1–5.8, 0.9–5.5, 0.5–1.8, and 1.5–5.2, respectively, for the stated consecutive components. It is observed that the relative

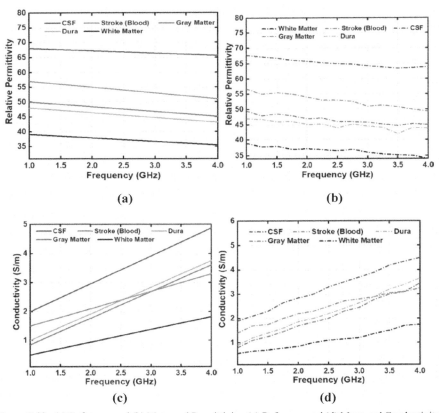

Figure 7.30. (a) Reference and (b) Measured Permittivity; (c) Reference and (d) Measured Conductivity.

Table 7.7. Electrical properties of the fabricated head phantom.

Properties	CSF				Dura			
	Data 1	Data 2	Data 3	Mean	Data 1	Data 2	Data 3	Mean
Permittivity	65.15	61.26	58.33	61.58	44.04	41.35	42.56	42.56
Conductivity	5.26	2.78	5.45	5.16	2.15	1.84	2.34	2.11

Properties	Gray Matter				White Matter			
	Data 1	Data 2	Data 3	Mean	Data 1	Data 2	Data 3	Mean
Permittivity	46.32	41.62	39.85	42.59	36.73	31.54	35.84	34.04
Conductivity	2.06	1.88	2.24	2.06	1.15	0.95	1.23	1.11

Properties	Stroke (Blood)			
	Data 1	Data 2	Data 3	Mean
Permittivity	52.81	49.65	55.87	52.78
Conductivity	2.55	2.78	2.12	2.48

permittivity and conductivity values are almost identical for the measured and reference parameters. Besides, the calculated mean value of relative permittivity and conductivity also show identical results compared to the reference values. Therefore, the measured phantom shows more realistic characteristics of real human head tissue that evaluates the effectiveness of the electromagnetic head imaging system.

It is important to evaluate the effectiveness of the fabricated tissue-mimicking head phantom as the brain's tissue distribution is more complicated in the real imaging scenario. The fabricated phantom elements are kept for the second and third consecutive measurements after four days and seven days in a week. The changes in the electrical properties are observed, and then the image reconstruction technique is performed to analyze timely effects.

The relative permittivity and conductivity of the fabricated phantom components are measured after four days, which is depicted in Figure 7.31(a) and Figure 7.31(b). A slight decrement is observed in the electrical properties compared to the first-day measurements. A decrement in measured electrical properties is also found after seven days, which is presented in Figures 7.31(c) and 7.31(d). This happens due to the evaporation of water from the phantom components over time. As the phantom components need to be measured at room temperature, they are expected to evaporate over time. Besides, preserving the phantom components below the freezing point can also make the water content evaporate. The better way to preserve the phantom components is to keep them in a less aired bag that reduces evaporation.

7.7 Em imaging results

This section presents the imaging results from the developed EM head imaging system associated with the nine developed metamaterial loaded antenna array setups and tissue-mimicking head phantoms. The reconstructed images with and without the presence of fabricated head phantoms are investigated in this section. Besides, the reconstructed images from CCSRR loaded 3D antenna setup are analyzed in this section to validate the efficiency of the system in detecting the brain stroke.

210 *Metamaterial for Microwave Applications*

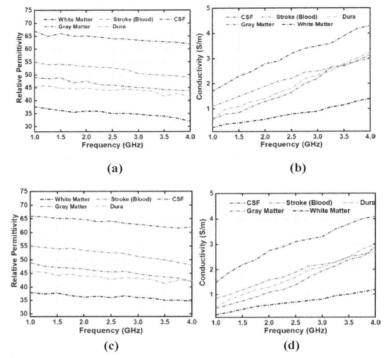

Figure 7.31. (a) Relative Permittivity and (b) Conductivity after four days; (c) Relative Permittivity and (d) Conductivity after seven days.

The tissue-mimicking phantom elements are analyzed after four and seven days. The sensitivity towards the image reconstruction and detection of brain stroke is examined in this section.

Initially, the scattering parameter of a single metamaterial loaded antenna with a realistic head model is analyzed to check the antenna's performance towards an appropriate return loss. The metamaterial loaded antenna connected with the head model is depicted in Figure 7.32(a). Figure 7.32(b) shows the measured scattering parameters compared with the simulation results. It is noticeable that the scattering parameters that remain below −10 dB show a good agreement from both simulation and measurement perspectives. These results imply the initial validation of the metamaterial loaded antenna towards the head imaging setup. The developed EM head imaging system is depicted in Figure 7.33. The system consists of nine metamaterial loaded antennas, a portable stand, RF switch, stepper motor, RF coaxial cables, and PNA. The tissue-mimicking phantom is placed inside the realistic 3D head skull. The skull is placed in the center of the nine antenna setup. A personal computer associated with the image processing unit is connected with the PNA using a USB-GPIB port. Figure 7.34 represents the front view of the developed EM head imaging system to demonstrate the portability and compatibility towards the clinical trial implementations.

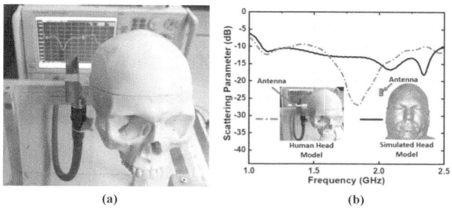

Figure 7.32. (a) Metamaterial loaded antenna with fabrication head phantom inside 3D skull (b) Scattering parameters on the simulated and measured setup.

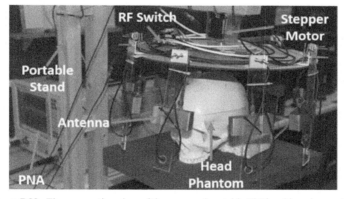

Figure 7.33. The perspective view of the proposed portable EM head imaging system.

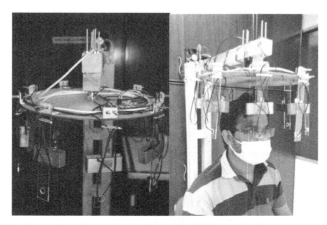

Figure 7.34. The front view of the proposed portable EM head imaging system with human trial compatibility.

212 *Metamaterial for Microwave Applications*

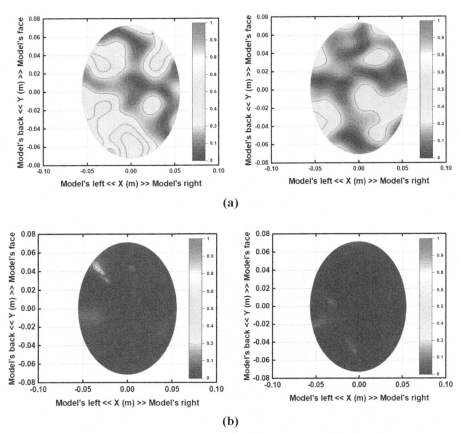

Figure 7.35. Reconstructed images of the head model without brain stroke (blank data) using (a) DMAS and (b) IC-CF-DMAS algorithm.

It is crucial to analyze the image reconstruction technique in free space to perform the comparison between the scattering map intensity. Figure 7.35 depicts the blank images reconstructed by applying the conventional DMAS and proposed IC-CF-DMAS algorithms without the placement of any head model inside the system. The observed fact is that the reconstructed images show very low noises within the regional area.

7.7.1 *Imaging results with CCSRR loaded 3D antenna setup*

This section presents the reconstructed images for brain stroke identification using the IC-CF-DMAS algorithm for both proposed metamaterial loaded 3D antenna array setups and tissue-mimicking phantoms. The reconstructed images using the conventional delay-multiply-and-sum (DMAS) algorithm are also investigated in this section. The reconstructed images using CCSRR loaded 3D antenna are presented in Figure 7.36. The conventional DMAS and proposed IC-CF-DMAS algorithms are applied to evaluate the imaging performances. The blue rectangular mark represents

Microwave Head Imaging and 3D Metamaterial-inspired Antenna 213

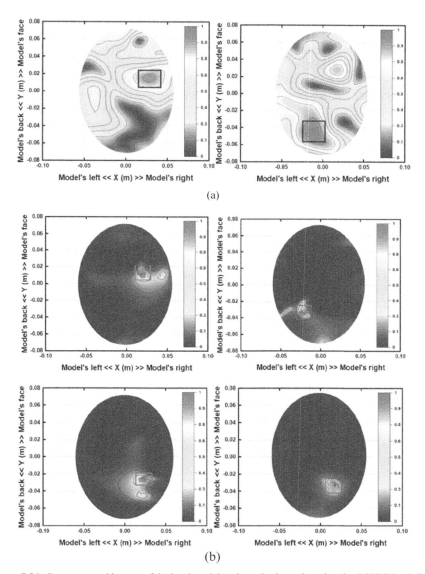

Figure 7.36. Reconstructed images of the head model to detect brain stroke using the CCSRR loaded 3D antenna at different positions with (a) DMAS and (b) IC-CF-DMAS algorithm.

the stroke object in the DMAS imaging domain, whereas the red rectangular mark represents the stroke object in the IC-CF-DMAS algorithm. It is noticeable from Figure 7.36(a) that the highest multiple contrast is formed within the imaging domain that reflects the localization error. This occurs potentially due to the underestimation of the average dielectric constant of the imaging domain.

Moreover, some "ghosting" or "halo effect" type distinct clutter is created due to the multiple reflections from the stroke object in conventional DMAS. This kind of clutter creates complications in detecting the actual stroke object. The IC-CF-DMAS is applied to enhance the performance of the conventional DMAS. Figure 7.36(b) represents the reconstructed images with brain stroke detection by applying the IC-CF-DMAS algorithm. Multiple stroke locations are applied to the head model to examine and evaluate the antenna performance and reconstructed images. In addition, the iterative correction converges after only a minor variation in the delay. Significant noise reduction is noticed for the generated images. IC-DMAS performs well by eliminating noise and ghosting and successfully detecting the stroke after removing localization errors. Using both proposed metamaterial loaded 3D antenna, the developed EM head imaging system can detect and locate the brain stroke positions inside the head model that verifies the system's capability.

7.8 Sensitivity analysis

The effectiveness of the phantom elements also depends on the image reconstruction results. The components are placed in the 3D skull after four and seven days to perform the image reconstruction process. The placement process is similar to Figure 7.37. The reconstructed images after four days and seven days are depicted in

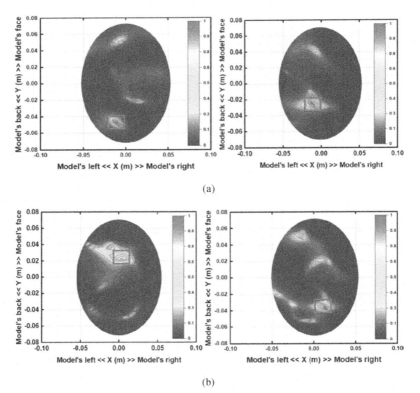

Figure 7.37. Reconstructed images of the human head model with stroke at different positions after (a) four days, and (b) seven days.

Figure 7.37(a–b). It is noticeable that the system can effectively identify the stroke from the phantom elements, although some noises are created due to the slight decrement of watery content from the phantom elements.

7.8.1 *Internet of things based image transfer*

The first approach, a relatively straightforward method, has been implemented to demonstrate the potential IoT framework strategy regarding portable EM head imaging systems. Refer to Figure 7.38; the process starts from the image processing unit after GPIB interfacing. Herein the PC used for image processing received the data from the image reconstruction system described earlier. This contour data for the fetch to MATLAB command is to generate a contour plot with the brain hemorrhage location. Now, the cloud operation starts at this stage, and we have adopted the MATLAB cloud platform for the image storage and remote access system.

The results presented in this section clearly indicate that the CCSRR loaded 3D antenna is compatible with the EM head imaging system. The measurement results obtained from both metamaterial loaded 3D antennas show good agreement with the simulation results. An acceptable reflection coefficient is obtained from both antennas when they are placed near the head model. The SAR analysis indicates the acceptability of the antennas due to the low radiation effect. The tissue mimicking phantom demonstrates good sensitivity performances over time. The IC-CF-DMAS algorithm can effectively detect and reconstruct brain stroke images. The brain stroke images are transferred through the cloud for the continuous monitoring process. Overall, the developed EM head imaging system can effectively detect the brain stroke images that represent the system's capabilities towards further clinical trials.

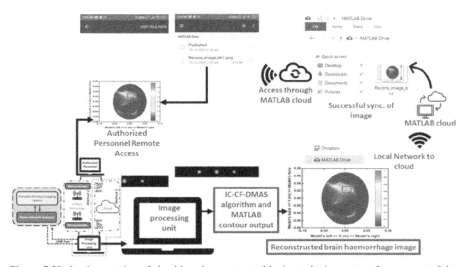

Figure 7.38. Implementation of cloud-based reconstructed brain stroke image transfer as a part of the IoT enabled framework.

7.9 Summary

The chapter describes the creation of a wideband EM head imaging system that is non-ionizing, non-invasive, inexpensive, portable, and compact that can be used to monitor patients in real time by detecting brain strokes. The field of EM imaging systems benefits greatly from this strategy.

An EM head imaging system is required for the identification of brain strokes, as shown by the context and underlying motive. The wideband portable EM head imaging system can be used to overcome the constraints of the current medical imaging technologies. Modern medical imaging systems may be seen to use simplified head phantoms and to have been examined numerically in most cases. Only a small portion of the imaging systems that are prototyped and tested using the realistic tissue-mimicking head phantom have advanced to preclinical testing. Despite the benefits of the conventional wideband EM head imaging system described in this chapter, the system faces a number of difficulties since it depends on so many different fields. While developing the EM head imaging system for brain stroke diagnosis, it is necessary to take into account a variety of fields, including electromagnetics, signal and image processing, instrumentation, material engineering, software development, and others. It is extremely difficult to design and manufacture effective and small wideband antennas for EM imaging systems. In EM imaging systems, the antenna is crucial since its sensing components are what determine the system's overall performance. It is also difficult to improve the antenna's performance in terms of bandwidth, gain, efficiency, and radiation pattern. We exhibit two 3D antennas loaded with metamaterial. The front and back of the antenna are made up of 33 metamaterial unit cell structures. The metamaterial is positioned between the parasitic and primary radiating patches.

In all scenarios, the antenna achieves good performance in terms of bandwidth, gain, and efficiency. Additionally, over a broad operating spectrum, the folded parasitic element aids in achieving directional radiation patterns. Furthermore, it is noted that when the metamaterial loaded antenna is positioned close to the head model, it performs well in terms of realized gain, radiation efficiency, and directional radiation patterns. Additionally, the examined SAR value shows that the metamaterial loaded antenna decreases the SAR value since the metamaterial is placed next to the antenna radiating element.

The performance and safety of EM head imaging systems must be quantitatively and explicitly validated in order for technology to advance. This inspired the creation of the convincing tissue-mimicking head phantom that is the subject of this thesis. According to the literature, the majority of the phantom materials utilized in EM head imaging systems are either solid, liquid, or semi-liquid based. The semi-solid phantom materials have received very little attention in the literature, but, since they are few, they still have trouble simulating the head phantom. This implies the manufacture of the realistic tissue-mimicking phantom, which is shown with the ingredient list and step-by-step instructions for addition to the 3D head model. The electrical properties of the phantom elements are also measured and tested over time to validate the sensitivity.

We describe a portable, nine antenna array-based EM head imaging system. Full-wave EM simulations are run to confirm the imaging system's security. Nine antenna array elements make up the system, with one serving as a transmitter and the remaining eight as receivers. The antenna receivers are used with the switching matrix, and fifty offset rotations are done sequentially. The VNA collects the reflected scattering EM signals, which are then post-processed using MATLAB for signal and image processing. The imaging method post-processes the variations in antenna performance with and without the presence of the head phantom. In addition, the antennas function in free space close to the head phantom. As a result, numerous calibration procedures and tests have been carried out to map the intensity of scattering from the inside of the cranium. The application of the IC-CF-DMAS image reconstruction algorithm in this chapter yields notable results in the recognition of brain strokes.

References

[1] Frykberg, R. G., L. A. Lavery, H. Pham, C. Harvey, L. Harkless et al. 1998. Role of neuropathy and high foot pressures in diabetic foot ulceration. *Diabetes Care*, 21(10): 1714–1719.
[2] Fernando, M. E., R. G. Crowther, P. A. Lazzarini, K. S. Sangla, P. Buttner et al. 2016. Gait parameters of people with diabetes-related neuropathic plantar foot ulcers. *Clinical Biomechanics*, 37: 98–107.
[3] Boulton, A. 2004. The diabetic foot: from art to science. The 18th Camillo Golgi lecture. *Diabetologia*, 47(8): 1343–1353.
[4] Farahpour, N., A. Jafarnezhad, M. Damavandi, A. Bakhtiari, P. Allard et al. 2016. Gait ground reaction force characteristics of low back pain patients with pronated foot and able-bodied individuals with and without foot pronation. *Journal of Biomechanics*, 49(9): 1705–1710.
[5] Sacco, I. C., A. P. Picon, D. O. Macedo, M. K. Butugan, R. Watari et al. 2015. Alterations in the lower limb joint moments precede the peripheral neuropathy diagnosis in diabetes patients. *Diabetes Technology & Therapeutics*, 17(6): 405–412.
[6] Uazman Alam, David R. Riley, Ravinder S. Jugdey, Shazli Azmi, Satyan Rajbhandari, Kristiaan D'Aou't and Rayaz A. Malik. 2017. Diabetic neuropathy and gait: a review. *Diabetes Therapy*, 8(6): 1253–1264.
[7] Corbee, R., H. Maas, A. Doornenbal and H. Hazewinkel. 2014. Forelimb and hindlimb ground reaction forces of walking cats: assessment and comparison with walking dogs. *The Veterinary Journal*, 202(1): 116–127.
[8] Armitano, C., S. Morrison and D. Russell. 2017. Upper body accelerations during walking are altered in adults with ACL reconstruction. *Gait & Posture*, 58: 401–408.
[9] Feigin, V. L., B. Norrving and G. A. Mensah. 2017. Global burden of stroke. *Circulation Research*, 120(3): 439–448.
[10] Johnson, W., O. Onuma, M. Owolabi and S. Sachdev. 2016. Stroke: a global response is needed. *Bulletin of the World Health Organization*, 94(9): 634.
[11] Vilela, P. and H. A. Rowley. 2017. Brain ischemia: CT and MRI techniques in acute ischemic stroke. *European Journal of Radiology*, 96: 162–172.
[12] Stroke Statistics [Online]. http://www.strokecenter.org/patients/about-stroke/stroke-statistics/ (accessed 2021).
[13] Pierre-Henri Tournier, Marcella Bonazzoli, Victorita Dolean, Francesca Rapetti and Frederic Hecht. 2017. Numerical modeling and high-speed parallel computing: new perspectives on tomographic microwave imaging for brain stroke detection and monitoring. *IEEE Antennas and Propagation Magazine*, 59(5): 98–110.
[14] Mohammed, B. J., A. M. Abbosh, S. Mustafa and D. Ireland. 2013. Microwave system for head imaging. *IEEE Transactions on Instrumentation and Measurement*, 63(1): 117–123.

[15] Fhager, A., S. Candefjord, M. Elam and M. Persson. 2018. Microwave diagnostics ahead: Saving time and the lives of trauma and stroke patients. *IEEE Microwave Magazine*, 19(3): 78–90.
[16] Mario R. Casu, Marco Vacca, Jorge A. Tobon, Azzurra Pulimeno and Imran Sarwar. 2017. A COTS-based microwave imaging system for breast-cancer detection. *IEEE Transactions on Biomedical Circuits and Systems*, 11(4): 804–814.
[17] Chandra, R., H. Zhou, I. Balasingham and R. M. Narayanan. 2015. On the opportunities and challenges in microwave medical sensing and imaging. *IEEE Transactions on Biomedical Engineering*, 62(7): 1667–1682.
[18] Shao, W. and T. McCollough. 2020. Advances in microwave near-field imaging: Prototypes, systems, and applications. *IEEE Microwave Magazine*, 21(5): 94–119.
[19] Islam, M., M. Mahmud, M. T. Islam, S. Kibria, M. Samsuzzaman et al. 2019. A low cost and portable microwave imaging system for breast tumor detection using UWB directional antenna array. *Scientific Reports*, 9(1): 1–13.
[20] Alqadami, A. S., A. Trakic, A. E. Stancombe, B. Mohammed et al. 2020. Flexible electromagnetic cap for head imaging. *IEEE Transactions on Biomedical Circuits and Systems*, 14(5): 1097–1107.
[21] Rezaeieh, S. A., A. Zamani and A. Abbosh. 2014. 3-D wideband antenna for head-imaging system with performance verification in brain tumor detection. *IEEE Antennas and Wireless Propagation Letters*, 14: 910–914.
[22] Mobashsher, A. and A. Abbosh. 2016. Compact 3-D slot-loaded folded dipole antenna with unidirectional radiation and low impulse distortion for head imaging applications. *IEEE Transactions on Antennas and Propagation*, 64(7): 3245–3250.
[23] Rokunuzzaman, M., A. Ahmed, T. C. Baum and W. S. Rowe. 2019. Compact 3-D antenna for medical diagnosis of the human head. *IEEE Transactions on Antennas and Propagation*, 67(8): 5093–5103.
[24] Salleh, A., C. Yang, T. Alam, M. Singh, M. Samsuzzaman et al. 2020. Development of microwave brain stroke imaging system using multiple antipodal vivaldi antennas based on raspberry Pi technology. *J. Kejuruterran*, 32: 1–6.
[25] Mohammadreza F. Imani, Jonah N. Gollub, Okan Yurduseven, Aaron V. Diebold and Michael Boyarsky. 2020. Review of metasurface antennas for computational microwave imaging. *IEEE Transactions on Antennas and Propagation*, 68(3): 1860–1875.
[26] Guohua Zhai, Xi Wang, Rensheng Xie, Jin Shi, Jianjun Gao, Bo Shi and Jun Ding. 2019. Gain-enhanced planar log-periodic dipole array antenna using nonresonant metamaterial. *IEEE Transactions on Antennas and Propagation*, 67(9): 6193–6198.
[27] Zada, M., I. A. Shah and H. Yoo. 2019. Metamaterial-loaded compact high-gain dual-band circularly polarized implantable antenna system for multiple biomedical applications. *IEEE Transactions on Antennas and Propagation*, 68(2): 1140–1144.
[28] Rezaeieh, S. A., M. A. Antoniades and A. M. Abbosh. 2016. Bandwidth and directivity enhancement of loop antenna by nonperiodic distribution of mu-negative metamaterial unit cells. *IEEE Transactions on Antennas and Propagation*, 64(8): 3319–3329.
[29] Musa, A., M. L. Hakim, T. Alam, M. H. Baharuddin, M. S. J. Singh et al. 2021. Dual-band metamaterial absorber for Ka-band satellite application. pp. 151–155. In: *2021 7th International Conference on Space Science and Communication (IconSpace)*, IEEE.
[30] Saddam, M. N. C., M. H. Baharuddin, M. L. Hakim, T. Alam, A. Musa et al. 2021. Single negative matamaterial absorber of K-band application. pp. 239–243. In: *2021 7th International Conference on Space Science and Communication (IconSpace)*, IEEE.
[31] Hakim, M. L., T. Alam, N. M. Sahar, N. Misran, M. F. Mansor et al. 2021. Elliptical slot metasurface high gain microstrip line antenna for sub-6 GHz 5G wireless communication. pp. 156–160. In: *2021 7th International Conference on Space Science and Communication (IconSpace)*, IEEE.
[32] Ahmad Musa, Mohammad Lutful Hakim, Touhidul Alam, Mohammad Tariqul Islam and Ahmed S. Alshammari. 2022. Polarization independent metamaterial absorber with anti-reflection coating nanoarchitectonics for visible and infrared window applications. *Materials*, 15(10): 3733.
[33] Hakim, M., T. Alam, M. Baharuddin and M. Islam. 2022. Frequency integration of dual-band hexagonal metamaterial resonator antenna for Wi-Fi and 5G wireless communication. In: *Journal of Physics: Conference Series*, 2250(1): IOP Publishing, p. 012001.

[34] Mohammad Lutful Hakim, Touhidul Alam, Md Shabiul Islam, M. Salaheldeen, Sami H. A. Almalki and Mohd Hafiz Baharuddin. 2022. Wide-oblique-incident-angle stable polarization-insensitive ultra-wideband metamaterial perfect absorber for visible optical wavelength applications. *Materials*, 15(6): 2201.
[35] Hakim, Mohammad Lutful, Abu Hanif, Touhidul Alam, Mohammad Tariqul Islam and Haslina Arshad. 2022. Ultrawideband polarization-independent nanoarchitectonics: a perfect metamaterial absorber for visible and infrared optical window applications. *Nanomaterials*, 12(16): 2849.
[36] Rahman, A., M. T. Islam, M. J. Singh, S. Kibria, M. Akhtaruzzaman et al. 2016. Electromagnetic performances analysis of an ultra-wideband and flexible material antenna in microwave breast imaging: To implement a wearable medical bra. *Scientific Reports*, 6(1): 1–11.
[37] Sohani, Behnaz, Banafsheh Khalesi, Navid Ghavami, Mohammad Ghavami, Sandra Dudley et al. 2020. Detection of haemorrhagic stroke in simulation and realistic 3-D human head phantom using microwave imaging. *Biomedical Signal Processing and Control*, 61: 102001.
[38] Rodriguez-Duarte, D. O., J. A. T. Vasquez, R. Scapaticci, L. Crocco, F. Vipiana et al. 2020. Brick-shaped antenna module for microwave brain imaging systems. *IEEE Antennas and Wireless Propagation Letters*, 19(12): 2057–2061.
[39] Alqadami, A. S., N. Nguyen-Trong, A. E. Stancombe, K. Bialkowski, A. Abbosh et al. 2020. Compact flexible wideband antenna for on-body electromagnetic medical diagnostic systems. *IEEE Transactions on Antennas and Propagation*, 68(12): 8180–8185.
[40] Hossain, A., M. T. Islam, M. E. Chowdhury and M. Samsuzzaman. 2020. A grounded coplanar waveguide-based slotted inverted delta-shaped wideband antenna for microwave head imaging. *IEEE Access*, 8: 185698–185724.
[41] Semenov, S., B. Seiser, E. Stoegmann and E. Auff. 2014. Electromagnetic tomography for brain imaging: From virtual to human brain. pp. 1–4. In: *2014 IEEE Conference on Antenna Measurements & Applications (CAMA)*, IEEE.
[42] Xiang, J., Y. Dong and Y. Yang. 2020. Multi-frequency electromagnetic tomography for acute stroke detection using frequency-constrained sparse Bayesian learning. *IEEE Transactions on Medical Imaging*, 39(12): 4102–4112.
[43] Alqadami, A. S., N. Nguyen-Trong, B. Mohammed, A. E. Stancombe, M. T. Heitzmann et al. 2019. Compact unidirectional conformal antenna based on flexible high-permittivity custom-made substrate for wearable wideband electromagnetic head imaging system. *IEEE Transactions on Antennas and Propagation*, 68(1): 183–194.
[44] Mobashsher, A. T., K. S. Bialkowski and A. M. Abbosh. 2016. Design of compact cross-fed three-dimensional slot-loaded antenna and its application in wideband head imaging system. *IEEE Antennas and Wireless Propagation Letters*, 15: 1856–1860.
[45] Rokunuzzaman, M., M. Samsuzzaman and M. T. Islam. 2016. Unidirectional wideband 3-D antenna for human head-imaging application. *IEEE Antennas and Wireless Propagation Letters*, 16: 169–172.
[46] Mobashsher, A. T., A. M. Abbosh and Y. Wang. 2014. Microwave system to detect traumatic brain injuries using compact unidirectional antenna and wideband transceiver with verification on realistic head phantom. *IEEE Transactions on Microwave Theory and Techniques*, 62(9): 1826–1836.
[47] Alqadami, A. S., K. S. Bialkowski, A. T. Mobashsher and A. M. Abbosh. 2018. Wearable electromagnetic head imaging system using flexible wideband antenna array based on polymer technology for brain stroke diagnosis. *IEEE Transactions on Biomedical Circuits and Systems*, 13(1): 124–134.
[48] Razzicchia Eleonora, Pan Lu, Wei Guo, Olympia Karadima, Ioannis Sotiriou, Navid Ghavami et al. 2021. Metasurface-enhanced antennas for microwave brain imaging. *Diagnostics*, 11(3): 424.
[49] Razzicchia, E., I. Sotiriou, H. Cano-Garcia, E. Kallos, G. Palikaras et al. 2019. Feasibility study of enhancing microwave brain imaging using metamaterials. *Sensors*, 19(24): 5472.
[50] Luukkonen, O., S. I. Maslovski and S. A. Tretyakov. 2011. A stepwise Nicolson–Ross–Weir-based material parameter extraction method. *IEEE Antennas and Wireless Propagation Letters*, 10: 1295–1298.

[51] Hakim, M. L., T. Alam, A. F. Almutairi, M. F. Mansor, M. T. Islam et al. 2021. Polarization insensitivity characterization of dual-band perfect metamaterial absorber for K band sensing applications. *Scientific Reports*, 11(1): 1–14.
[52] Hakim, M. L. et al. 2022. Polarization insensitive symmetrical structured double negative (DNG) metamaterial absorber for Ku-band sensing applications. *Scientific Reports*, 12(1): 1–18.
[53] Hakim, M. L., T. Alam, M. T. Islam, M. H. Baharuddin, A. Alzamil et al. 2022. Quad-band polarization-insensitive square split-ring resonator (SSRR) with an inner jerusalem cross metamaterial absorber for Ku-and K-band sensing applications. *Sensors*, 22(12): 4489.
[54] Bahl, I. J. 2003. *Lumped Elements for RF and Microwave Circuits*. Artech House.
[55] Bahl, I. J. 1977. A Designer's Guide to Microstrip Line.
[56] Garg, R., P. Bhartia, I. J. Bahl and A. Ittipiboon. 2001. *Microstrip Antenna Design Handbook*. Artech House.
[57] Bahl, I. and P. Bhartia. 2003. *Microwave Solid State Circuit Design*. Wiley-Interscience.
[58] Balanis, C. A. 2016. *Antenna Theory: Analysis and Design*. John Wiley & Sons.
[59] Hoque, A., M. Tariqul Islam, A. F. Almutairi, T. Alam, M. Jit Singh et al. 2018. A polarization independent quasi-TEM Metamaterial absorber for X and Ku band sensing applications. *Sensors*, 18(12): 4209.
[60] Kibria, S., M. Samsuzzaman, M. T. Islam, M. Z. Mahmud, N. Misran et al. 2019. Breast phantom imaging using iteratively corrected coherence factor delay and sum. *IEEE Access*, 7: 40822–40832.
[61] IEEE Standards Coordinating Committee. 1992. IEEE standard for safety levels with respect to human exposure to radio frequency electromagnetic fields, 3 kHz to 300 GHz. *IEEE C95*.

Chapter 8

Metamaterial Inspired Stacked Antenna Based Microwave Brain Imaging

Mohammad Tariqul Islam[1,*] and *Amran Hossain*[2]

8.1 Introduction

Globally, brain-related diseases such as brain tumors are an enormous burden for people and healthcare systems. A **brain tumor** is one of the severest reasons for death because it damages the central tissues of the brain. A brain tumor is a mass growth of abnormal cells formed inside the brain that transforms into brain cancer [1]. Brain cancer can be a threat to human life and critically affects longevity. Brain cancer possibilities have grown over time due to the uncontrolled growth of brain tumors, and it is the 9th leading cause of death for women, men, and children [2]. Brain tumors might be fatal, crucially affecting the quality of longevity and changing the dynamics of life for patients and their families. The brain tumor is mainly classified into two categories, benign and malignant [3]. The number of brain tumor cases around the globe is growing at a frightening rate. According to a 2022 survey of the National Brain Tumour Society (NBTS), 88,970 people in the USA are living with a primary brain tumor diagnosis, off which 63,040 are **benign** and 25,930 are **malignants** [2]. The benign tumors are non-cancerous cells with a homogeneous structure and regular shape, and they do not invade neighbouring tissues or spread to other parts of the body. In contrast, malignant tumours are cancerous cells with heterogeneous structures and irregular shapes. In addition, benign tumours have a

---

[1] Department of Electrical, Electronic and Systems Engineering, Universiti Kebangsaan Malaysia, Bangi, Selangor, Malaysia.
[2] Department of Computer Science and Engineering, Dhaka University of Engineering & Technology, Gazipur, Bangladesh.
Email: amran_duet38@duet.ac.bd
* Corresponding author: tariqul@ukm.edu.my

222 *Metamaterial for Microwave Applications*

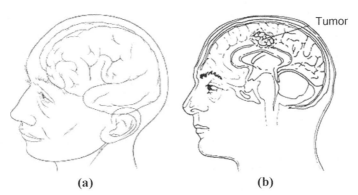

Figure 8.1. Human brain (a) Healthy brain (b) Brain with a tumor.

slower growth rate, and malignant tumours can grow uncontrollably. The schematic representation of the healthy brain and brain tumour is illustrated in Figure 8.1.

The death rate percentage increases due to the invasive features of the tumors. The **symptoms** of a brain tumor vary greatly and depend on the brain tumor's location, size, and rate of growth. The general symptoms caused by brain tumors may include a change in the pattern of headaches, unexplained nausea, vomiting, vision problems, gradual loss of sensation, speech difficulties, difficulty in decision making and personality or behavior changes. However, **early detection**, classification, monitoring, and proper investigation can reduce the death rate and increase the human survival rate. In addition, automatic classification of brain tumors from reconstructed images is significant for clinical evaluation and treatment preparation of brain cancers. Brain tumor analysis, classification, and detection are important issues for radiologists and medical doctors. Accurate and timely investigation of brain cancer is imperative for appropriate treatment. Brain tumor classification can be a vital technique in medical imaging applications that classify the specific tumor from the head images. Therefore, automated detection and classification of brain tumors utilizing deep learning-based microwave brain imaging is crucial for timely treatment and recovery of the afflicted patient.

Currently, different types of imaging technologies: PET (positron emission tomography), MRI (magnetic resonance imaging), ultrasound screening, X-ray screening, and CT (computed tomography) screening are utilized to diagnose brain tumors in modern healthcare facilities. These standard imaging techniques help physicians and radiologists identify different types of health-related diseases like brain cancer. Motivated by the mentioned limitations, it is desirable to implement a brain imaging system integrated with deep learning techniques that can be applied to automatically detect, segment, and classify brain tumors from the reconstructed brain images. **Microwave Brain Imaging** (MBI) is nowadays an emerging technology for medical diagnostic systems that has drawn significant interest among researchers all over the world because of its cost-effective, non-ionizing, non-invasive, and portable features. This technology can reconstruct the brain images, hence it is essential to detect and classify brain tumors from the reconstructed brain images. In this situation, deep learning techniques can be reliable in detecting, segmenting, and classifying the target tumors from the reconstructed microwave brain (RMB) images.

Deep Learning (DL) is a type of **Machine Learning** (ML) modality that can use the convolutional neural network (CNN) model to classify and detect target objects. Deep learning has gained massive popularity in scientific computing, image processing, and data analysis, and its algorithms are widely used by medical applications to solve complex problems. Deep learning uses artificial neural networks (ANN) to perform sophisticated computations on large amounts of data, such as images. Deep learning algorithms train machines by learning from input samples. During the training process algorithms use unknown elements in the input distribution to extract features, group objects and discover useful data patterns. The schematic diagram of the deep learning model is depicted in Figure 8.2. The CNN, also known as ConvNets, consists of multiple layers and is mainly used for image processing and object detection. The deep learning model is primarily divided into four parts: (i) Input image as a dataset, (ii) Feature extraction, (iii) Fully connected layer (FCL), and (iv) Output layer as a classification or detection layer. The images as a dataset are fed to the input of the model. Then, the feature extraction layer extracts the upcoming image samples.

The extraction layer consists of many convolutional layers for feature extractions with several filters to perform the convolution operation. CNNs have a rectified linear unit (ReLU) layer to perform operations on elements. It is a non-linear activation function that operates in layers and produces the output as a rectified feature map. The rectified feature map next feeds into a pooling layer. Max-pooling is a down-sampling operation that reduces the dimensions of the feature map. The pooling layer then flattens the resulting two-dimensional arrays from the pooled feature map into a single, long, continuous and linear vector. A **fully connected layer** forms when the flattened matrix from the pooling layer is fed as an input, classifying and identifying the images as the desired output. However, deep learning techniques with CNNs have become the state-of-the-art approaches for brain tumor detection and segmentation in a microwave imaging system. On the other hand, image classification is an essential role of microwave image analysis, in which deep convolutional neural networks (DCNNs) are used to classify brain tumors. The image classification identifies whether the target object or disease is present or not in the image under investigation image. Thus, there is a demand to design and implement a deep learning model that can automatically segment and classify brain tumors from the RMB images in the MBI system.

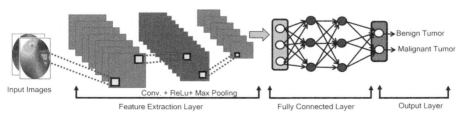

Figure 8.2. Basic architecture of a deep learning model.

8.2 Importance of deep learning in current brain imaging technologies

Currently, different categories of imaging technologies: Computed Tomography (CT) scanning, X-ray screening, Magnetic Resonance Imaging (MRI), Biopsy, Positron Emission Tomography (PET), and Ultrasound screening are used to diagnose brain tumors in modern medical healthcare facilities. These imaging standards help physicians and radiologists identify different types of health-related diseases like brain cancer. However, these techniques still have some drawbacks for the diagnostic process. The major drawback of these imaging technologies is the growing cancer risk because of the high radiation dose [4], the danger for pregnant women and old patients [5] high ionization of brain cells [6] the risk and high expense of pacemakers and implantable patient cardioverters [3, 7], high time consumption, and less susceptibility [4]. **Microwave imaging** (MWI) demonstrated excellent potential to researchers for medical applications due to attractive features such as non-ionizing radioactivity, low power penetration capability, non-invasive nature, absence of ionization risk for the human body and low profile cost effectiveness [8–12]. Recently, researchers have been using microwave imaging technology to overcome the drawbacks of the traditional medical imaging modalities. A microwave brain imaging (MBI) system has been utilized to identify the brain tumors inside the brain [8, 10, 12–16]. The MBI system consists of an **antenna array**, mechanical devices, and an image processing unit. The antenna is an essential piece of equipment, and its characteristics are a significant factor in producing the desired image. The MBI system demand wideband compact antennas with unidirectional radiation characteristics for efficient and significant data acquisition within the frequency range of 1 GHz to 4 GHz. A single antenna transmits the microwave signals towards the region of interest, and receiver antenna(s) receive the backscattering signals. Different antennas have been offered to develop a MBI system to detect brain tumours in **reconstructed microwave brain** (RMB) images. Most of them lack compactness, low bandwidth, higher gain, and directionality towards the head. For a microwave brain imaging system, the antenna should have the following properties: high gain, small size, directive radiation, wide bandwidth with higher efficiency, model simplicity, compatible penetration into head tissue, and low specific absorption rate (SAR) [9, 17, 18].

However, the main limitations of the current MBI systems are (i) Noisy, blurry, and low-resolution reconstructed microwave brain (RMB) images created by the system, (ii) Complicated identification of the tumor with location from RMB images for incompetent physicians and radiologists (iii) extremely time-consuming manual detection, classification, and segmentation of tumors (iv) Difficulty in locating tumor regions in RMB images by automatic detection, and (v) Difficulty in classifying the tumor into benign and malignant from RMB images due to its pattern, size, and shape.

So, to overcome such issues, researchers have been applying deep learning techniques in microwave imaging applications [9, 19–22]. Different types of deep learning models such as YOLOv3, ResNet50, DenseNet201, Unet, faster R-CNN, segmentation and classification models have been used to identify, segment, localize, and classify tumors from MRI images. However, all models use deeper architectures for analysing the MRI images. Thus, it is natural that these networks require longer training and inference durations and are not suitable for portable device deployment. According to a recent study on MBI systems, no deep learning model has been applied to automatically detect and classify tumors from the RMB images. Thus, there is a strong demand to implement a MBI system and design a lightweight deep learning-based classification and segmentation model that can automatically segment the tumor region from the reconstructed microwave brain (RMB) images in the MBI system. It is also, necessary to design and implement a lightweight classification model to classify the tumors from the RMB images with an improved performance classification.

8.3 Metamaterial loaded stacked antenna

This chapter provides brief details of the design and analysis procedure of the net-shaped triple split ring resonator (NTSRR) based metamaterial loaded stacked antenna along with its specifications. The antenna's geometry and design evolution analysis are presented. Moreover, the parametric analysis also analyses the evolution diagrams and their simulated scattering parameters, realized gain, and efficiency. In addition, the radiation directivity with stacked antenna pattern and measurement outcomes are also explained.

8.3.1 Stacked antenna structure design and analysis

The schematic diagram of the **metamaterial** (MTM) loaded **stacked antenna's** structure is depicted in Figure 8.3. The antenna consists of three layers: Top Layer (TL), Middle Layer (ML), and Bottom Layer (BL). The MTM array unit cells are used in three layers. A 2 mm air gap thickness is considered between the TL and ML and the ML and BL. The design mechanism of the antenna starts with printing on low loss Rogers RT5880 and RO4350B substrates. The Rogers RT5880 substrate (Relative permittivity $\varepsilon_r = 2.2$, loss tangent $\delta = 0.0009$, thickness $h_1 = h_2 = 1.575$ mm) material is used in the TL and MLs, whereas the Rogers RO4350B substrate (Relative permittivity $\varepsilon_r = 3.66$, loss tangent $\delta = 0.0037$, thickness $h_3 = 1.524$ mm) material is used in the BL. The ML is attached with the TL, and the BL is attached with the ML by using a 2 mm thick double-sided foam tape. The main **radiating patch** and feed line are designed

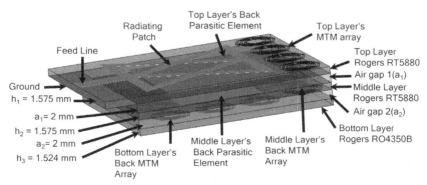

Figure 8.3. Overall design structure of the stacked antenna.

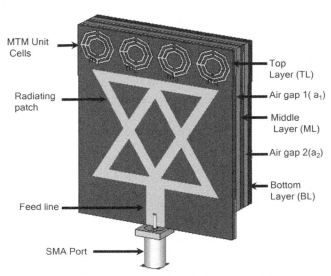

Figure 8.4. MTM loaded stacked antenna model.

on the TL. A 50Ω SMA port is connected to the feeding line that feeds the signal to the antenna. The antenna's feed line width is responsible for the 50Ω impedance matching. The NTSRR MTM array is employed on the TL, ML, and BL to enhance the antenna's performance. A 1 × 4 MTM array (m_1 to m_4) is used on the top layer and backside of the middle layer, and a 3 × 2 MTM array (m_1 to m_6) is used on the backside of the bottom layer to enhance the **radiation directionality**, gain, and bandwidth of the antenna. The proposed antenna's overall layer-based stacked structure layout is illustrated in Figure 8.3, and its simulated model is illustrated in Figure 8.4. The three layers of the antenna, including two air gaps (a_1 and a_2) make it a **stacked antenna** structure. The stacked structure is responsible for the high performance of the antenna.

8.3.2 Stacked antenna geometry and design evolution analysis

The wideband antenna is required in microwave brain imaging (MBI) systems with special features such as higher gain, efficiency, and directional radiation characteristics, where the antenna operates in the frequency range of 1 GHz to 4 GHz. The MBI system requires sufficient **signal penetration** to detect abnormalities in the brain. An NTSRR MTM loaded stacked **wideband antenna** is designed, which achieves microwave brain imaging features. The geometry and evolution of the proposed antenna are depicted in Figure 8.5. The antenna consists of three main substrate layers: Top layer (TL), Middle layer (ML), and Bottom layer (BL), including two air gaps. The air gaps are considered between the TL and ML & ML and BL. The **optimized parameters** with their values are presented in Table 8.1.

The antenna is printed on cost-effective Rogers substrate materials. The TL and ML are printed on the Rogers RT5880 substrate, and the BL is printed on the Rogers RO4350B substrate. In the initial step, a main radiating patch and feed line are designed on the TL. Two triangle-shaped patches are attached in opposite directions

Table 8.1. The geometric parameters of NTSRR loaded stacked antenna.

Parameters	Value (mm)	Parameters	Value (mm)	Parameters	Value (mm)
L	50	C	10.68	l_4	7
W	40	D	14.05	l_5	24
L_1	43	F	10.50	l_6	18
W_1	40	f_w	4.80	g	0.50
L_2	43	L	32.91	a_1	2.00
W_2	40	l_1	28	a_2	2.00
A	16.45	l_2	27	...	...
B	10.68	l_3	40	...	...

to make the main radiating patch. The dimensions of the TL are 50 mm × 40 mm (L × W). The partial ground ($l_3 \times l_4$ mm²) and a rectangle-shaped parasitic element ($l_1 \times l_2$ mm²) are attached to the backside of the TL. The lowest operating frequency (f_l), patch width (l), and feed line width (f_w) are calculated by the following equations:

$$f_w = \frac{7.48 \times h}{e^{\left(z_0 \frac{\sqrt{\varepsilon_r + 1.41}}{87}\right)}} - 1.25 \times t \quad (8.1)$$

$$\varepsilon_{eff} = \frac{(\varepsilon_r + 1)}{2} + \frac{(\varepsilon_r - 1)}{2}\left[1 + \frac{10 \times h}{l}\right]^{-0.5} \quad (8.2)$$

$$f_l = \frac{c}{2 \times l \sqrt{\varepsilon_{eff}}} \quad (8.3)$$

where, f_r denotes the resonance frequency, h denotes the thickness of the substrate, t denotes the copper thickness, c denotes the speed of light in free space and ε_{eff} denotes the substrate's effective dielectric constant. The **CST (Computer simulation Tool) simulator** software is used to optimize the geometric parameters for achieving the required band and special features of the antenna. The feed line width is responsible for the 50Ω **impedance matching**.

Figure 8.5 depicts the evolution of the proposed antenna. The simulated reflection coefficient $|S_{11}|$, efficiency, and realized gain curves of the stacked antenna's evolution are illustrated in Figures 8.6(a) and 8.6(b), respectively.

Figure 8.5(a) shows the top view of Antenna-1 (TL without MTM). It is observed from Figure 8.6, that the frequency band of Antenna-1 is 1.94 GHz to 2.94 GHz with a resonance frequency at 2.29 GHz, and the maximum realized gain is 3.36 dBi at 2.96 GHz with 79.13 percent efficiency. Thereafter, when a 1 × 4 MTM array is applied at the top of the patch in Antenna-2 (TL with MTM), the frequency band increases towards both the lower and upper frequencies due to the MTM characteristics and achieves a frequency band of 1.89 GHz to 3.04 GHz with a resonance at 2.24 GHz. The attained maximum **realized gain** of Antenna-2 is 3.69 dBi at 2.96 GHz with 81.32 percent efficiency.

228 *Metamaterial for Microwave Applications*

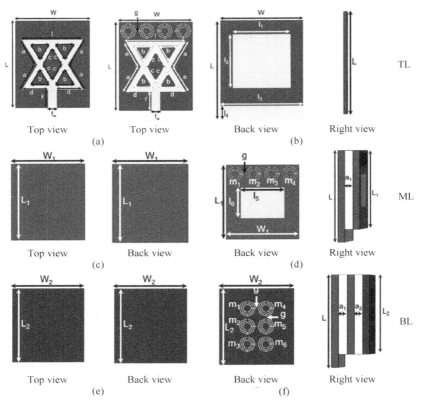

Figure 8.5. Design evaluation of NTSRR MTM loaded stacked antenna.

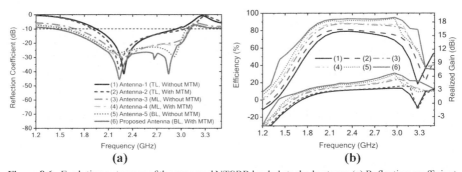

Figure 8.6. Evolution outcomes of the proposed NTSRR loaded stacked antenna (a) Reflection coefficient (b) Efficiency and realized gain.

In the second step, the ML (Antenna-3) without a MTM array is placed 2 mm away and attached with Antenna-2 by using a double-sided foam tape. The air gap (a_1) between Antenna-2 and Antenna-3 is 2 mm. The dimensions of the ML are 43 mm × 40 mm ($L_1 \times W_1$). The observed operating band, gain, and efficiency of antenna-3 are 1.69 GHz to 3.10 GHz, 5.38 dBi at 2.96 GHz, and 87.84 percent, respectively. When another 1 × 4 MTM array (m_1–m_4) and a rectangular-shaped ($l_5 \times l_6$) parasitic element are attached to the backside of the ML (Antenna-4), as shown in Figure 8.5(d), the gain increases. In addition, the radiation directivity and bandwidth also increase. The operating frequency band is 1.61 GHz to 3.10 GHz, with two resonances at 2.21 GHz and 2.92 GHz. The recorded simulated realized gain is 5.68 dBi at 2.96 GHz and the efficiency is 88.59. The right view of the two-layered stacked antenna (Antenna-4) is shown in Figure 8.5(d). In the third step, the BL (Antenna-5) without a MTM array is positioned 2 mm away from the ML, which is shown in Figure 8.5(e). The air gap (a_2) between antenna-4 and antenna-5 is 2 mm. The dimensions of BL are 43 mm × 40 mm ($L_2 \times W_2$). The achieved frequency band, realized gain, and efficiency of antenna-5 are 1.55 GHz to 3.06 GHz, 5.54 dBi, and 93.89 percent at 2.96 GHz, respectively. Finally, when a 3 × 2 MTM array (m_1–m_6) is applied at the backside of the BL (proposed antenna), it gives better outcomes than other antennas in terms of **bandwidth, radiation directionality**, realized gain, and efficiency. The resultant simulated operating band of the proposed antenna is 1.43 GHz to 3.12 GHz, with a fractional bandwidth (FBW) of 74.45 percent for the 2.27 GHz central frequency. It also produces three resonances at 2.24 GHz, 2.66 GHz, and 2.84 GHz. The back and right view of the three layer-based proposed antenna is illustrated in Figure 8.5(f). The recorded maximum realized gain and efficiency are 6.74 dBi and 95.06 percent at 2.96 GHz, respectively, shown in Figure 8.6. The overall performance summary of the antenna evolution is presented in Table 8.2.

In the antenna design, it is noticeable that the overall performance of the antenna is enhanced due to the MTM array element effectiveness in all layers and stacked architecture. The **negative permittivity** characteristics of MTM array elements enhance the overall bandwidth. Also, it produces extra resonances and improves radiation efficiency by resulting in an additional **electromagnetic coupling** between radiating patches and layers. In addition, MTM unit cells increase the electrical length and enhance the current flow on the layer surface. As a result, it produces a strong electrical coupling between the antenna layers which leads to improved gain, radiation directivity, and efficiency.

Table 8.2. Simulated performance of NTSRR MTM loaded stacked antenna.

Evaluation Steps	Operating Frequency (GHz)	Bandwidth (GHz)	Resonance Frequency (GHz)	Realized Gain (dBi)
Antenna-1	1.94 ~ 2.94	1.00	2.29	3.36
Antenna-2	1.89 ~ 3.04	1.15	2.24	3.86
Antenna-3	1.69 ~ 3.10	1.41	2.29	5.38
Antenna-4	1.61 ~ 3.10	1.49	2.27, 2.92	5.68
Antenna-5	1.55 ~ 3.06	1.51	2.21, 2.91	6.54
Proposed	1.43 ~ 3.12	1.69	2.24, 2.66, 2.84	6.74

8.3.3 Parametric analysis of MTM loaded stacked antenna

In Table 8.1, there are 22 parameters are presented, which are used to demonstrate the overall antenna structure. The antenna consists of three layers with a feeding line attached to the top layer. The ground plane is attached to the backside of the top layer. Especially, the length of the ground plane (l_4), two air gaps (a_1 and a_2), and used MTM array elements in the layers are responsible for getting the desired frequency band. The **parametric analysis** based on the variations in the length of the ground plane, air gaps, and MTM unit cell is presented in Figure 8.7. When the length of the ground plane is varied, and the remaining parameters remain constant, the resulting reflection coefficient is shown in Figure 8.7(a). When $l_4 = 6$ mm, the achieved operating band is 1.71 GHz to 3.08 GHz with a resonance at 2.18 GHz, whereas when $l_4 = 8$ mm, the achieved operating band is 1.87 GHz to 3.09 GHz with a resonance at 2.33 GHz. However, if the length is set at, $l_4 = 7$ mm, then the antenna shows better outcomes than others, and the resultant band is 1.43 GHz to 3.12 GHz with three resonances at 2.24 GHz, 2.66 GHz, and 2.84 GHz, respectively. The gaps (a_1 and a_2) between layers are significant parameters to give the antenna a stacked structure.

The reflection coefficient due to changes in air gaps between the layers is illustrated in Figure 8.7(b). When no air gaps ($a_1 = 0$ and $a_2 = 0$) are considered, then the antenna shows a narrow band with a low reflection coefficient. If the

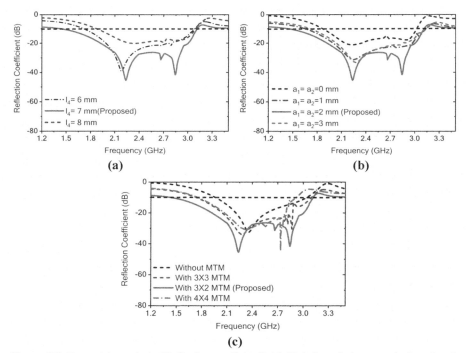

Figure 8.7. Parametric analysis "Reflection coefficient" (a) Variation of the ground plane length (b) Variations of the air gaps between layers (c) Variations of the MTM array elements in the BL.

middle air gap is gradually increased, the operating band increases due to the effect of the substrate and MTM unit cell, but it gets saturated at a certain value. When $a_1 = a_2 = 1$ mm, then the operating frequency band of the antenna is 1.64 GHz to 3.13 GHz, with two resonances at 2.20 GHz and 2.90 GHz providing a low reflection coefficient. When $a_1 = a_2 = 2$ mm, the antenna, with a 3 mm air gap, achieved the highest **operating band**. The MTM unit cell is an important element for the antenna because the MTM array elements can change and enhance the antenna performance and increase the bandwidth, radiation directivity, and gain. This analysis, shows the MTM cell effects when used in the bottom layer of the antenna.

The **reflection coefficient** of the antenna with the MTM arrays in BLs is depicted in Figure 8.7(c). Initially, when no MTM unit cell is used in the antenna layers, it shows a resonance at 2.35 GHz, and the functioning band is 1.96 GHz to 3.04 GHz. When 3 × 3 MTM array elements are applied to the backside of the bottom layer, the antenna's operating frequency is 1.73 GHz to 3.09 GHz, with two resonances at 2.27 GHz and 2.86 GHz, respectively, giving a high reflection coefficient. The operating frequency shifts towards the lower frequencies due to MTM effects, but the resonances are distorted at the upper frequencies. The **distorted resonances** are also observed when the 4 × 4 MTM array elements are applied and show an operating band narrower than others. However, if 3 × 2 MTM array elements are applied, then the antenna shows a distortion free frequency band, which covers a bandwith of 1.43 GHz to 3.12 GHz with three resonances at 2.24 GHz, 2.66 GHz, and 2.84 GHz, respectively.

8.4 Stacked antenna prototype fabrication and performance analysis

After the completion of the design and analysis of the metamaterial-loaded stacked antennas, fabrication and measurement of the designed prototypes is required. The geometric parameters of the metamaterial-loaded antenna are collected from the CST simulation software for further processing with **fabrication**. Furthermore, the **reflection coefficient,** realized gain, efficiency, and radiation patterns are measured after completing the fabrication process. A power network analyzer (PNA) is the typical equipment used to measure antenna performance. This PNA is a form of radiofrequency (RF) network analyser widely used for RF designs and applications. It enables the user to measure the RF performance of the microwave devices to be characterized as scattering parameters. The Agilent P-series PNA (Agilent N5227A) is used to collect the measured results from the two equipped measurement ports. Figure 8.8 illustrates the different pictorial views of the fabricated stacked antenna prototype. The fabricated antenna consists of three layers, including two air gaps. Initially, the top, middle, and bottom layers are printed on the proposed substrate materials. The middle layer is attached 2 mm away from the top layer using double-sided foam tape. Then the bottom layer is attached 2 mm away from the middle layer by using double-sided foam tape. The double-sided foam tape is 2 mm thick. The **prototype measurement** is performed using PNA (Power Network Analyzer, PNA-N5227A).

232 Metamaterial for Microwave Applications

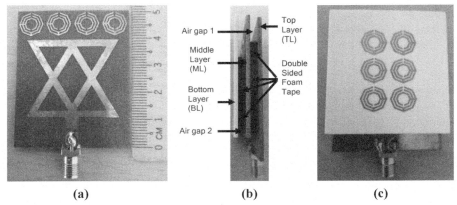

Figure 8.8. Fabricated prototype of the stacked antenna (a) Top view (b) Side view (c) Bottom view.

The simulated and measured reflection coefficient is depicted in Figure 8.9. From Figure 8.9, it is observed that the simulated operating band is 1.43 GHz to 3.12 GHz with 74.45% **fractional bandwidth** (FBW), whereas the measured operating band is 1.37 GHz to 3.16 GHz with 79.20% FBW. It is seen in the measured result that the antenna generates three resonances at 2.39 GHz, 2.57 GHz, and 2.81 GHz under −35 dB. The first resonance has shifted towards the upper frequency from 2.24 GHz to 2.39 GHz (150 MHz). The second resonance has shifted towards a lower frequency from 2.66 GHz to 2.57 GHz (90 MHz) and the third resonance is approximately the same as the simulated resonance. However, the measured and simulated outcomes show good agreement. In addition, the Satimo near the filed chamber is utilized to perform the measurement of the **radiation pattern**, realized gain, and efficiency of the fabricated stacked antenna prototype. Figure 8.10 illustrates the antenna's measured and simulated realized gain. Figure 8.10 shows that the measured maximum

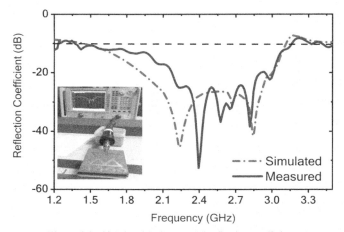

Figure 8.9. Simulated and measured reflection coefficient.

Metamaterial Inspired Stacked Antenna Based Microwave Brain Imaging 233

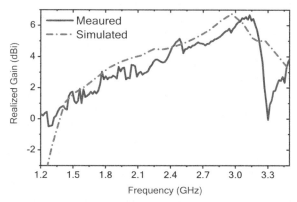

Figure 8.10. Measured and simulated realized gain.

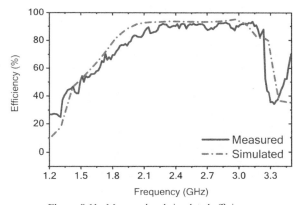

Figure 8.11. Measured and simulated efficiency.

realized gain is 6.67 GHz at 3.13 GHz, whereas the simulated maximum realized gain is 6.74 GHz at 2.96 GHz. Also, as investigated in Figure 8.11, the maximum measured **efficiency** is 93.0 percent, whereas the simulated maximum efficiency is 95.06 percent. The simulated and measured outcomes exhibit good agreement in both gain and efficiency between them.

8.5 Radiation characteristic analysis of the stacked antenna

It is noticeable that the directional radiation characteristic of the antenna is required in microwave brain imaging applications. Thus, the far-field and **near-field radiation** patterns of the proposed antenna are investigated. It is also notable that the antenna must operate in the far-field region when it works in the brain imaging system. Thus, the **far-field radiation** investigation is necessary because of the high permittivity of the human brain. The simulated 2D far-field radiation patterns in the E-plane ($\varphi = 0°$) and H-plane ($\varphi = 90°$) of layer-wise antenna evolution (Ant.-1, Ant.-2, Ant.-3, Ant.-4, Ant.-5, and proposed antenna) at 2.24 GHz, 2.66 GHz, and

234 *Metamaterial for Microwave Applications*

2.84 GHz are illustrated in Figure 8.12(a–i) It is observed from Figure 8.12, the back lobe is reduced due to the use of the MTM on the stacked layers of the antenna, as no side loop is created. As a result, the front-to-back ratio (FBR), gain, radiation efficiency, and **directivity** of the antenna are increased. In addition, the radiation directivity at a **higher frequency** is more directive rather than the lower frequency. Also, the radiation pattern in the H-plane of the antenna exhibits successively wider beamwidths towards the boresight than the E-plane. That means the antenna mostly

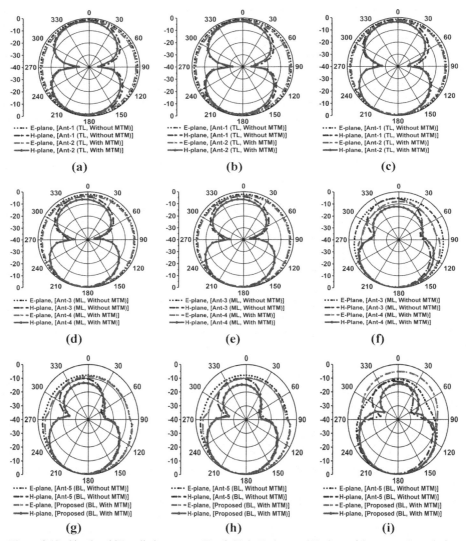

Figure 8.12. Simulated 2D radiation pattern (Far-field) in E-plane and H-plane of the antenna's evolution for: (a) Ant-1 and Ant-2 at 2.24 GHz (b) Ant-1 and Ant-2 at 2.66 GHz (c) Ant-1 and Ant-2 at 2.84 GHz (d) Ant-3 and Ant-4 at 2.24 GHz (e) Ant-3 and Ant-4 at 2.66 GHz (f) Ant-3 and Ant-4 at 2.84 GHz (g) Ant-5 proposed at 2.24 GHz (h) Ant-5 proposed at 2.66 GHz (i) Ant-5 proposed at 2.84 GHz.

Metamaterial Inspired Stacked Antenna Based Microwave Brain Imaging 235

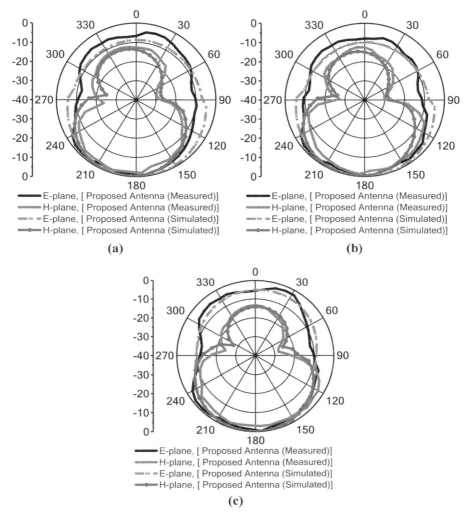

Figure 8.13. Simulated and measured radiation patterns in E and H planes of the proposed antenna at (a) 2.24 GHz (b) 2.66 GHz (c) 2.84 GHz.

radiate towards the − Z direction. The simulated and measured far-field radiation patterns of the stacked antenna at 2.24 GHz, 2.66 GHz, and 2.84 GHz are portrayed in Figure 8.13(a–c), respectively.

The measured and simulated radiation patterns showed good agreement with each other in the entire operating band. It is seen that the antenna shows **directional radiation characteristics**, which satisfy the imaging system requirements. Moreover, the antenna is placed near the head model for imaging purposes; hence verification of the near field characteristics of the stacked antenna is required. Thus, the antenna is measured in the **near-field region**. Figure 8.14 presents the near-field radiation measurement in the H-plane scenario. It is observed that the proposed prototype

236 *Metamaterial for Microwave Applications*

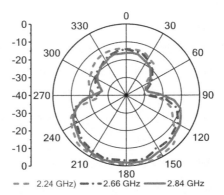

Figure 8.14. Measured near-field in H-plane at 2.24 GHz, 2.66 GHz, and 2.84 GHz.

demonstrates a directional characteristic for 2.24 GHz, 2.66 GHz, and 2.84 GHz, respectively, and shows an almost symmetrical pattern in both the far and near fields.

8.6 Performance analysis of the stacked antenna with head model

Different views and cutting plane views of the Hugo model with a single stacked antenna are depicted in Figure 8.15. A six-layered **tissue-mimicking head phantom** model with different layers and tumor(s) is fabricated to investigate the antenna's performance. The tumors are placed at different positions in the model. Figure 8.16(a) represents the head model with a single benign tumor, and Figure 8.16(b) represents the head model with a benign and a malignant tumor. The simulated and measured S-parameters of the head model with and without tumour using a single stacked antenna are presented in Figure 8.16(c). The antenna showed a good agreement between simulated and measured outcomes. The reflection coefficients are slightly decreased with resonance frequencies and shifted towards **higher frequency** due to the head tissues and tumour dielectric properties but remain stable within the required operating band because of the effectiveness of the MTM array in different layers.

Furthermore, the microwave signal penetration to the head is also investigated to evaluate the antenna effectiveness. Figure 8.17 illustrates the E and H field penetration inside the head model. The E-field and H-**field distributions** in the

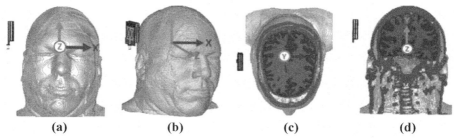

Figure 8.15. Different views and planes of Hugo Head model with a single stacked antenna (a) Front view (b) Perspective view (c) Transverse plane (d) Coronal plane.

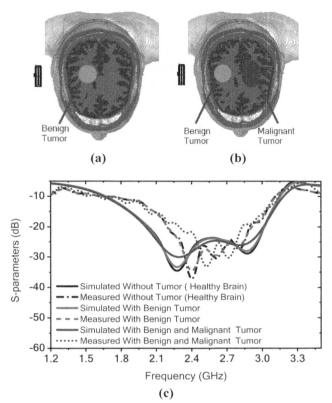

Figure 8.16. Simulated and measured results with head model including tumors (a) Head model with a benign tumor (b) Head model with a benign and a malignant tumor (c) S-parameters.

xz-plane at 2.24 GHz and 2.84 GHz are shown in Figure 8.17(a–b) and Figure 8.17(c–d), respectively. Besides, the E-field and H-field distributions in the yz-plane at 2.24 GHz and 2.84 GHz are shown in Figure 8.17(e–f) and Figure 8.17(g–h), respectively. It is observed from Figure 8.17 that the antenna continues to show directionality to the model, and the signal is able to penetrate the head tissues covering two-thirds of the head. Figure 3.18(a) illustrates the simulated nine-antenna array set-up, where antenna one acts as a **transmitter** and the remaining eight antennas act as **receivers**. The received S-parameters without tumour (i.e., Healthy brain), with a single benign tumor, and a single benign and malignant tumor are depicted in Figure 8.18(b–d), respectively.

It is examined from the S-parameters that there is **significant distortion** of the **backscattered signals** for the benign and malignant tumors. These distortion differences happened due to the presence of high dielectric properties of the tumors and the peak resonance frequencies algorithm. However, the nine-antenna array setup covers the complete area of the head, which carries all the sufficient information for **reconstructing the brain images**.

238 *Metamaterial for Microwave Applications*

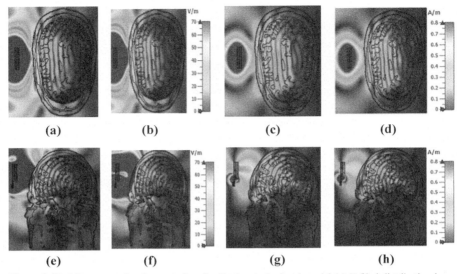

Figure 8.17. Microwave signal penetration distribution to the head model (a) E-filed distribution in xz plane at 2.24 GHz (b) E-filed distribution in xz plane at 2.84 GHz (c) H-filed distribution in xz plane at 2.24 GHz (d) H-filed distribution in xz plane at 2.84 GHz (e) E-filed distribution in yz plane at 2.24 GHz (f) E-field distribution in yz plane at 2.84 GHz (g) H-filed distribution in yz plane at 2.24 GHz (h) H-filed distribution in yz plane at 2.84 GHz.

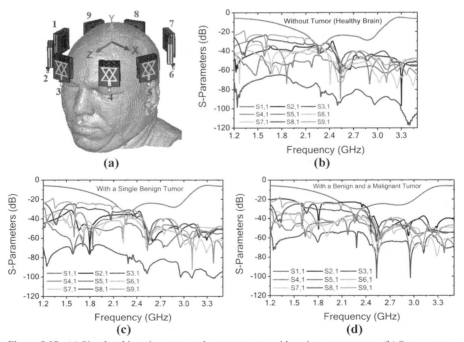

Figure 8.18. (a) Simulated imaging setup and measurements with a nine-antenna array (b) S-parameters without tumor (c) S-parameters with a single benign tumor (d) S-parameters with a benign and a malignant tumor.

8.7 Specific absorption rate (SAR) analysis of stacked antenna

This section explains the analysis of the specific absorption rate (SAR) because **microwave radiations** are extremely harmful to the human brain when it is exposed to them. So, the **SAR analysis** is an essential consideration for microwave brain imaging systems to ensure operational safety. The SAR is measured by the following formula:

$$SAR = \frac{|E_f|^2 . \sigma}{M_d} \qquad (8.4)$$

where, E_f represents the electric fields, M_d represents the mass density, and σ represents the conductivity of the human brain tissue. According to the standard regulations (IEEE radiation exposure standard regulations 1992), the highest SAR must not exceed 1.60 watt per kilogram for one gm of tissue, and two watt per kilogram for 10 gms of tissue. For investigating the SAR by utilizing the proposed stacked antenna, one milli watt power is applied as an input to the antenna (positions: 1, 3,5,7, and 8), and the SAR is observed at 2.24 GHz, 2.66 GHz, and 2.84 GHz, respectively, for 1 gm and 10 gm of tissue. The measured **SAR** values for the stacked antenna are presented in Table 8.3. The investigation shows that the observed maximum SAR

Table 8.3. SAR analysis of stacked antenna.

Antenna position	Frequency (GHz)	Without MTM 1 gm	Without MTM 10 gm	With MTM 1 gm	With MTM 10 gm
	2.24	0.0036	0.0023	0.0016	0.0013
	2.66	0.0040	0.0025	0.0018	0.0013
	2.84	0.0038	0.0021	0.0019	0.0012
	2.24	0.0025	0.0022	0.0012	0.0008
	2.66	0.0024	0.0020	0.0011	0.0007
	2.84	0.0023	0.0019	0.0011	0.0005
	2.24	0.0047	0.0031	0.0029	0.0018
	2.66	0.0046	0.0022	0.0020	0.0017
	2.84	0.0041	0.0026	0.0010	0.0005
	2.24	0.0030	0.0018	0.0012	0.0006
	2.66	0.0025	0.0015	0.0010	0.0004
	2.84	0.0025	0.0015	0.0011	0.0004
	2.24	0.0029	0.0014	0.0018	0.0012
	2.66	0.0030	0.0015	0.0017	0.0011
	2.84	0.0046	0.0022	0.0015	0.0008

value is 0.0047 watt per kilogram for one gm and 0.0031 watt per kilogram for 10 gms of tissue without MTM **array components** at 2.24 GHz. In contrast, when MTM array elements are applied to the different layers of the antenna, the SAR value decreases. Also, it is inferred from Table 3, the highest observed SAR value is 0.0029 watt per kilogram and 0.0018 watt per kilogram for one gm and 10 gms of tissue, respectively, for the proposed stacked antenna at 2.24 GHz. Therefore, it is concluded that the stacked antenna is applicable to microwave brain imaging systems to reduce the SAR.

8.8 Microwave brain imaging (MBI) system implementation method

This section presents the deep learning-based **microwave brain imaging** (MBI) system implementation method by utilizing NTSRR loaded stacked antennae. The nine-antenna array setup with fabricated head phantom and overall system analyses are also described in this section. The system is comprised of a portable stand, stepper motor, microcontroller, rotating platform, RF cables, RF switch, PNA transceiver, and GPU based image processing unit. Initially, the scattering parameters are received by the PNA transceiver and then transferred to the GPU-based image processing unit. The data is post-processed and the image reconstruction unit produces **reconstructed microwave brain** (RMB) **images**. Later, the images are pre-processed and augmented, and a training dataset is created. Then the training data is applied to the deep learning models to detect, segment, and classify the tumours in the RMB images. The detailed explanation is described in the following sections. Figure 8.19 presents the schematic diagram of the deep learning based MBI system.

The proposed MBI system consists of a stepper motor, a portable mounting stand, SP8T RF switch, rotating platform, motor driver, colling fan, microcontroller, a PNA E8358A transceivers, image reconstruction processing unit, and a high-performance GPU based personal computer for tumor detection and classification purposes. The stepper motor is attached to the portable rotating mounting stand, which rotates clockwise with a 7.2-degree angle at every step to cover the whole (360 degrees) area. A metamaterial loaded stacked antenna array of nine elements is installed on a rotatable circular ABS (Acrylonitrile butadiene styrene) plastic holder rotating platform. The rotatable circular holder consists of nine holders, which hold the antennas. The rotating platform is connected to the motor by the motor shaft. The **angular distance** from antenna to antenna is 40 degrees to cover the whole area of the system. The antenna position is set at the middle of the antenna holder to adjust the phantom head position. The PNA is connected with the computer through the GPIB port, port A is connected with the transmitting antenna, and Port B is connected to the RF switch for receiving the backscattered signals. A fabricated six-layered (i.e., DURA, CSF, Gary matter, White matter, Fat, and Skin) tissue-mimicking head phantom model is placed at the centre of the rotatable platform to verify the stacked antenna and system performance. The **fabricated tissue** mimicking head phantoms with the tumors are presented in Figure 8.20(a). The step-by-step process of tissue layering in the 3D skull is presented in Figure 8.20(b). The recipe for making the

Metamaterial Inspired Stacked Antenna Based Microwave Brain Imaging   241

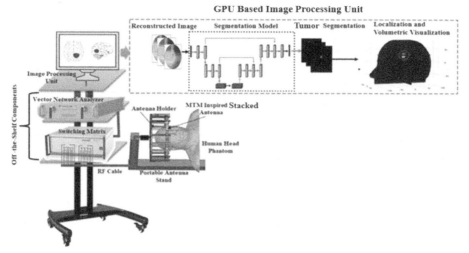

Figure 8.19. Schematic diagram of the deep learning based MBI system.

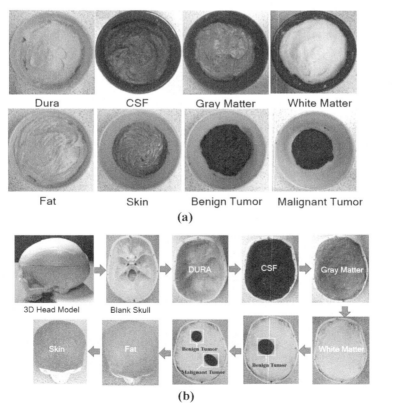

Figure 8.20. (a) Fabricated six tissue-mimicking phantom layers with tumours (b) Step by step phantom fabrication process.

242 *Metamaterial for Microwave Applications*

Table 8.4. Ingredients for 500 gm phantom/tissues fabrication.

Ingredients/Tissue	DURA	CSF	Gray Matter	White Matter	Benign Tumor	Malignant Tumor	Fat	Skin
Water (gm/ml)	361.90	418.75	403.25	353.35	409.85	405.85	54.50	240.00
Corn Flower (gm)	120.65	10.15	82.95	134.30	13.35	14.55	111.90	150.00
Gelatine(gm)	0.00	0.00	0.00	7.05	0.00	0.00	30.00	0.00
Agar (gm)	4.58	56.20	5.20	0.00	63.70	63.55	0.00	0.00
Sodium Azide (gm)	1.80	1.85	1.75	1.75	1.80	1.30	0.00	0.00
Propylene Glycol (gm)	9.65	7.45	4.60	3.55	4.55	5.30	0.00	0.00
NaCl (gm)	1.20	5.60	2.30	0.00	6.40	6.40	1.50	1.60
Sodium benzoate (gm)	0.00	0.00	0.00	0.00	0.00	1.55	1.50	2.00
N-Propanol (gm)	0.00	0.00	0.00	0.00	0.00	1.25	31.10	0.00
TX-151	0.00	0.00	0.00	0.00	0.00	0.00	0.00	25.00
Glycerin (gm)	0.00	0.00	0.00	0.00	0.00	0.00	100.00	0.00
Kerosene (gm)	0.00	0.00	0.00	0.00	0.00	0.00	79.50	46.00
Canola oil (gm)	0.00	0.00	0.00	0.00	0.00	0.00	75.00	0.00

tissue mimicking phantom is presented in Table 8.4. The phantom fabrication process is presented in [11].

The S-parameters data (S_{21}, S_{31}, S_{41}, S_{51}, S_{61}, S_{71}, and S_{81}) is collected by the PNA at each 7.2 degrees rotation and 50 equal points, covering 360 degrees. Thus, in-total 8 × 9 × 50 locations are scanned. Later, the collected signals are post-processed by the IC-DMAS **image reconstruction algorithm** [23] to reconstruct microwave brain (RMB) images to create an original RMB image database for further processing. Some samples of the six-layered tissue-mimicking phantom, different tumour scenarios and corresponding RMB images are depicted in Figure 8.21. For visual explanation purposes, the layout of the **phantom model** is shown in Figure 8.21(a). The tumors are placed in different locations in the model. Figure 8.21(a–f) represents the Non-tumor (NT) image (i.e., healthy brain), Single Benign Tumor (BT) image, Single Malignant Tumor (MT) image, two Benign Tumors (BBT) image, two Malignant Tumors (MMT) image, and one Benign Tumor and one Malignant Tumor (BMT) image, respectively. After that, four hundred (400) samples are collected to create an original RMB image dataset. Later, different pre-processing steps and augmentation techniques are applied to the images to train, validate, and assess the deep learning models.

8.9 Image data collection pre-processing and augmentation techniques

This section describes the image data collection procedure from the implemented MBI system, **image pre-processing** and **image augmentation** for the deep learning model. The Microwave Segmentation Network (MSegNet) is used for tumour segmentation and localizing it in an RMB image. The BrainImageNet (BINet) model is used for tumour classifications from the RMB images. The required input image

Metamaterial Inspired Stacked Antenna Based Microwave Brain Imaging 243

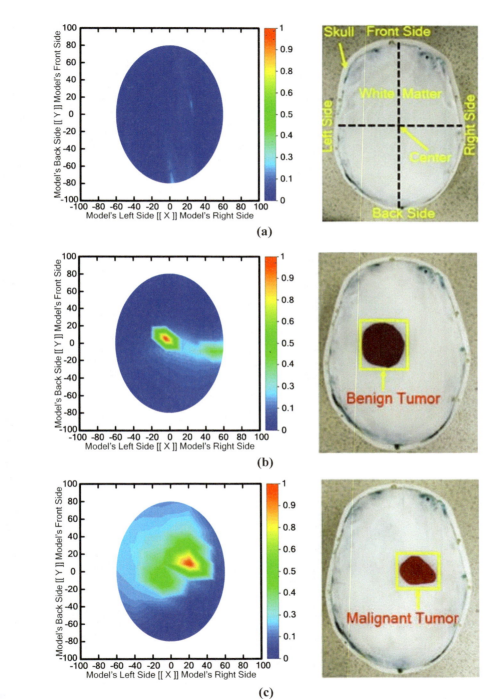

Figure 8.21 contd. ...

244 *Metamaterial for Microwave Applications*

...*Figure 8.21 contd.*

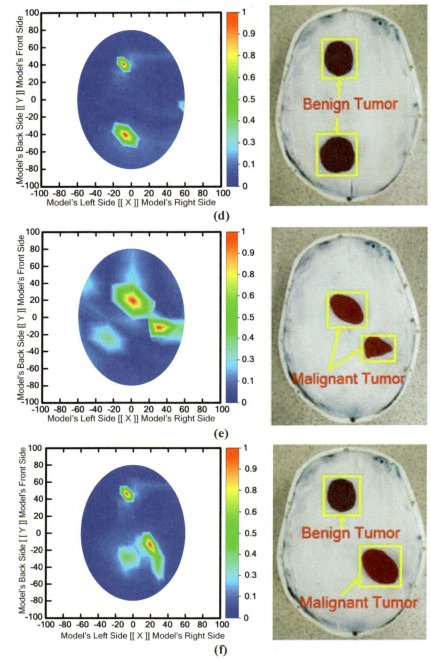

Figure 8.21. Some RMB image samples from stacked antenna-based MBI system (a) Non-tumour (NT) (b) Single benign tumour (BT) (c) Single malignant tumour (MT) (d) Two benign tumours (BBT) (e) Two malignant tumours (MMT) (f) Single benign and a single malignant tumour (BMT).

size and dataset are different for different models. So, these factors are described in the following sections.

8.9.1 Image data collection

In this chapter, the required data is collected from the implemented MBI system which is described in the previous section. The MBI system produces four hundred (400) RMB image samples. The original dataset consists of total 400 RMB image samples. The different models are used with different data from the original dataset for further processing.

8.9.2 Image pre-processing and input size requirement

Due to its input image size limitations, the image pre-processing technique serves as the foundation of a deep learning model. Different deep learning models have different input size requirements. As a result, prior to training the network model, the images are pre-processed. Initially, every image's head area region is cropped, and then resizing and **normalization techniques** are applied to the original sample dataset. The CNN network models, including the segmentation and classification models, have different input size requirements. Thus, before training the models, images are pre-processed (resized and normalized). The images are resized to 256×256 pixels for the investigation of nine Unet segmentation network models such as: (i) U-net, (ii) Modified Unet (M-Unet), (iii) MultiResUnet, (iv) Keras Unet (K-Unet), (v) Unet with ResNet50 backbone, (vi) Unet with DenseNet161 backbone, (vii) ResNet152 FPN, (viii) DenseNet121 FPN, and (ix) proposed MSegNet. On the other hand, for the raw and segmented brain tumor image classification purposes, the images are resized to 224×224 pixels for vanilla CNN, Self-ONN, and BINet. Using the mean (M) and standard deviation (SD) of all images of the original dataset the z-score normalization method is used to normalize the images.

8.9.3 Image augmentation technique

It is noticeable that the deep learning based MSegNet and BINet models require a large image dataset to train a model effectively to classify, segment, and detect the target tumor located in the RMB images. Since we have a small dataset, which is not suitable for training the deep models thus, the image augmentation technique is used to generate a large training dataset for training the models. Image augmentation can enhance the performance of the models by augmenting the existing image dataset rather than collecting new data or samples, and it can significantly increase the diversity of available data for training the models as well as to create a rich dataset from a small sample dataset, which helps to enhance the network performance. In addition, image augmentation may significantly increase the range of data available for the training model and turn a sparse image sample dataset for **image categorization** into a rich dataset. To create a training dataset, this study employs eight image augmentation techniques: rotation, scaling, translation, horizontal flipping, vertical flipping, width shifting, height shifting, and zooming. The rotation operation is done by rotating the images in the anticlockwise and clockwise directions with an angle

246 *Metamaterial for Microwave Applications*

between three to 50 degrees. As a result, the tumour objects are shifted at different locations in the images. Scaling is the reduction or magnification of the image size. Here, two percent to 15 percent image magnifications are used. The image translation technique translates the images vertically and horizontally by three to 10 percent to shift the tumour objects at different places in the images. Additionally, a probability factor of between 0.2 and 0.5 is used for vertical and horizontal flipping. Thus, the tumour objects change their position in the images. This study uses a five-fold **cross-validation** technique for training, validation, and testing purposes. To do five-fold cross-validation, 80 percent of the total images were used for training and 20 percent for testing. Furthermore, from 80 percent of the training dataset, 20 percent is utilized for validation to avoid overfitting. After augmentation, 4,400 images were created per fold for training the MSegNet and BINet models, respectively.

8.10 Deep learning based tumor segmentation and classification models

This section discusses the methodology, dataset description, and analysis of the proposed segmentation and classification model. The complete methodology of the segmentation and classification model is presented in Figure 8.22. This study used RMB images, which were collected from the implemented MBI system. The brain images, including non-tumour, tumour, and corresponding segmented tumour

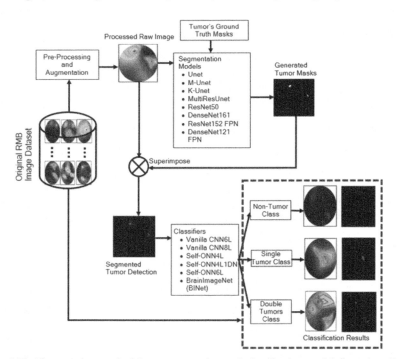

Figure 8.22. The complete methodology segmentation and classification model flow chart Dataset preparation.

region masks, are obtained. The study has mainly two different types of images such as (i) Healthy brain images (i.e., non-tumor images), (ii) Unhealthy brain images (i.e., tumor-based images). The unhealthy images are classified into two categories: (i) Single tumor images, and (ii) Double tumor images. This work first explored the proposed **lightweight** Microwave Segmentation Network (MSegNet) segmentation model along with eight other state-of-the-art segmentation models to investigate the segmentation performance of detecting tumour regions in the RMB images. At first, raw RMB image samples are collected, and then image pre-processing is applied. Besides, the corresponding **ground truth masks** are also created and then applied along with the image dataset. Thereafter, the tumour mask is superimposed on the raw RMB images to create a segmented tumour region-based image dataset. Then, the proposed lightweight BrainImageNet (BINet) classification model and five other CNN-based classification models are used to investigate the classification performances of the raw and segmented RMB images for three-class classification: Non-tumor, single tumor, and double tumors.

In this work, the RMB images and their corresponding ground truth masks are used as an original dataset. The original dataset consists of 400 RMB images, where 100 images are in the non-tumor (i.e., healthy brain) class, and three-hundred images are in the tumor (i.e., **unhealthy brain**) class, and corresponding ground truth masks are made available in the dataset. The tumor class is further divided into two subclasses: 150 images for single tumor, and another 150 for double tumors. Samples of the raw and **segmented RMB images** and their ground truth masks of the dataset are shown in Figure 8.23.

The complete dataset for the segmentation and classification model is presented in Table 8.5. Image pre-processing and augmentation techniques are applied to the raw RMB image and ground truth masks to create a **training dataset** for training the models. Some randomly selected **augmented samples** are depicted in Figure 8.24.

8.11 Microwave segmentation network (MSegNet)-brain tumor segmentation model

8.11.1 Architecture of MSegNet segmentation model

Brain tumor segmentation is done to segment the tumors from the RMB tumor images to identify the correct spatial location of a tumor in the images. Nowadays, U-net based deep learning architecture is popularly used to segment objects in medical imaging applications. The main benefit of this network is that it can precisely segment the target features and effectively process and evaluate the images. This thesis proposed a lightweight segmentation model, called MicrowaveSegmentationNetwork (MSegNet) for segmenting the tumor region in the RMB tumor images. The proposed MSegNet model architecture is illustrated in Figure 8.25. Typically, a U-net model has four encoding and decoding blocks and some skip connections. The MSegNet model used only two levels in both **encoding** and **decoding** to make it a lightweight network.

The model consists of a contracting path with two encoding blocks followed by an expanding path with two decoding blocks. Every block in the encoder and

248 *Metamaterial for Microwave Applications*

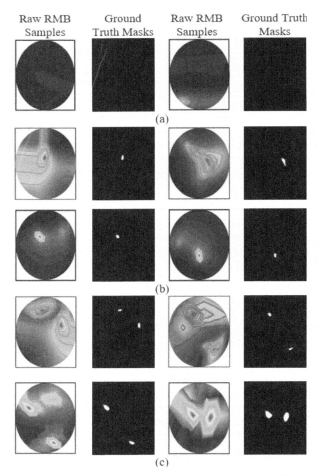

Figure 8.23. Randomly selected RMB brain image samples and their corresponding ground truth masks from the original dataset (a) Non-tumor (b) Single tumor (c) Double tumors.

Table 8.5. Dataset description for training, testing, and validation for segmentation model.

Dataset	Number of Original Images	Image Classes	Training Dataset			
			Number of Images Per Class	Augmented Train Images Per Fold	Testing Images Per Fold	Validation Image Per Fold
Raw RMB Image Samples	400	Non-Tumor	100	1000	20	16
		Single Tumor	150	1500	30	24
		Double Tumors	150	1500	30	24
Total			400	4000	80	64

Metamaterial Inspired Stacked Antenna Based Microwave Brain Imaging   249

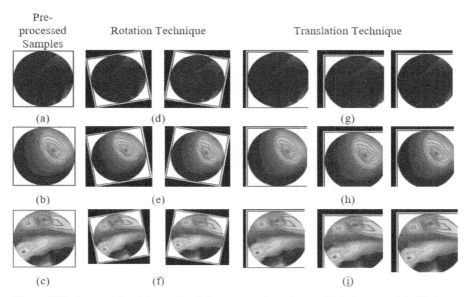

Figure 8.24. Augmented training set (a–c) Pre-processed non-tumor, single tumor, and double tumor images (d–f) Images after rotation by 20 degrees anticlockwise and clockwise for non-tumor, single tumor, and double tumors (g–i) Images after 3% horizontal, 5% vertical and horizontal, 5% horizontal and 3% vertical translation for non-tumor, single tumor, and double tumors.

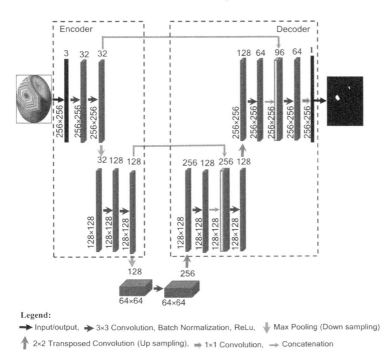

Figure 8.25. Proposed lightweight MSegNet model for tumor segmentation.

decoder consists of two 3 × 3 convolutional layers followed by a non-linear activation function. The input image (256 × 256) is fed into the encoder of the network. Every encoding block consists of two 3 × 3 **convolutional layers** consecutively followed by a 2 × 2 **max-pooling layer** for down sampling. In the decoder part, every decoding block consists of an up sampling, followed by one 3 × 3 convolutional layer, a concatenation layer, and one 3 × 3 convolutional layer. The decoder begins with a 2 × 2 transposed convolutional layer for upsampling. All convolutional layers in both encoder and decoder are followed by BN (Batch Normalization) and Rectified Linear Unit (ReLu) activation function. The contracting path from the encoder block is directly connected with the decoder blocks' concatenation layer to create a high-resolution segmentation features map. At the ending layer, 1 × 1 convolution is used to create the output map from the last decoding block to two-channel feature maps. A pixel-wise Softmax function is utilized in two-channel feature maps to map every pixel into a binary background or tumor class.

8.11.2 Training experiment of MSegNet model

This section presents the experimental training results of the segmentation model. Two sets of experiments (brain tumor segmentation, classification with raw and segmented RMB brain images) were carried out. For the experiment, nine different segmentation models, including the proposed model, were implemented on the anaconda distribution platform using the PyTorch library with the Python 3.7 version. The experiments were carried out on a 64-bit Windows 10 operating system with a 128GB RAM and 64-bit Intel(R) Xeon(R)W-2016 @ 3.30GHz CPU. In addition, a 11GB NVIDIA GeForce GTX 1082Ti GPU was used to accelerate the network training performances. The two sets of experimental analyses (brain tumor segmentation and RMB image classification) were explained in this section. Finally, the average of the performance metrics of the five folds was calculated. For experimental purposes, the proposed MSegNet model and the other eight models (as mentioned earlier) were trained and validated by using a five-fold cross-validation image dataset to evaluate tumor segmentation performance. The training was executed using a learning rate (LR) of 0.0005 for a maximum of 30 epochs, batch size of 8, and Adam optimizer for network optimization. During training, if no improvement was observed for ten successive epochs, the learning rate was decreased by a learning factor of 0.2, and the training was stopped if there was no improvement detected for 15 successive epochs. The detailed **hyperparameters** for all segmentation models are presented in Table 8.6. Moreover, the dice score (DSC) and loss plots for different epochs during the training of the proposed MSegNet model is presented in Figure 8.26. It is observed from Figure 8.26, the model was trained for 20 **epochs**, and the model's performance gets saturated after a few epochs in terms of DSC and loss. So, it is seen that the proposed model is not over-fitting and converges well and should segment the desired tumor regions in RMB images reliably.

Metamaterial Inspired Stacked Antenna Based Microwave Brain Imaging 251

Table 8.6. Hyperparameters for all segmentation models.

Parameter's name	Value	Parameter's name	Value
Input Channels	3	Output Channels	1
Batch Size	8	Optimizer	Adam
Learning Rate (LR)	0.0005	Loss Type	Dice Loss
Maximum Number of Epochs	30	Epochs Patience	10
Maximum Epochs Stop	15	Learning Factor	0.2
Initial Feature	32	Number of folds	5

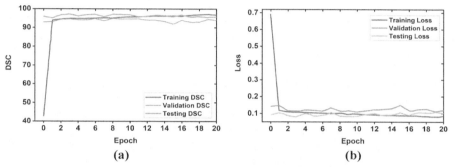

Figure 8.26. The training results graph (a) The DSC graph (b) Loss graph.

8.11.3 *Evaluation matrix for the MSegNet segmentation model*

After completion of the training and **validation phase**, the tumor segmentation performances of the different networks: MSegNet, U-net, M-Unet, K-Unet, MultiResUnet, ResNet50, DenseNet161, ResNet152 FPN, and DenseNet121 FPN for testing the RMB image dataset are evaluated. The performance evaluation matrices are Accuracy (A), Intersection-over-Union (IoU), and Dice score (DSC); and these are calculated by the following equations:

$$A = \frac{(N_{TP} + N_{TN})}{(N_{TP} + N_{FN}) + (N_{FP} + N_{TN})} \tag{8.5}$$

$$IoU = \frac{N_{TP}}{(N_{TP} + N_{FN} + N_{FP})} \tag{8.6}$$

$$Dice\,Score(DSC) = \frac{(2 \times N_{TP})}{(2 \times N_{TP} + N_{FN} + N_{FP})} \tag{8.7}$$

where, N_{TP} represents the number of tumor images identified as tumors, N_{FP} represents the number of images identified as tumor errors, N_{FN} represents the number of missed images that were identified as tumors, N_{TN} represents the number of missed images that were classified as tumors.

8.11.4 Brain tumor segmentation performances

This section discusses the brain tumor **segmentation performance** and compares it with those of different types of models. It is notable that, the main advantages of the MSegNet model are: (i) a lightweight architecture with only two layers in the encoding and decoding blocks, (ii) low training and inference time, (iii) can segment the desired tumor (small and large) regions precisely with a high-resolution image, (iv) shows high segmentation performances in terms of accuracy, IoU, and Dice Score (DSC) compared to other deeper segmentation networks. For experimental purposes, the proposed MSegNet model and the other eight segmentation models (U-net, M-Unet, K-Unet, MultiResUnet, ResNet50, DenseNet161, ResNet152 FPN, and DenseNet121 FPN) were used to investigate the tumor segmentation performances. The **tumor segmentation** performance results of the MSegNet model are shown in Figure 8.27 which illustrates the non-tumor, single tumor, and double tumor images, corresponding ground truth masks, generated masks, and resultant segmented tumor

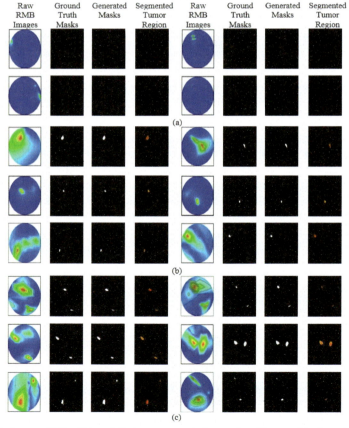

Figure 8.27. Proposed MSegNet model's tumor segmentation results with ground truth masks, generated masks, and resultant segmented tumor images for (a) Non-tumor class (b) Single tumor class (c) Double tumors class.

Table 8.7. Performance evaluation matrices of all segmentation models.

Network Model Name	Accuracy (%)	IoU (%)	Dice Score (%)	Loss
U-net	99.96	85.72	91.58	0.1127
Modified Unet (M-Unet)	99.96	86.47	92.20	0.1086
Keras Unet (K-Unet)	99.96	86.01	91.91	0.1156
MultiResUnet	99.96	86.55	92.20	0.1064
ResNet50	99.95	86.43	92.13	0.1121
DenseNet161	99.95	85.62	91.59	0.1145
ResNet152 FPN	99.94	82.86	89.58	0.1312
DenseNet121 FPN	99.95	83.30	89.91	0.1318
Proposed MSegNet	99.97	86.92	93.10	0.1010

regions of the raw RMB images. It is observed that the MSegNet model precisely segments the desired region of the tumor in the RMB images. The four evaluation performance matrices of the segmentation models are presented in Table 8.7. It is observed from Table 8, that the MSegNet model exhibited better performances compared to the other eight segmentation models. The achieved accuracy (A), IoU, dice score (DSC), and loss of the proposed model are 99.97, 86.92, and 93.10 percent, and 0.101, respectively. So, it is concluded that the high **accuracy**, dice score, and low **loss** ensure that the MSegNet model can clearly segment desired tumor regions in the raw RMB images, and it can be applied in the MBI system.

8.12 BrainImageNet (BINet)-brain tumor classification model

8.12.1 Mathematical analysis of the classification model

The deep learning-based classification model consists of many convolutional neural network (CNN) layers. The **CNN layer** weight is calculated based on the linear operation. Recently, the Operational Neural Network (ONN) based model was introduced in [24] to overcome the **linear nature** of CNNs. The ONN is a heterogeneous network that has demonstrated a promising performance in a number of applications, including image denoising and image restoration [25, 26]. It uses a fixed set of nonlinear operators to learn complicated patterns of any input. However, the fixed set of operator libraries limits the learning process of the ONN. To overcome this problem, Self-Organized ONN (Self-ONN) is proposed in [27]. Self-ONN inevitably learns the best set of operators throughout the training progression instead of a fixed set of operator libraries. This creates a more robust model which can handle more diverse situations and generalizes well in real-life cases. **Self-ONN networks** choose the best set of operators during the training process, which can be a combination of all standard functions or some other functions that we do not know. The output O_k^L at the kth neuron of the Lth layer of any ONN can be illustrated as following equation:

$$O_k^L = b_k^L + \sum_{i=1}^{N_{L-1}} \Psi_{ki}^L(w_{ki}^L, y_i^{L-1}) \tag{8.8}$$

where, b_k^L and w_{ki}^L are the biases and weights associated with that neuron and layer, y_i^{L-1} is the input from the previous layer, N_{L-1} is the kernel size in that layer, Ψ_{ki}^L is the nodal operator of that neuron and layer. If Ψ_{ki}^L is linear then the equation simply corresponds to a conventional CNN. In ONN, the **composite nodal operator** Ψ can be constructed using a set of standard equation functions as follows:

$$\Psi(w, y) = w_1 \sin(w_2 y) + w_3 \exp(w_4 y) + \ldots + w_q y \qquad (8.9)$$

Here, w denotes the q-dimensional array of parameters which is composed of internal parameters of the individual functions and weights. Instead of a fixed set of operators, the composite nodal operator Ψ can be constructed using a **Taylor series** approximation. The Taylor series approximation of a function $f(x)$, near point, $x = a$ is stated by the following equation:

$$f(x) = f(a) + \frac{f'(a)}{1!}(x-a) + \frac{f''(a)}{2!}(x-a)^2 + \frac{f'''(a)}{3!}(x-a)^3 + \ldots + \frac{f^n(a)}{n!}(x-a)^n \qquad (8.10)$$

Equation 8.10 can be used to construct the nodal operator as follows:

$$\Psi(w, y) = w_0 + w_1(y-a) + w_2(y-a)^2 + \ldots + w_q(y-a)^q \qquad (8.11)$$

Here, $w_q = \dfrac{f^{(n)}(a)}{q!}$ denotes the qth parameter of the qth order polynomial. In Self-ONN, *tanh* has been used as an activation function that is bounded at the range of [−1, 1]. So, for *tanh*, a is equal to zero in Equation 8.11.

8.13 Architecture of BINet classification model

A new lightweight classification model called **BrainImageNet** (BINet) is proposed to classify the raw and segmented brain tumor images. The BINet is designed using self-organized operational neural network (Self-ONN) architecture. The detailed architecture of the BINet classification model is shown in Figure 8.28. As illustrated in Figure 8.28, the BINet has six Self-ONN layers, where the first four layers have eight neurons, and the other two have 16 neurons, respectively. Through the self-organization of its nodal operators, it can accomplish the required **non-linear transformations** to extract the optimal features from the brain tumor images. The

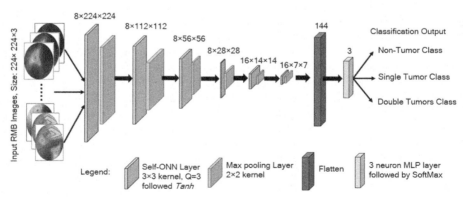

Figure 8.28. Proposed BrainImageNet using Self-ONN.

kernel sizes are set as 3 × 3 for the Self-ONN layer and 2 × 2 for the max-pooling layer, respectively.

Moreover, the Q value is set to 3 as the order of the qth order polynomial for all operational layers. The input image of dimension 224 × 224 is fed to the input layer. The images are propagated through the Self-ONN and max pooling layers and their features are extracted into different feature maps. A flattened layer with 144 neurons is used to convert the output of the convolutional layer into a one-dimensional **feature vector** and applied to the final dense layer. The dense layer is the final classifier of the network that uses a three neuron MLP layer followed by the **SoftMax activation function** and then classifies the upcoming images into non-tumor, single tumor, and double tumor classes.

8.14 Training experiment of the BINet classification model

In this section, two classification experiments (classification with raw and segmented RMB brain images) were carried out. In the experiment, classification models were implemented on the anaconda distribution platform using the PyTorch library with the Python 3.7 version. The experiments were carried out on a 64-bit Windows 10 operating system with a 128GB RAM and 64-bit Intel(R) Xeon(R)W-2016 @ and 3.30GHz CPU. In addition, a 11GB NVIDIA GeForce GTX 1082Ti GPU was used to accelerate the classification network training performances. Finally, the average of the performance metrics of the five folds for classification was calculated. Two classification experiments to investigate the classification performances of the networks are: (i) Classification using the raw RMB images (non-segmented) and (ii) Classification using the segmented RMB images. However, the proposed BINet model and three variations of the Self-ONN based model, such as two Self-ONN models with 4 operational layers and one with 6 operational layers (Self-ONN4L, Self-ONN4L1DN, and Self-ONN6L) as well as two **vanilla CNN models** with 6 and 8 layers (Vanilla CNN6L and Vanilla CNN8L) were investigated and compared the results separately by using the raw (non-segmented) and segmented RMB tumor images. In the model names, "4L" means the model consists of four layers, "6L" means the model consists of six layers, and "1DN" means the model consists of one dense layer in the final stage. The training was executed using a **learning rate** (LR) of 0.0005 for a maximum of 30 epochs, batch size of 16, utilized Adam optimizer for network optimization, and set stop criteria based on training loss. The Q order value is a significant factor during training Q = 1 is set to train the two vanilla CNNs, and Q = 3 is set for the Self-ONN and BINet models. The hyperparameters for the classification models are presented in Table 8.8. The training accuracy and loss investigation curve of the BINet model are illustrated in Figure 8.29. It is observed that the model was trained for 30 epochs, and the model's performance gets saturated after a few epochs in terms of accuracy and loss. So, it can be seen that the proposed model is not over-fitting and converges well and should classify the RMB images reliably.

Table 8.8. Hyper-parameters for all classification models.

Parameter's name	Assigned value	Parameter's name	Assigned value
Input Channels	3	Q order	1 for CNN, 3 for Self-ONNs
Batch Size	16	Optimizer	Adam
Learning Rate (LR)	0.0005	Stop criteria	Loss
Maximum Number of Epochs	30	Epochs Patience	5
Maximum Epochs Stop	10	Learning Factor	0.2
Image size	224	Number of folds	5

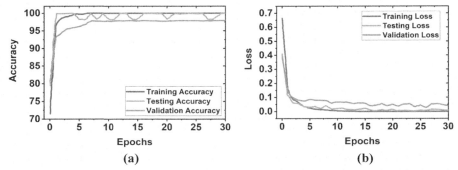

Figure 8.29. The training results graph (a) Accuracy graph (b) Loss graph.

8.15 Evaluation matrix for the BINet classification model

The classification performance of the various CNN and Self-ONN models is evaluated by the five evaluation matrices, namely: (i) Overall accuracy (A), (ii) Weighted Recall or **sensitivity** (R), (iii) Weighted **specificity** (S) (iv) Weighted **precision** (P) and (v) Weighted F1-score (Fs). The evaluation metrics for classification are calculated by using the following equations:

$$A = \frac{(N_{TP} + N_{TN})}{(N_{TP} + N_{FN}) + (N_{FP} + N_{TN})} \quad (8.12)$$

$$R = \frac{N_{TP}}{(N_{TP} + N_{FN})} \quad (8.13)$$

$$S = \frac{N_{TN}}{(N_{FP} + N_{TN})} \quad (8.14)$$

$$P = \frac{N_{TP}}{(N_{TP} + N_{FP})} \quad (8.15)$$

$$F_s = \frac{(2 \times N_{TP})}{(2 \times N_{TP} + N_{FN} + N_{FP})} \quad (8.16)$$

where, N_{TP} represents the number of tumor images identified as tumors, N_{TN} represents the number of non-tumor images identified as non-tumors, N_{FP} represents the number of images incorrectly identified as tumors, N_{FN} represents the number of images with tumor(s) that were missed by the network.

8.16 Brain images classification performances of the BINet model

This section, explains the three Self-ONNs (Self-ONN4L, Self-ONN4L1DN, and Self-ONN6L), two vanilla CNNs (vanilla CNN6L and vanilla CNN8L), and proposed BINet classification models to investigate the classification effectiveness by applying the raw and segmented RMB images. The classification models classify the images into three classes: non-tumor, single tumor, and double tumor classes. The main advantages of the BINet model are: (i) a lightweight architecture that uses a non-linear operation to boost up the **network diversity** along with the classification effectiveness, (ii) ability to optimize the learning weight of each layer during the training process, and (iii) attains superior classification performances while significantly reducing the computational complexity compared to conventional CNNs models. All classification models were trained by using the raw RMB tumor images. The comparative classification performance outcomes of the models for the raw RMB tumor images are presented in Table 8.9. It was investigated that the conventional deeper CNN networks have achieved lower performances than the three Self-ONNs models, but the BINet model was the best model among all networks and achieved the highest performances. The BINet has exhibited an accuracy, precision, recall, specificity, and **F1 score** of 89, 88.74, 88.67, 94.33, and 88.61 percent, respectively for the raw RMB images. Then it investigated the classification performances of all the mentioned classification models for segmented RMB images. All models were trained by utilizing the resultant segmented RMB tumor images to verify the classification efficacy. The comparative classification performances of the models for classifying the segmented RMB tumor images into the three classes are presented in Table 8.10. It was observed that the conventional deeper CNN networks and Self-ONN models improved performances but they were poorer than the BINet model performance. However, the BINet model was the best among all the networks and achieved the best performance. The attained accuracy, precision, recall, specificity, and F1 score of the BINet model are 98.33, 98.35, 98.33,

Table 8.9. Classification results of all models for the raw RMB images.

Image Type	Network Model Name	Overall	Weighted			
		Accuracy (A) (%)	Precession (P) (%)	Recall (R) (%)	Specificity (S) (%)	F1 Score (F_s) (%)
Raw RMB Images	Vanilla CNN6L	84.33	84.17	84.33	92.17	84.06
	Vanilla CNN8L	85.33	85.62	85.33	92.67	85.14
	Self-ONN4L	85.00	84.91	85.00	92.50	84.87
	Self-ONN4L1DN	87.00	87.05	87.00	93.50	86.95
	Self-ONN6L	87.00	86.85	87.00	93.50	86.82
	BrainImageNet (BINet)	89.33	88.74	88.67	94.33	88.61

258   *Metamaterial for Microwave Applications*

Table 8.10. Classification results of all models for the segmented RMB images.

Image Type	Network Model Name	Overall Accuracy (A) (%)	Weighted Precession (P) (%)	Recall (R) (%)	Specificity (S) (%)	F1 Score (F_s) (%)
Segmented RMB Images	Vanilla CNN6L	95.00	94.98	95.00	97.50	94.96
	Vanilla CNN8L	95.67	95.77	95.67	97.83	95.65
	Self-ONN4L	94.00	93.96	94.00	97.00	93.96
	Self-ONN4L1DN	96.33	96.41	97.00	98.17	97.00
	Self-ONN6L	96.67	96.79	96.67	98.33	96.66
	BrainImageNet (BINet)	98.33	98.35	98.33	99.17	98.33

99.17, and 98.33 percent, respectively. Therefore, it is concluded that the proposed classification model exhibited better performance for the segmented RMB images classified into three classes: Non-tumor, single tumor, and double tumor classes.

8.17 Receiver operating characteristic of BINet model

The **receiver operating characteristic** (ROC) curve is an essential performance metric used in multi-class classification problems. The ROC is used to visualize the performance of a classification model across all thresholds. It also shows the capability for distinguishing classes. The ROC curve with the **area under the curve** (AUC) classification of all classification models across all thresholds is depicted in Figure 8.30. Figure 8.30 presented the ROC with the AUC for segmented RMB image classification and it was observed that the proposed BINet model performed better. The calculated AUC for vanilla CNN6L, Vanilla CNN8L, Self-ONN4L, Self-ONN4L1DN, Self-ONN6L, and BINet models are 98.15, 99.22, 98.38, 99.73, 99.32, and 99.94 percent respectively.

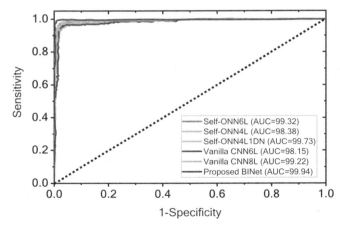

Figure 8.30. Receiver operating characteristic (ROC) curve with AUC for three class classifications by the BINet model.

8.18 Miss classification performance analysis of BINet model

It was concluded from the classification performances in Tables 9 and 10 that the best classification model was BINet for classifying the raw and segmented RMB images. The overall classification accuracy was 89.33 percent and 98.33 percent for the raw and segmented images, respectively. For the classification results, the BINet model **confusion matrix** for the raw RMB images is illustrated in Figure 8.31(a). It is shown that there is a total of thirty-four **misclassified** images during model testing. For instance, eight misclassified images are illustrated in Figure 8.32. It is observed from Figure 8.31(a) that three non-tumor and fourteen double tumor images were misclassified as having single tumors. Three double and six single tumor images

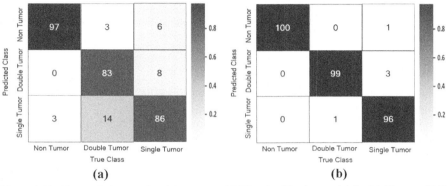

Figure 8.31. The confusion matrix of the proposed BINet classification model for (a) The raw RMB images (b) The segmented RMB images.

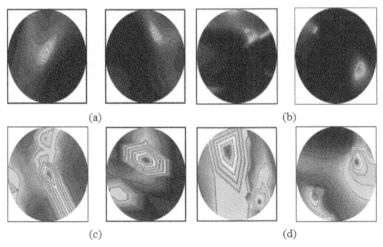

Figure 8.32. Some misclassified images by the BINet model for the raw RMB images (a) Non-tumor images were misclassified as a single tumor class (b) Single tumor images were misclassified as a non-tumor class (c) Single tumor images were misclassified as a double tumor class (d) Double tumor images were misclassified as a single tumor classification.

were misclassified as a non-tumor class as well as eight single tumor images were misclassified as a double tumor class. In contrast, after segmenting the tumors, the confusion matrix of the BINet classification model is shown in Figure 8.31(b). It is observed from Figure 8.31(b) that only five tumor images were misclassified and none of the non-tumor images were misclassified. One double tumor image was misclassified as a single tumor class. Also, one single tumor was misclassified as a non-tumor, and three single tumor images were misclassified as a double tumor class.

Through the training of Self-ONNs, the optimum non-linear parameters can be learned to exploit the **learning performance** and attain a superior classification performance in terms of non-tumor and tumor images. However, the proposed model performed better and presented satisfactory outcomes for the segmented tumor images rather than the raw RMB tumor images. Finally, it is concluded that the segmentation technique abetted the classification model for improving the **classification performance** and is applicable to the portable microwave brain imaging system.

8.19 Summary

Automated brain anomalies such as tumor detections from the reconstructed microwave brain images and image classification are essential for the investigation, surgical planning, and monitoring of the progression of brain disease. Since the manual identification, classification, and segmentation of tumors are extremely time-consuming, expensive, and crucial tasks due to the tumor's pattern, size, location, and shape are involved. Thus, an automatic, interactive, and reliable detection and classification technique is required in microwave brain imaging systems. This chapter proposes a metamaterial inspired stacked antenna-based microwave brain imaging system (MBI) to detect brain tumors using a deep learning model. A new lightweight segmentation model called MicrowaveSegNet (MSegNet) was utilized to segment the brain tumors in the reconstructed microwave brain (RMB) images. Then, a seven-layer lightweight classifier called BrainImageNet (BINet) using a self-organized operational neural network (Self-ONN) model was used to classify the RMB images. The MSegNet model consists of two encoding and decoding layers, making it a lightweight network that reduces inference time while maintaining performance. Initially, four hundred RMB brain image samples were collected from the stacked antenna-based MBI system to create an original dataset. Then, image preprocessing and augmentation techniques were applied to make 4,000 training images per fold for 5-fold cross-validation. Later, the proposed MSegNet was compared with eight state-of-the-art segmentation models (Unet, M-Unet, K-Unet, MultiresUnet, Unet with ResNet50 and DenseNet161 backbones, Feature Pyramid Network (FPN) with ResNet152, and DenseNet121 backbones for brain tumor segmentation. In addition, the proposed BINet model is investigated and compared with three Self-ONN models and two conventional Convolutional Neural Network (CNN) models for three-class classification using raw and segmented RMB images. The proposed MSegNet has achieved an **Intersection over Union** (IoU), and **Dice score** (DSC) 86.92 percent, and 93.10 percent, respectively for tumors segmentation. The BINet has achieved an accuracy, precision, recall, F1-score, and specificity of 89.33, 88.74, 88.67, 88.61, and 94.33 percent, respectively for a three-class classification using raw RMB images, whereas, 98.33, 98.35, 98.33, 98.33, and 99.17 percent, respectively for segmented RMB images. The outcomes ensure that the classification performance of the proposed model for the segmented tumor images is better than the original raw images. Therefore, the cascaded MSegNet and BINet deep learning models can be used for reliably detecting and classifying the tumor(s) from the RMB brain images in the portable MBI system.

References

[1] Kavli, F. The Kavli Foundation [Online]. Available: https://kavlifoundation.org/tags/brain-diseases-disorders.

[2] Chaturia Rouse, H. G., Quinn T. Ostrom, Carol Kruchko, Jill S. Barnholtz-Sloan. (2022, 17/06/2022). *Quick brain tumor facts*. Available: https://braintumor.org/brain-tumor-information/brain-tumor-facts/.

[3] Tariq, M., A. A. Siddiqi, G. B. Narejo and S. Andleeb. 2019. A cross sectional study of tumors using bio-medical imaging modalities. *Current Medical Imaging*, 15: 66–73.

[4] Kerlikowske, K., C. C. Gard, B. L. Sprague, J. A. Tice, D. L. Miglioretti et al. 2015. One versus two breast density measures to predict 5-and 10-year breast cancer riskbreast density measures to predict breast cancer risk. *Cancer Epidemiology, Biomarkers & Prevention*, 24: 889–897.

[5] Cazzato, R. L., J. Garnon, B. Shaygi, G. Koch, G. Tsoumakidou et al. 2018. PET/CT-guided interventions: Indications, advantages, disadvantages and the state of the art. *Minimally Invasive Therapy & Allied Technologies*, 27: 27–32.

[6] Adamson, E. B., K. D. Ludwig, D. G. Mummy and S. B. Fain. 2017. Magnetic resonance imaging with hyperpolarized agents: methods and applications. *Physics in Medicine & Biology*, 62: R81.

[7] Jones, K. M., K. A. Michel, J. A. Bankson, C. D. Fuller, A. H. Klopp et al. 2018. Emerging magnetic resonance imaging technologies for radiation therapy planning and response assessment. *International Journal of Radiation Oncology* Biology* Physics*, 101: 1046–1056.

[8] Alqadami, A. S., K. S. Bialkowski, A. T. Mobashsher and A. M. Abbosh. 2018. Wearable electromagnetic head imaging system using flexible wideband antenna array based on polymer technology for brain stroke diagnosis. *IEEE Transactions on Biomedical Circuits and Systems*, 13: 124–134.

[9] Hossain, A., M. T. Islam, M. S. Islam, M. E. Chowdhury, A. F. Almutairi et al. 2021. A YOLOv3 deep neural network model to detect brain tumor in portable electromagnetic imaging system. *IEEE Access*, 9: 82647–82660.

[10] Islam, M. S., M. T. Islam, A. Hoque, M. T. Islam, N. Amin et al. 2021. A portable electromagnetic head imaging system using metamaterial loaded compact directional 3D antenna. *IEEE Access*, 9: 50893–50906.

[11] Mobashsher, A. and A. Abbosh. 2014. Three-dimensional human head phantom with realistic electrical properties and anatomy. *IEEE Antennas and Wireless Propagation Letters*, 13: 1401–1404.

[12] Mobashsher, A. T., A. M. Abbosh and Y. Wang. 2014. Microwave system to detect traumatic brain injuries using compact unidirectional antenna and wideband transceiver with verification on realistic head phantom. *IEEE Transactions on Microwave Theory and Techniques*, 62: 1826–1836.

[13] Alam, M. M., M. S. Talukder, M. Samsuzzaman, A. I. Khan, N. Kasim et al. 2022. W-shaped slot-loaded U-shaped low SAR patch antenna for microwave-based malignant tissue detection system. *Chinese Journal of Physics*.

[14] de Oliveira, A. M., A. M. de Oliveira Neto, M. B. Perotoni, N. Nurhayati, H. Baudrand et al. 2021. A fern antipodal vivaldi antenna for near-field microwave imaging medical applications. *IEEE Transactions on Antennas and Propagation*, 69: 8816–8829.

[15] Rezaeieh, S. A., A. Zamani and A. Abbosh. 2014. 3-D wideband antenna for head-imaging system with performance verification in brain tumor detection. *IEEE Antennas and Wireless Propagation Letters*, 14: 910–914.

[16] Rodriguez-Duarte, D. O., J. A. T. Vasquez, R. Scapaticci, L. Crocco, F. Vipiana et al. 2020. Brick-shaped antenna module for microwave brain imaging systems. *IEEE Antennas and Wireless Propagation Letters,* 19: 2057–2061.
[17] Hossain, A., M. T. Islam, M. E. Chowdhury and M. Samsuzzaman. 2020. A grounded coplanar waveguide-based slotted inverted delta-shaped wideband antenna for microwave head imaging. *IEEE Access,* 8: 185698–185724.
[18] Jalilvand, M., X. Li, L. Zwirello and T. Zwick. 2015. Ultra wideband compact near-field imaging system for breast cancer detection. *IET Microwaves, Antennas & Propagation,* 9: 1009–1014.
[19] Gerazov, B. and R. C. Conceicao. 2017. Deep learning for tumour classification in homogeneous breast tissue in medical microwave imaging. pp. 564–569. *In: IEEE EUROCON 2017-17th International Conference on Smart Technologies.*
[20] Khoshdel, V., M. Asefi, A. Ashraf and J. LoVetri. 2020. Full 3D microwave breast imaging using a deep-learning technique. *Journal of Imaging,* 6: 80.
[21] Shah, P. and M. Moghaddam. 2017. Super resolution for microwave imaging: A deep learning approach. pp. 849–850. *In: 2017 IEEE International Symposium on Antennas and Propagation & USNC/URSI National Radio Science Meeting.*
[22] Shao, W. and Y. Du. 2020. Microwave imaging by deep learning network: Feasibility and training method. *IEEE Transactions on Antennas and Propagation,* 68: 5626–5635.
[23] Islam, M. T., M. Samsuzzaman, S. Kibria, N. Misran and M. T. Islam. 2019. Metasurface loaded high gain antenna based microwave imaging using iteratively corrected delay multiply and sum algorithm. *Scientific Reports,* 9: 1–14.
[24] Kiranyaz, S., T. Ince, A. Iosifidis and M. Gabbouj. 2020. Operational neural networks. *Neural Computing and Applications,* 32: 6645–6668.
[25] Kiranyaz, S., J. Malik, H. B. Abdallah, T. Ince, A. Iosifidis et al. 2021. Self-organized operational neural networks with generative neurons. *Neural Networks,* 140: 294–308.
[26] Malik, J., S. Kiranyaz and M. Gabbouj. 2020. Operational vs convolutional neural networks for image denoising. *arXiv preprint arXiv:2009.00612.*
[27] Malik, J., S. Kiranyaz and M. Gabbouj. 2021. Self-organized operational neural networks for severe image restoration problems. *Neural Networks,* 135: 201–211.

Chapter 9

Lower UHF Metamaterial Antenna for Nanosatellite Communication System

Mohammad Tariqul Islam[1,*] and *Touhidul Alam*[2]

9.1 Introduction

With the advent of modern technology, nanosatellites are a flourishing new dimension in **space communication.** In the last decade, nanosatellite launches for low earth orbit missions have increased dramatically due to a continual reduction in development costs and building time. The communication between the nanosatellite and Earth significantly depends upon antenna systems. There is a strong demand for compact, lightweight and stable performing antenna since it is a requirement for smooth operation of the nanosatellite mission. Deployable antennas are widely used in limited volume in 1U and 2U nanosatellites. The deployable antennas for **nanosatellite** have to be deployed mechanically. However, mechanical deployment is quite sophisticated, and this might increase the chances of mission failure. Several nanosatellite missions have failed due to antenna deployment complexity. In contrast to the deployable antenna a patch antenna provides a low profile. However, this type of antenna has a frequency shift issue due to a finite ground plane and metallic satellite body effect. Designing ultra-high frequency (UHF) antennae for nanosatellites has become a big challenge for researchers. To overcome these problems, the EMNZ structure has been developed and included in the ground plane of the antenna to stabilize the resonant frequency and to improve efficiency and impedance matching over the desired frequency range when embedded in the nanosatellite structure.

---

[1] Department of Electrical, Electronic and Systems Engineering, Universiti Kebangsaan Malaysia, Bangi, Selangor, Malaysia.
[2] Institute of Climate Change, Universiti Kebangsaan Malaysia, Bangi, Selangor, Malaysia.
Email: touhidul@ukm.edu.my
* Corresponding author: tariqul@ukm.edu.my

264 *Metamaterial for Microwave Applications*

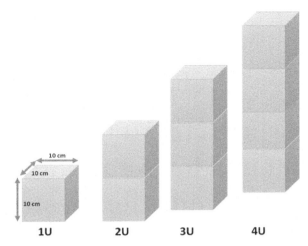

Figure 9.1. Nanosatellites are classified by the number of cubic units (U) and are typically built in 1U, 1.5U, 2U, or 3U sizes.

Nanosatellite or CubeSat is a smaller form of the **SmallSat family**. Typical nanosatellites are classified based on their mass and size. An illustration of various nanosatellite structures is given in Figure 9.1, where the nanosatellites are presented in a one cubic centimeter unit. One unit (1U) of a nanosatellite has a size of 10 × 10 × 10 cm^3, and they have historically been built in 1U, 1.5U, 2U, or 3U sizes. Moreover, nanosatellites can be categorized by mass, which is 1 to 10 Kg (Swartwout, 2013). A nanosatellite is an optimal balance between cost and mission lifetimes to benefit from space.

The development of nanosatellites or **CubeSat standards** provides a unique platform and real-world nanosatellite experience to researchers. The nanosatellite standard states that the size of a single unit of nanosatellite should be 10 × 10 × 10 cm^3 with a total mass of no more than 1 Kg [1, 2]. According to another nanosatellite standard, the weight of a 1U nanosatellite should no more than 1.33 Kg, and similarly, for a **2U nanosatellite** weight should be no more than 2.66 Kg [3]. The standardization of weight and volume must be maintained for the universal launch system called the Poly-Picosatellite Orbital Deployer (P-POD) and JAXA-Picosatellite Orbital Deployer (J-POD). The standards are developed considering some aspects. For example, the size of the components, the dimensions and features of the Poly-Picosatellite Orbital Deployer (P-POD), safety standards and launch vehicle environment should be commercial off-the-shelf [1].

The operating frequency of the nanosatellite depends on its requirements and applications. The UHF-band is highly desirable for nanosatellite research, because, the International Telecommunication Union has allocated the UHF 420–450 MHZ band as the International amateur satellite frequency band [4]. Moreover, VHF and UHF band frequencies are also chosen specifically for **tracking, telemetry, and commands (TT&C)** in nanosatellites.

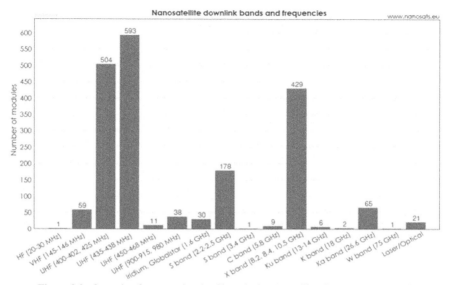

Figure 9.2. Operating frequency bands of launched nanosatellite (Source: nanosats.eu).

Besides the **lower UHF** (401–402 MHz) band is also widely used in the nanosatellite communication system, shown in Figure 9.2. In addition, VHF, UHF, and S-bands are mostly used for communication with the ground station. For deep space exploration X-band is highly preferred over other bands [5]. To establish inter-satellite communication, a higher operating frequency is preferred, specifically S-band (2.45 GHz) [6]. The operating frequency bands of launched nanosatellite are presented in Figure 9.2. Recently, the nanosatellite operating frequency bands are moving towards higher frequencies like Ku, K, W and V-band due to power efficient and high-speed data requirements as well as RF payload size miniaturization. Besides, nanosatellite missions have been launched to explore quantum communication, which is an optical communication between the Earth and space using the optical field frequency band [7, 8].

9.2 Antennas for nanosatellite

The data communication system is one of the most important aspects of a satellite, in which the antenna is a key element for establishing a communication link between the nanosatellite and the earth or from satellite to satellite. The inherent relationship between antenna performances and size compels antenna researchers to compromise with **link quality** for compliance with the nanosatellite standards. Maintaining good performance with specific nanosatellite requirements represent a major mechanical and RF challenge. The key challenges of the **nanosatellite antenna** are shown in Table 9.1 [9]. A variety of antenna architectures have been studied and summarized into two categories: deployable and non-deployable antenna.

Wire antennas are widely used in nanosatellites for HF, VHF and UHF applications. Monopole, dipoles, Yagi-Uda arrays and helical antennas are different

266 *Metamaterial for Microwave Applications*

Table 9.1. Key challenges of nanosatellite antenna designs [9].

Design challenges	Reasons
High reliability	Having mission failure risk
Mechanically robust	To survive launch
Compact, efficient and cost	Small volume with low mass and cost-effective
Space quality material	Must comply with space environment (high/low temperature)
Nanosatellite structure consideration	The interaction between the nanosatellite structure and antenna must be taken into account including other subsystems, because it can also radiate
Mutual EM coupling	Mutual coupling between the antenna and other components can degrade antenna performances.

types of wire antennas, which are typically placed on the external face of the nanosatellite structure to facilitate space for other electronic components. However, mechanical deployment of this type of antenna is quite sophisticated and is liable to increase the chance of **mission failure** [10]. Moreover, dipole antennas are very popular antenna systems for lower operating frequency applications due to their omnidirectional radiation characteristics. A composite tape-spring dipole antenna has been designed for nanosatellite communication system in [11], where the antenna achieves an impedance bandwidth of about 8% with a gain of 2.0 dBi. A 437 MHz half-wave crossed dipole antenna has been developed for NUTS CubeSat [12], where the element length was considered to be 172 mm. The antennas were made of measuring tape and wrapped around the satellite structure. Nichrome-wire deployment mechanism was utilized for deploying the antennas. Moreover, HIT-SAT, a Japanese nanosatellite has also used dipole antennas for communication and altitude estimation [13].

Reflector antennas have drawn attention to the antenna and nanosatellite research due to fine resolution and high gain possibilities, though they come with a complex deployment mechanism. One of the first implementations of deployable reflector antennas was the Aeneas mission [14], which was a S-band umbrella structured reflector. It is shown from the literature that most of the reflector antenna have been designed for higher frequencies [15]. Horn antennas and waveguide antennae can be a viable option for the future nanosatellite communication systems. These antennas are suitable for higher frequencies with higher gain applications. The feasibility of horn antennae has been investigated for the last few years [16, 17].

The Planar Inverted F Antenna (PIFA) addresses the problems associated with the limited volume and surface area for the solar panel and antenna placement of the nanosatellite technology. Several PIFA antennas have been designed for nanosatellite communication systems. In [18], an Inverted-F antenna has been developed for the IITMSAT nanosatellite communication system. The antenna can operate at 435 MHz with a gain of 2.33 dB. An S-band PIFA antenna has been used in the Picpot nanosatellite [19]. Moreover, in [20], circular polarized PIFA array antennae have been developed for S-band nanosatellite communication systems. In addition, PIFA antennae have been used in the ESTCube-2 nanosatellite mission [21].

In contrast to the deployable antenna, patch antennae provide a low profile and improve the mission reliability; and makes the patch antennae a good replacement for wire antennas. However, a UHF patch antenna occupies a large amount of the nanosatellite body surface and introduces complexity to integrate sufficient solar cells. In 1994, Tanaka et al. proposed a microstrip antenna for microsatellites for the very first time [22]. They designed a 2.225 GHz microstrip patch antenna with a solar cell on the surface to mitigate power scarcity of the microsatellite system. Later, in 2001, He & Arichandran focused on a microsatellite patch antenna, operated at 10.74 GHz with a high gain of 6.5 dBi [23]. In [24], a folded shorted patch antenna has been demonstrated for UHF 400 MHz microsatellite applications. The developed antenna offered CP performance with a 130 mm × 130 mm ground plane. Consequently, Podilchak et al. developed a multi-layered shorted patch antenna for UHF 400 MHz microsatellite applications in 2017 [25]. The antenna prototype is illustrated in Figure 2.15. The antenna achieved a gain of 0.4 dBiC with overall antenna dimensions, 150 mm × 150 mm × 37 mm. Slotted patch technique is one of the earliest techniques for antenna miniaturization and dual-band achievement. In 2001, a miniaturized dual port slotted patch antenna was presented for the Mars mission [26]. The antenna diameter was 120 mm and 7.6 mm thick. Transparent antenna are potential antennae for nanosatellite missions. This type of antenna is integrated with the nanosatellite structure and placed above its solar panels. The main design consideration of this type of antenna is maintaining minimal blockage of the solar panel. Meshed structures are used to make a transparent patch and ground plane.

Metamaterials contribute a lot to the field of antenna design. The reactive impedance surface based U-slot patch antenna is one of them [27], where the antenna achieved lower UHF band (410–485 MHz) with antenna size reduction. However, the area of the antenna is $220 \times 220 \times 20$ mm^3, and the antenna is incompatible with the CubeSat structure. In [28], to reduce the conventional RIS antenna size, a two-layer mushroom-like RIS is presented for 400 MHz UHF wireless communication, where the size is $66 \times 66 \times 11.2$ mm^3. Though the technique reduced the antenna size, higher antenna height remains a critical issue for nanosatellite communication. High impedance surface is another form of metamaterial, which is used to reduce the antenna size and weight as well as antenna gain and for efficiency enhancement. In [29], a 3×3 HIS array has been utilized to enhance antenna gain and efficiency of UHF frequency bands from 470 to 790 MHz.

Thus, designing an ultra-high frequency (UHF) antenna that is strategically integrated with the solar panel and that does not require **mechanical deployment** has become a major challenge for nanosatellite and antenna researchers. The motivation of this chapter is directed design, analysis and prototyping of lower EMNZ metamaterial-based UHF patch antenna for nanosatellite application. Attention is given to antenna size reduction and achieving vital features demanded by standard nanosatellite communication systems.

9.3 EMNZ metamaterial design and characterization

There have been tremendous research efforts since the previous decades in the field of artificially engineered materials that show infrequent properties and do not readily exist in nature [30–33]. The unique properties of metamaterials such as negative permittivity, permeability, refractive index or double negative characteristic have been utilized to improve antenna characteristics. Now, researchers show interest in another type of engineered material known as near-zero-metamaterials. This is the type of metamaterial whose characteristics are near zero like **epsilon-near-zero (ENZ)**, **mu-near-zero (MNZ)**, or both **epsilon-and-mu-near-zero (EMNZ)** [34, 35]. Metamaterial with individual ENZ or MNZ show impedance mismatches in free space, which result in high reflectance, high impedance and high loss [36]. However, EMNZ has low loss since the impedance is matched with free space. This type of metamaterial has been efficiently used in the field of antennae and wave propagation for enhancing the radiation efficiency, antenna size miniaturizing, coupling effect reduction, or for modifying the radiation patterns [36–39]. In this research, an EMNZ metamaterial structure has been designed. The geometric layout of the designed structure is depicted in Figure 9.3 and the corresponding design parameters are listed in Table 9.2. The metamaterial structure is composed of thin metallic arms and modified split resonant rings (SRRs), where the outer metallic arms are utilized to realize electric resonance and effective ε-near-zero (ENZ), and the interconnected split ring resonators are used to realize magnetic resonance and effective μ-near-zero (MNZ). So, the designed structure shows the properties of an impedance-matched near-zero-index metamaterial with ENZ and MNZ simultaneously.

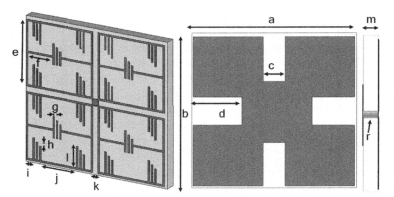

Figure 9.3. Geometric layout of the designed EMNZ metamaterial structure.

Table 9.2. Design parameters of the metamaterial structure.

Parameters	Value (mm)	Parameters	Value (mm)
a	21.20	g	0.25
b	20.50	i	0.75
c	2.80	j	5.25
d	6.00	k	0.75
e	9.50	l	3.00
f	3.75	m	1.57

Lower UHF Metamaterial Antenna for Nanosatellite Communication System 269

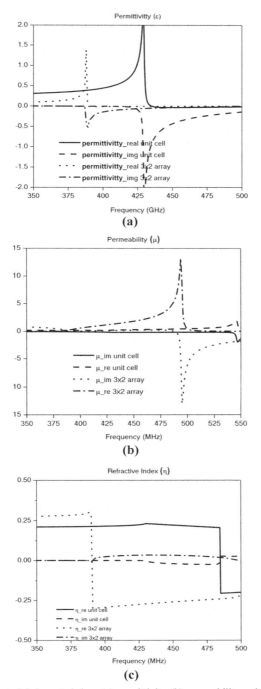

Figure 9.4. Metamaterial characteristics - (a) permittivity, (b) permeability and (c) refractive index.

270 *Metamaterial for Microwave Applications*

Table 9.3. EMNZ properties of the proposed metamaterial at 450 MHz.

Configuration	Permittivity	Permeability	Refractive index
Unit Cell	−0.035	0.20	0.21
3 × 2-unit cell array	−0.032	1.78	−0.22

The electromagnetic parameters of the proposed metamaterial have been retrieved using the following equation of the robust metamaterial parameter extraction method [40, 41]. In Figure 9.4, the retrieved metamaterial parameters for the unit cell are depicted, where it can be observed that the proposed structure shows near zero **permittivity, permeability and refractive index** values. The metamaterial's constitutive parameters show near-zero permittivity, near-zero permeability, and near-zero refractive index. The summary of the study is enlisted in Table 9.2. According to Table 9.3, the designed metamaterial structure shows near zero index characteristics at desired frequencies from 400 MHz to 450 MHz.

9.4 EMNZ inspired UHF antenna

The meander-line antenna concept is to fold the conductors back and forth to reduce the antenna's physical size. The resonant frequency of the conventional meander line antenna can be achieved by increasing the width of the meander line which leads the capacitance to the ground plane or increases the meander line section to generate effective **self-inductance** for lower frequency adjustment. In this design, the main parameters have been optimized by tuning line width, number of folds and the distance between folds to reduce the antenna size. As the physical dimensions of the antenna are reduced, the antenna radiation efficiency and bandwidth also decrease. To enhance radiation performance, EMNZ metamaterial structure has been used in the ground plane of the antenna. The performance of the proposed antenna has been characterized using the **CST microwave studio** software. The design layout of the proposed antenna is illustrated in Figure 9.5. The antenna consists of a modified meander line patch and a partial ground plane with a 3 × 2 near-zero index metamaterial (NZIM) array. The patch is designed on a Rogers Duroid 5880 substrate with a thickness of 1.575 mm. The overall size of the antenna is 80 × 40 × 1.575 mm³. The antenna is fed by a 50 Ω **MMCX connector**. A slot L_3 has been etched on the upper arm of the meander line to tune the resonating frequency. The NZIM structure concept has been adopted to address the resonant **frequency shift issues** and efficiency reduction while the antenna is placed in close proximity to the metallic structure [42, 43]. The array configuration of the metamaterial unit cell has been optimized until the desired antenna reflection coefficient and efficiency have been achieved. Moreover, the antenna has an omnidirectional radiation pattern, which has also been accomplished by the defected ground structure of EMNZ metamaterial array.

The **reflection coefficient** of the proposed antenna is illustrated in Figure 9.6. The final antenna resonates at 450 MHz. The resonating frequency can be controlled by shifting the L_3 slot position, depicted in Figure 4.22. It is shown from Figure 9.7 that when the slot moves upwards, the L_2 value decreases and the L_4 value increases. The resonant frequency moves to a higher value with a decreasing L_2 value. The **realized**

Lower UHF Metamaterial Antenna for Nanosatellite Communication System   271

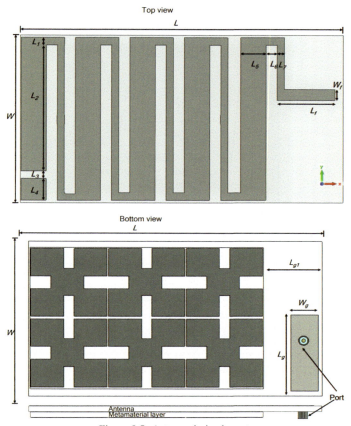

Figure 9.5. Antenna design layout.

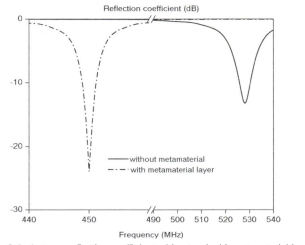

Figure 9.6. Antenna reflection coefficient without and with metamaterial layer.

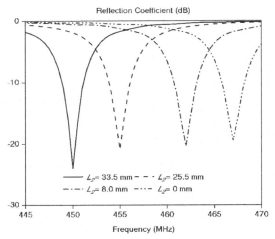

Figure 9.7. Reflection coefficient response of L_2 parameter.

Table 9.4. Overall antenna design parameters.

Parameters	Value (mm)	Parameters	Value (mm)
L	80	L_6	2.72
L_1	2	L_7	1.8
L_2	33.5	L_f	14.38
L_3	2	W	45
L_4	6.18	W_f	3
L_5	6.3		

gain and **total efficiency** remain nearly constant. The overall design parameters are listed in Table 9.3.

The antenna has been fabricated according to the optimized parameters listed in Table 9.3 and Table 9.4, shown in Figure 9.8. The reflection coefficient of the proposed antenna has been measured and attached the antenna with a 2U nanosatellite structure. The antenna S_{11} performance has been investigated with a 2U nanosatellite body, presented in Figure 9.9. The simulated and measured results are in good agreement for metamaterial antenna with a 2U nanosatellite, though a little mismatch is observed. The main reasons for the mismatch between the two results are **fabrication tolerance** and deficient soldering effects of the MMCX connector. Moreover, this disagreement may have occurred due to the RF feeding cable, which was not considered in the simulation.

The **radiation characteristics** of the antenna with the 2U nanosatellite structure have also been investigated using the Satimo **Nearfield Measurement** System. The simulated and measured radiation patterns of the antenna with nanosatellite structure are found to be in good agreement in the H-plane and the maintained approximate **omnidirectional radiation**. However, the measured **cross-polarization** value has increased and that might have occurred due to the presence of the metallic structure.

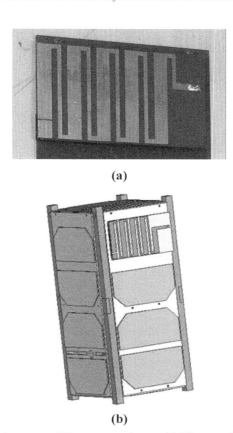

Figure 9.8. (a) Fabricated antenna and (b) antenna attachment with 2U nanosatellite structure (simulation).

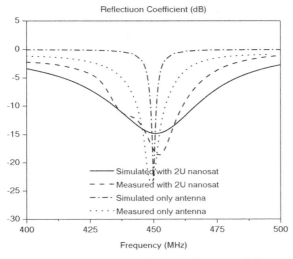

Figure 9.9. Simulated and measured reflection coefficient.

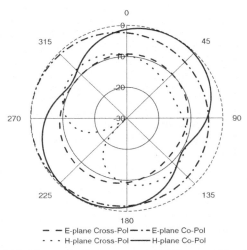

Figure 9.10. Measured radiation pattern with 2U nanosatellite structure.

Similarly, the E-plane radiation pattern shows little disagreement in measurement. Radiation efficiency and gain of the antenna while mounted on the nanosatellite body have also been investigated. The antenna obtained 69% radiation efficiency with the nanosatellite structure with a peak gain of about 2.5 dB.

The **Free Space Path Loss (FSPL)** using variable attenuation has been conducted to estimate the maximum signal propagation of the proposed antenna with active nanosatellite. The FSPL has been calculated considering LEO orbit nanosatellite (400 km). The free space path loss has been calculated using the Friis transmission equation 9.1 [44]. The antenna has been tuned at 467 MHz and mounted on a 2U nanosatellite structure that can only measure FSPL at 467 MHz due to measurement limitations. Extra attenuation required to achieve the signal level was calculated to be 89.85 dB. The **maximum attenuation** level was increased until demodulation was no longer possible. It is seen from Table 9.5 that the proposed antenna facilities 9 dB extra attenuation for signal demodulation.

$$FSPL = 20log_{10}(d) + 20log_{10}(f) + 20log_{10}\left(\frac{4\pi}{c}\right) - G_{Tx} - G_{Rx} \qquad (9.1)$$

where, d is the distance between the receiver and transmitter ends, f is the operating frequency, G_{Tx} is the transmitter antenna gain and G_{Rx} is the receiver antenna gain.

The active satellite and communication board have been placed on the anechoic chamber's turntable and the proposed antenna (Tx) transmitting signal is shown in Figure 9.11. The receiving antenna (Rx) is oriented horizontally and connected to the Receiver through a variable attenuator. Attenuation increases gradually until the demodulation of the transmitted signal is no longer possible. The maximum attenuation value at which the receiver can demodulate the signal represents the **signal strength** of the transmitter antenna, which can address the maximum path loss. The SS power has been calculated by integrating over a 100 kHz bandwidth. Care was taken to ensure that the RF power at the receiver input never exceeds

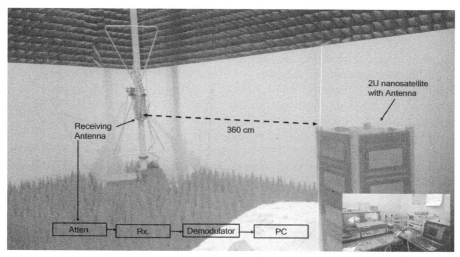

Figure 9.11. Free Space Path Loss (FSPL) measurement system.

Table 9.5. Maximum attenuation level until demodulation was no longer possible.

Nanosatellite rotation angle (Degree)	Max. attenuation Value (dB)
0	99
20	105
80	99
100	99
180	99
200	99
280	99
300	99

over −30 dBm and the RF power received by the demodulator doesn't exceed over −50 dBm for extended time periods to protect the devices. The FSPL analysis and signal modulation system using the proposed antenna is tabulated in Table 9.5.

The design, simulation and measurement result analysis of a UHF metamaterial antenna have been presented for a **nanosatellite communication** system. The simulated results are in good agreement with the measurement results. The advantages of the proposed metamaterial antenna are wide impedance bandwidth with high gain, compact size, easy fabrication and flexibility in antenna resonant frequency tuning. The proposed antenna has the potential to be used in nanosatellites for command functions, tracking and data downlink applications.

9.5 Summary

This chapter focuses on the simulation and measurement of EMNZ metamaterial inspired printed microstrip planar antenna for nanosatellite communication systems. The design, a conventional meander line planar antenna with **EMNZ metamaterial** ground plane is presented for lower UHFs, capable of establishing communication between the **LEO nanosatellite** and the Earth. Several parametric studies have been performed for understanding the operation mechanism. The prototype of the proposed antenna was fabricated and experimentally verified the simulated results The Free Space Path Loss (FSPL) experiment has been performed with an active nanosatellite and the experimental results confirm that the antenna can be a potential candidate for UHF nanosatellite communication systems.

References

[1] Puig-Suari, J., R. Coelho, R. Munakata and A. Chin. 2008. The CubeSat: The picosatellite standard for research and education. pp. 9–11. *In: AIAA Space 2008 Conference and Exhibition, San Diego.*

[2] Pignatelli, D. 2014. Poly Picosatellite Orbital Deployer Mk. III Rev. E User Guide. *The CubeSat Program, California Polytechnic State University, CP-PPODUG-1.0-1, San Luis Obispo,* vol. 93407.

[3] Lee, S., A. Hutputanasin, A. Toorian, W. Lan, R. Munakata, J. Carnahan, D. Pignatelli and Arash Mehrparvar. 2014. CubeSat design specification, rev. 13: The CubeSat program. *San Luis Obispo, California Polytechnic State University.*

[4] Council, N. R. 2007. *Handbook of Frequency Allocations and Spectrum Protection for Scientific Uses.* National Academies Press.

[5] Yu, X. and J. Zhou. 2014. CubeSat: A candidate for the asteroid exploration in the future. pp. 261–265. *In: 2014 International Conference on Manipulation, Manufacturing and Measurement on the Nanoscale (3M-NANO),* IEEE.

[6] Budianu, A., T. J. W. Castro, A. Meijerink and M. J. Bentum. 2013. Inter-satellite links for cubesats. Presented at the 2013 IEEE Aerospace Conference, Big Sky, MT, USA, March 2013.

[7] Arvizu, A., J. Santos, E. Domínguez, R. Muraoka, J. M. Núñez, J. Valdes and F. J. Mendieta. 2015. ATP subsystem for Optical Communications on a CubeSat. pp. 1–5. *In: 2015 IEEE International Conference on Space Optical Systems and Applications (ICSOS),* IEEE.

[8] Kingsbury, R., D. Caplan and K. Cahoy. 2015. Compact optical transmitters for CubeSat free-space optical communications. *In: Free-Space Laser Communication and Atmospheric Propagation XXVII,* 9354: 93540S: International Society for Optics and Photonics.

[9] Gao, S., Y. Rahmat-Samii, R. E. Hodges and X.-X. Yang. 2018. Advanced antennas for small satellites. *Proceedings of the IEEE,* 106(3): 391–403.

[10] Ernest, A. J., Y. Tawk, J. Costantine and C. G. Christodoulou. 2015. A bottom fed deployable conical log spiral antenna design for CubeSat. *IEEE Transactions on Antennas and Propagation,* 63(1): 41–47.

[11] Costantine, J., Y. Tawk, C. G. Christodoulou, J. Banik, S. Lane et al. 2012. CubeSat deployable antenna using bistable composite tape-springs. *IEEE Antennas and Wireless Propagation Letters,* 11: 285–288.

[12] Birkeland, R., T. A. Stein, M. Tømmer, M. Beermann, J. Petrasch et al. 2015. The NUTS cubesat project: spin-offs and technology development. *In: Proceedings of the 22nd ESA Symposium on European Rocket and Balloon Programmes and Related Research,* Tromso, Norway.

[13] Sato, T., R. Mitsuhashi and S. Satori. 2009. Attitude estimation of nano-satellite "hit-sat" using received power fluctuation by radiation pattern. pp. 1–4. *In: 2009 IEEE Antennas and Propagation Society International Symposium,* IEEE.

[14] Aherne, M., T. Barrett, L. Hoag, E. Teegarden, R. Ramadas et al. 2011. Aeneas--Colony I meets three-axis pointing. Presented at the 25th Annual AIAA/USU Conference on Small Satellite, Utah State Universit, August 2011.

[15] Chahat, N. J. Sauder, M. Thomson, R. Hodges, Y. Rahmat-Samii et al. 2015. CubeSat deployable Ka-band reflector antenna for deep space missions. pp. 2185–2186. *In: 2015 IEEE International Symposium on Antennas and Propagation & USNC/URSI National Radio Science Meeting*, IEEE.

[16] Lucente, M., T. Rossi, A. Jebril, M. Ruggieri and S. Pulitano. 2008. Experimental missions in W-band: A small LEO satellite approach. *IEEE Systems Journal*, 2(1): 90–103.

[17] Vourch, C. J. and T. D. Drysdale. 2014. Inter-CubeSat communication with V-band "Bull's eye" antenna. pp. 3545–3549. *In: The 8th European Conference on Antennas and Propagation (EuCAP 2014)*, IEEE.

[18] Gulati, A., N. Sivadas, D. Kannapan, D. Koilpillai, H. Ramachandran et al. 2013. IITMSAT, An efficient nano-satellite bus design for a large payload. Presented at the 5th Nano-Satellite Symposium, University of Tokyo, Japan, Novemver 20–22, 2013.

[19] Reyneri, L., C. Sansoè, D. Del Corso, C. Passerone, S. Speretta et al. 2007. PicPot: a small satellite with educational goals. pp. 1–4. *In: Proc. 18th EAEEIE Conf. Innov. Edu. Elect. Inf. Eng.*

[20] Kurnia, G. F., B. S. Nugroho and A. D. Prasetyo. 2014. Planar inverted-F antenna (PIFA) array with circular polarization for nano satellite application. pp. 431–432. *In: 2014 International Symposium on Antennas and Propagation Conference Proceedings*, IEEE.

[21] Dalbins, J., H. Ehrpais, E. Ilbis, I. Iakubivskyi and E. Oro. 2017. ESTCube-2 integrated platform for interplanetary missions. *In: European Planetary Science Congress*, vol. 11.

[22] Tanaka, M., R. Suzuki, Y. Suzuki and K. Araki. 1994. Microstrip antenna with solar cells for microsatellites. *In: Proceedings of IEEE Antennas and Propagation Society International Symposium and URSI National Radio Science Meeting*, 2: 786–789: IEEE.

[23] He, Y. and K. Arichandran. 2001. The design of X band dual feed aperture coupled patch antenna for microsatellites. *In: IGARSS 2001. Scanning the Present and Resolving the Future. Proceedings. IEEE 2001 International Geoscience and Remote Sensing Symposium (Cat. No. 01CH37217)*, 6: 2784–2786: IEEE.

[24] Podilchak, S. K., M. Caillet, D. Lee, Y. M. M. Antar and L. Chu. 2012. Compact antenna for microsatellite using folded shorted patches and an integrated feeding network. pp. 1819–1823. *In: 2012 6th European Conference on Antennas and Propagation (EUCAP)*, IEEE.

[25] Podilchak, S. K., A. P. Murdoch and Y. M. Antar. 2017. Compact, Microstrip-Based Folded-Shorted Patches: PCB antennas for use on microsatellites. *IEEE Antennas and Propagation Magazine*, 59(2): 88–95.

[26] Huang, J. 2001. Miniaturized UHF microstrip antenna for a Mars mission. *In: IEEE Antennas and Propagation Society International Symposium. 2001 Digest. Held in conjunction with: USNC/URSI National Radio Science Meeting (Cat. No. 01CH37229)*, 4: 486–489: IEEE.

[27] Sarabandi, K., A. M. Buerkle and H. Mosallaei. 2006. Compact wideband UHF patch antenna on a reactive impedance substrate. *IEEE Antennas and Wireless Propagation Letters*, 5: 503.

[28] Wu, J. and K. Sarabandi. 2014. Reactive impedance surface TM mode slow wave for patch antenna miniaturization [AMTA corner]. *IEEE Antennas and Propagation Magazine*, 56(6): 279–293.

[29] Mohamed-Hicho, N. M., E. Antonino-Daviu, M. Cabedo-Fabrés and M. Ferrando-Bataller. 2015. A novel low-profile high-gain UHF antenna using high-impedance surfaces. *IEEE Antennas and Wireless Propagation Letters*, 14: 1014–1017.

[30] Landy, N. I., S. Sajuyigbe, J. Mock, D. Smith, W. Padilla et al. 2008. Perfect metamaterial absorber. *Physical Review Letters*, 100(20): 207402.

[31] Enoch, S., G. Tayeb, P. Sabouroux, N. Guérin, P. Vincent et al. 2002. A metamaterial for directive emission. *Physical Review Letters*, 89(21): 213902.

[32] Zhang, K., Y. Yuan, D. Zhang, X. Ding and B. Ratni. 2018. Phase-engineered metalenses to generate converging and non-diffractive vortex beam carrying orbital angular momentum in microwave region. *Optics Express*, 26(2): 1351–1360.

[33] Zhang, K., X. Ding, D. Wo, F. Meng, Q. Wu et al. 2016. Experimental validation of ultra-thin metalenses for N-beam emissions based on transformation optics. *Applied Physics Letters*, 108(5): 053508.

[34] Mahmoud, A. M. and N. Engheta. 2014. Wave–matter interactions in epsilon-and-mu-near-zero structures. *Nature Communications*, 5: 5638.
[35] Dehbashi, R., K. S. Bialkowski and A. M. Abbosh. 2017. Size reduction of electromagnetic devices using double near-zero materials. *IEEE Transactions on Antennas and Propagation*, 65(12): 7102–7109.
[36] Liu, Y., X. Zhou, Z. Zhu and X. Zhao. 2016. Broadband impedance-matched near-zero-index metamaterials for a wide scanning phased array antenna design. *Journal of Physics D: Applied Physics*, 49(7): 075107.
[37] Soric, J. C., N. Engheta, S. Maci, and A. Alu. 2013. Omnidirectional metamaterial antennas based on ε-near-zero channel matching. *IEEE Transactions on Antennas and Propagation*, 61(1): 33–44.
[38] Yang, J. J., Y. Francescato, S. A. Maier, F. Mao, M. Huang et al. 2014. Mu and epsilon near zero metamaterials for perfect coherence and new antenna designs. *Optics Express*, 22(8): 9107–9114.
[39] Ullah, M. H., M. J. Uddin, T. A. Latef, W. N. L. Mahadi and M. T. Islam. 2016. ZIM cover for improvement of the bandwidth and gain of patch antenna. *Current Applied Physics*, 16(8): 837–842.
[40] Chen, X., T. M. Grzegorczyk, B.-I. Wu, J. Pacheco Jr, J. A. Kong et al. 2004. Robust method to retrieve the constitutive effective parameters of metamaterials. *Physical Review E*, 70(1): 016608.
[41] Islam, M. T., F. Bin Ashraf, T. Alam, N. Misran, M. S. J. Singh et al. 2018. Investigation of left-handed meta-atom for radio frequency shielding application. *Science of Advanced Materials*, 10(11): 1582–1587.
[42] Qing, X. and Z. N. Chen. 2007. Proximity effects of metallic environments on high frequency RFID reader antenna: Study and applications. *IEEE Transactions on Antennas and Propagation*, 55(11): 3105–3111.
[43] Ukkonen, L., L. Sydanheimo and M. Kivikoski. 2005. Effects of metallic plate size on the performance of microstrip patch-type tag antennas for passive RFID. *IEEE Antennas and Wireless Propagation Letters*, 4(1): 410–413.
[44] Saakian, A. 2011. *Radio Wave Propagation Fundamentals*. Artech House.

Index

+3 dB Bandwidth 61
2 Theta Angle Range 149
2U nanosatellite 263, 264, 272–274
3D antenna 176, 178, 184–189, 200–207, 209, 212–216

A

abnormal 22, 23
Absorbers 87
Accuracy 251–253, 255–260
ADS Simulator 51
Advance Design System (Ads) 161
Algorithm 177, 195–200, 212–215, 217
Analogous 161
Angular Distance 240
Annealed Copper Finite Integration Technique 45
Antenna Array 176, 177, 194, 197, 202, 209, 212, 217, 224, 237, 238, 240
Antenna Gain 94, 99, 101, 104, 105
Antenna Miniaturization 143
Antenna Randoms 87
Antenna Technology 28
Antennas 65–67, 69, 74, 76, 83–90, 123–125, 127, 128, 130, 133–139, 175–178, 184–190, 194–197, 200–207, 209–217
Antiparallel Surface Current 113
Area Under the Curve 258
Array Components 240
Array MTM structures 78, 79
Asymmetric Structure 98
Augmented Samples 247
Axial Ratio (AR) 87

B

Backscattered Signals 237, 240
Bandwidth 49, 61, 65, 74, 79, 80, 83, 85–89, 123–125, 127–129, 132, 134, 176–178, 189, 198, 205, 206, 216, 224, 226, 229, 231, 232
Benign Tumor 221, 236–238, 242
Biomedical 146

Biomedical Instrument 124
Biosensing 14, 16, 22–24
Blood 175, 176, 190, 191, 193, 208, 209
Blood/stroke 190, 191
Boundary condition 9, 14, 31
Brain injuries 199
Brain Stroke 177, 194, 195, 199, 200, 209, 210, 212–216
Brain tissue 177, 200
Brain Tumor 221–224, 245, 247, 250, 252–254, 260
Brain Tumor Segmentation 247, 250, 252, 260
Broadcast Pulses 138

C

Calibration Kit 42
Capacitance 5, 7–9, 179, 181, 182, 185, 187
Characterization 4, 18, 20
Chelating Agent 147
Chemical Reactions 57
Classification Performance 247, 255–257, 259–261
Clean Engine Oil 38, 41, 43, 44
CNN Layer 253
Co and Cross Polarization 117
Coaxial probe 193, 194, 208
Coherence Factor 178, 196–198
Component 124, 131
Composite Nodal Operator 254
Compositional Ratio 158, 172
Computer Simulation Technology 30, 71, 73, 88
Computer Simulation Tool 227
concentric crossed line split ring resonator (CCSRR) 178
Conductivity 192, 200, 208–210
Confusion Matrix 259, 260
Conventional 132
Conventional Substrate Materials 145
Converging 21
Convolutional Layers 223, 250
Copper Sputtering 161

Cost-Effective 62
Coupling Metallic Stubs 102
Cross Section View 134
Cross-Polarization 135, 272
Cross-Validation 246, 250, 260
Crystallite Size 149–151
CSF 190, 191, 193, 208, 209
CST microwave studio 270
CST Simulator 161
CubeSat standard 264
Current 175, 183–187, 189, 196, 216
Current Flow 72, 80

D

Data extraction 16
Decoding 247, 250, 252, 260
Deep Learning 222–225, 240–242, 245–247, 253, 260, 261
Defective Ground Structure 84
Dice Score 250–253, 260
Dielectric Assessment Kit (Dak) 148, 154, 171
Dielectric Materials 27, 37, 54
Dielectric Probe 153, 154
Dielectric Probe Kit 39–41
Dielectric Properties 191, 193, 202, 208
Differential Equations 37, 38
Diffraction Angle 149
Diffraction Peaks 149
Dipole Moment 51
Directional Radiation Characteristics 226, 235
Directivity 225, 229, 231, 234
Discrepancies 135
Distorted Resonances 231
Double Negative 67, 68, 76, 78, 88
Double Negative Property 33, 34
Double-Negative (DNG) Characteristics 163
double-positive (DPS) 4
Double-Positive (DPS) Medium 167
Double-Positive Materials 68
Dura 190, 191, 193, 208, 209

E

Early Detection 222
Effective Medium Ratio 30, 57
Effective Parameters 33, 34, 47, 49, 160, 163–165, 168, 169
Effective Permeability 160, 163, 168
Effective Permittivity 160, 163, 168, 180, 181
Efficiency 177, 178, 188–190, 202, 203, 206, 209, 216, 225–229, 231–234
Electric Conductor 70
Electric Dipole 80
Electric Field (E) 165, 167, 179, 181, 183, 190, 193, 194

Electric Field Density (D) 165
Electric Field Distribution 31
Electric Permittivity 94
Electrical Dimension 170, 171
Electrical noise 198
Electrodynamical 11
Electromagnetic 1–4
Electromagnetic Coupling 229
Electromagnetic field 179
Electromagnetic Properties 28, 145–148, 165, 171, 172
Electromagnetic Waves 68–70, 79, 86, 87, 176, 180, 181, 198
Em Head Imaging 194–196, 199, 202, 209–211, 214–217
EM head imaging system 194–196, 199, 202, 209–211, 214–217
EMNZ metamaterial 267, 268, 270, 276
EMR (Effective Medium Ratio) 170
Encoding 247, 250, 252, 260
Energy Harvesting 145, 147
E-Plane Radiation Pattern 207
Epochs 250, 251, 255, 256
Epsilon Negative (ENG) 83
epsilon-and-mu-near-zero (EMNZ) 268
Epsilon-near-zero (ENZ) 12, 268
Equivalent circuit 6, 8, 14, 18, 181, 182, 185–187
Experimental 4, 14, 16, 19, 24
Extended Guided Wave 42, 43
Extra Virgin Coconut Oils 40, 41, 43, 61
Extra virgin olive oil 40, 42–44

F

F1 Score 256–258, 260
Fabricated Prototype 55
Fabricated Tissue 240
Fabrication 231, 241, 242
Fabrication tolerance 272
Far-Field Radiation 233, 235
Feature Vector 255
Federal Communications Commission 123, 127, 128
Ferromagnetic Resonance (Fmr) 157, 158
Fidelity Factor 136, 139
Field Analysis 79
Field Distribution 236–238
Field Emission Scanning Electron Microscope (FESEM) 151
Fifth Generation (5G) 65
finite-integration technique 9, 18
Flame Retardant-4 30
Flexible Antenna 142–147, 171
Flexible Displays 146
Flexible Electronic Gadgets 144

Flexible Electronics Market 144, 145
Flexible Metamaterial 142, 143, 145–147, 158, 162, 163, 165, 168, 170–172
Flexible Sensors 143, 146
Flexible Substrates 144–146, 153, 154, 158, 161, 171, 172
Fluctuating Magnetic Field 49
Fractional Bandwidth 229, 232
Free Space Path Loss (FSPL) 274–276
Frequency Bands 47
Frequency domain algorithm 195
Frequency Range 123, 126, 128, 129, 132–136
Frequency Selective Surface (FSS) 87
Frequency shifting issues 263, 270
Frequency spectrum 28, 43, 51
Frequency-Dependent Variables 165
Fully Connected Layer 223

G

Gain 67, 74, 76, 83, 84, 86, 88–90
Gain Variation 39
Gap Distances 79
Geometrical 2, 3, 7, 13, 17, 24, 134
glutamate concentration 21–24
Grain Size 153, 154
Gray matter 190, 191, 193, 208, 209
Ground Truth Masks 247, 248, 252

H

Head Imaging 175–178, 194–196, 199, 202, 209–211, 214–217
Head Model 176–178, 190, 200–205, 210, 212–216
Head phantom 176, 177, 190, 192–194, 203, 204, 208, 209, 211, 216, 217
Head Tissue 177, 190, 202, 204, 209
High Absorption 101, 119
High Dielectric Constant 113
higher density 9
Higher Frequency 234, 236, 265, 266
Highly Concentrated 50
High-Speed Wireless Devices 139
Horn Antennas 55
H-Plane 203, 207, 208
Hyperparameters 250, 251, 255

I

Image Augmentation 242, 245
Image Categorization 245
Image Pre-Processing 242, 245, 247
Image Reconstruction Algorithm 242
Images 176–178, 194–200, 202, 209, 210, 212–217

Images Of The Human Head Model With Stroke 214
Imaginary and Real Values 74
Imaging domain 195, 197, 213
Impedance 9, 10
Impedance Matching 226, 227
Incident Wave 37, 45
Inductance 35, 36, 51–53
Inductively Tuned 97, 98, 101
Interrelationship 50
Intersection Over Union 251, 260
Investigates 125, 127, 128, 136, 139
IoT framework 199, 215
ISM band 24
isotropic materials 4

K

Kapton Polyimide 144, 145
Key challenges 265, 266
Key Effects 53
Koop's Phenomenological Theory 154
Ku-band 12, 24

L

Lattice Coordination Number 150
Learning Performance 260
Learning Rate 250, 251, 255, 256
Left-Handed 66, 68
LEO nanosatellite 276
Lightweight 225, 247, 249, 252, 254, 257, 260
Linear 8
Link quality 265
Liquid Under Test 39
Long Wavelength 29
Loss 222, 225, 250, 251, 253, 255, 256
Loss Tangents 30, 39, 41, 58, 142, 145, 147, 153–156, 171
Lower Frequency 71, 79
Lower UHF 263, 265, 267, 276
lumped components 5, 18

M

Machine Learning 223
Magnetic and Electric Fields 75
Magnetic Conductivity 145
Magnetic Conductor 70, 83
Magnetic Field (H) 156–158, 165, 167, 179, 181, 183, 190, 202
Magnetic Field Density (B) 165
Magnetic Properties 156, 158, 172
Magnetic Response 32
Malignant Tumor 236–238, 242
Manufactured 124, 135

282 Metamaterial for Microwave Applications

Massive MIMO 66
Material Characterizations 59, 60
Mathematical Modelling 36, 38
Mathematically 4
MATLAB Software 163
Maximum attenuation 274, 275
Max-Pooling Layer 250, 255
Maxwell's Equations 38, 49, 50
Maxwell-Wagner Model 154
Measured Magnitude 43
Measurement Error 153
Measurement Setup 55, 56, 59
Measurement System 29
Mechanical deployment 263, 266, 267
Medical Imaging 175, 216
Metallic Layer 43
Metamaterial 1–7, 9–16, 20, 22, 24, 93–102, 104–110, 112, 113, 115, 119, 120, 175–180, 187, 189, 194, 202, 204, 209–212, 214–216, 221, 225, 231, 240, 260, 263, 267–272, 275, 276
Metamaterial Array 52
Metamaterial structure 178, 179, 189
Metamaterial Unit Cells 158–164, 167–170
Metamaterial-loaded antenna 189, 194
Metasurface 66, 67, 86, 87
M-H Hysteresis Loops 157
Microcontroller 194, 199, 200
Microfluidic Sensors 61
micro-ring resonator 10
Microstrip feeding 184
Microstrip transmission line 187
Microwave Brain Imaging 221, 222, 224, 226, 233, 239, 240, 260
Microwave Dielectric 142, 154, 170
Microwave head imaging 175, 176, 178
Microwave Imaging 133, 223, 224
Microwave Radiation 239
Microwave Sensing 26–29, 31, 61
Microwave signals 177
Miniaturization 66, 68, 84, 87–89
Miniaturized Antenna 84, 87
Mirror Symmetric 105, 106, 108, 119
Misclassified 259, 260
Mission failure 263, 266
MMCX connector 270, 272
Mm-Wave 65, 69, 70, 79–83
modeling 5
Molecular Weight 150
Mu Negative (MNG) 83
Multiband 76, 83, 86
Multiband Antenna 119
Multiband Resonances 104
mu-near-zero (MNZ) 268

N

Nanoparticles 145–151, 153, 154, 156–158, 161–163, 170–172
Nanosatellite 263–267, 272–276
Nanosatellite antenna 265, 266
Nanosatellite communication 263, 265–267, 275, 276
Nanosomic Structures 125
Near Zero Permeability 82
Nearfield Measurement 272
Near-Field Radiation 233, 235
Near-Field Region 235
Near-Zero Index (NZI) 67
Negative Bandwidth 165, 168
negative effective parameter 4
Negative Index Frequency 132, 136
Negative Permeability 26
Negative Permittivity 26, 229
Negative Real Values 74, 76, 83, 90
Negative Refractive Index 145, 163
Network Diversity 257
Nicolson-Ross-Weir (NRW) 19
Nicolson–Ross–Weir Technique 47
non-invasive 12
Non-Linear Transformations 254
Normal Impedance 160
Normal Incidence 47
Normalization Techniques 245
NRW (Nicolson-Ross-Weir) Method 157, 160, 163, 168
Numerical Simulation 95, 109

O

Oil Samples 37–39, 41–43, 61
Omnidirectional Manner 135
Omnidirectional radiation 266, 270, 272
Open Add Spaces 31
Operating Band 228–232, 235, 236
Optical Bandgap Energy (Eg) Parameters 155
Optimized 13, 15, 17
Optimized Parameters 226
Optimum focal point 196
Overall Performance 53
Oxide-Based Flexible Microwave Composites 146

P

Parametric Analysis 225, 230
Partial Ground 134, 135
Particle Size Histograms 151, 153
Peak Absorption 101, 110–112, 114, 118, 120
Perfect Electric Conductor 28, 31, 45

Permeability 66–69, 73–75, 78, 80, 82, 86–88, 90, 156, 160, 161, 163, 165, 167, 168, 268–270
Permittivity 66–69, 73–75, 78, 80, 82, 86, 88, 90, 123, 125, 126, 130, 132, 268–270
Perpendicular 125, 130
Phantom 176–178, 190–194, 198, 203, 204, 208–212, 214–217
Phantom Model 236, 240, 242
Phenomena 126
Photoluminescence (Pl) Spectra 156
Physical Structure 28, 36
Pivotal 128
Planar Inverted F Antenna (PIFA) 266
Polar Angle 160
polarization 9, 14, 15, 19
Polytetrafluroethene 15
Polyvinyl Alcohol (PVA) 158, 161
Portable platform 178
Power Analysis 113
Power Loss Distribution 115
Precision 256, 257, 260
Propagate 131
propagation constant 7, 8
Propagation Direction 37
Prototype Measurement 231
Prototypes 4, 5, 13, 14, 17, 18, 24, 73, 83, 85
Pulse Distortions 136, 138

Q

Q-factor 1, 3
Quad Resonance Frequencies 48
Quality Factor 27, 61, 62
Quasi-State Theory 159
quasi-TEM 14, 15

R

Radar Cross Section 14
Radiating Patch 134, 225–227, 229
Radiation characteristics 266, 272
Radiation Directionality 226, 229
Radiation Efficiency 124, 135, 136, 142, 144, 202, 203, 216
Radiation Pattern 146, 189, 190, 202, 206–208, 216, 231–235
Realized gain 177, 178, 188–190, 202, 203, 206, 216, 225, 227–229, 231–233
Receiver 224, 237, 258
Receiver Operating Characteristic 258
Reconfigurable 87
Reconstructed Microwave Brain 223–225, 240, 260
Reconstructed Microwave Brain Images 260
Reconstructing the Brain Images 237

Reconstruction algorithms 196, 217
Reflected Wave 45
Reflection 9–12, 14, 15, 18–20
Reflection Coefficient 75, 76, 85, 86, 95, 99, 100, 109, 117, 118, 131–136, 180, 182, 188–190, 193, 200, 205, 206, 215, 227, 228, 230–232, 236, 270–273
Reflection Parameter 163, 168
Reflector antennas 266
Refractive Index 28, 30, 33, 34, 47–49, 67, 69, 74–76, 78, 80, 82, 83, 88, 90, 145, 160, 161, 163, 165, 168, 180, 181, 183, 268–270
Relative permittivity 154, 159, 177, 181, 192, 208–210
Relative Permittivity and Permeability 112
Resonance 1, 2, 5–9, 11, 14, 16, 18–23, 71–73, 75, 76, 79, 80, 86–90, 157–159, 163, 165, 167, 168, 170, 171
Resonance Frequency 27, 29, 31, 32, 34–36, 38, 46, 47, 50–54, 57–61, 95, 97, 107, 108, 111, 119
Resonance Phenomena 33
Resonances Shifted 59
Resonator 1–5, 8, 10–14, 16, 17, 24
Resonator Patch 53
Resonator Width 35, 36, 52–54
Retrieval 21
Right Hand 67–69
Ring Copper 79
Robust Approach 73
Rogers RO4350B 55, 58, 60
Rogers RT5880 45, 55, 58, 60

S

S21 71, 75, 76, 79–81
Sample Holder 30, 31, 35, 42, 43
sandwich 7, 9, 12, 15, 24
Sandwiched 58, 61
Sar 224, 239, 240
Sar Analysis 239
S-band 78
Scattering intensity 195, 197, 198
Scattering parameters 176–178, 182, 188–190, 195, 200, 201, 204, 210, 211
Segmentation Performance 247, 250–252
Segmented RMB Images 247, 255, 257–261
Self-inductance 270
Self-Onn Networks 253
Sensing 1–7, 9, 11, 12, 14–18, 21–24
Sensitivity 29, 30, 38, 39, 61, 62, 256
Sensor Layer 30–32, 43, 59, 60
Shot noise 198
Signal Penetration 226, 236, 238
Signal strength 274

Signal-To-Signal Correlation 136
Simulated 10, 14, 15, 19, 21
Simulated Magnitude 43
Simultaneous Negative Permittivity 123
Single Element 133
Single-Negative MTM 74, 76
Six-Generation (6G) 65
Size 65, 66, 68, 83–89
SmallSat family 264
Soft Magnetic Material 157
Softmax Activation Function 255
Solid-State Components 29
Space communication 263
S-parameter 6, 9, 18, 19, 21, 38, 47, 48, 70, 72, 73, 160, 161, 164, 169
Specific Absorption Rate 176, 204
Specificity 256–258, 261
Spectrophotometer 155
Spectrum 124, 127–130, 133
Spinel Oxide Materials 145, 147
Split Gap 36, 51–53, 79
Split Ring Resonators (SRR) 75, 158
Square SRR 45–47, 51, 52, 55–58, 61
Stacked Antenna 221, 225–233, 235, 236, 239, 240, 244, 260
Stacking 16, 17
Stop Band 89
Straight Metallic 126
Stripline 47, 51, 52
Strokes 175–177, 190, 191, 193–195, 199–202, 209, 210, 212–217
Structural Flexibility 147
Structural Morphology 151
Structural Symmetry 94, 97
Sub-6 GHz 74, 79, 80
Superstrate 94, 99–101, 104, 105, 119
Surface Current 72, 80, 81, 87, 183
Surface Current Distribution 31–33, 165, 166
symmetric 5
Symmetric Split Rings 79
Symptoms 222

T

Tapered Log Periodic Dipole (TLPD) 99
Taylor Series 254
telemetry and command (TT&C) 264
Thickness Sensing 57–59, 61
Tissue-Mimicking Head Phantom 236, 240
Total efficiency 272
Tracking 264, 275
Training Dataset 240, 245–248
Transmission Coefficient 5, 7, 9, 18, 22, 71, 75, 79, 80, 180, 187, 190
transmission line 3, 5, 6, 8, 9, 12, 17, 18

Transmission Parameter 163, 168
Transmission Response 31, 34–36, 38, 43, 44
Transmission Spectra 96, 102
Transmitters 129, 237
Transverse Electric (TE) Or Transverse Magnetic (TM) 116
Traumatic brain 21
traveling wave 8
Tri Circle 30, 35, 61
Tumor Segmentation 247, 249–252, 260
Tuning 52
Tuning Effect 107
Tunning Metallic Stubs 106, 107, 109, 110, 120
Two-Axes Symmetrical Structure 102

U

Ultra-Wideband 85
Unhealthy Brain 247
Unit Cell 3–7, 9, 10, 12–22, 24, 30, 45, 47, 48, 50–53, 55–57, 66–77, 79, 83, 84, 87–90, 130–134, 136, 177–184, 187–189, 202, 216
Unrefined Oil 40

V

Vanilla CNN Models 255
Vector Network Analyzer (VNA) 42, 55, 153, 157, 160, 161, 193, 194
Vibrating Sample Magnetometer (VSM) 148, 171
Vibration noise 198
Voltage 185–187, 190
Voltage Standing Wave Ratio (VSWR) 135

W

Wave Propagation 68–70, 74, 76, 78, 79, 90
Wave Vector 161
waveguide 5, 9, 11, 12, 14, 15
Waveguide Adapter 117
Waveguide Ports 36, 42, 55, 58
Wearable Computers 142, 143
White matter 190–193, 208, 209
Wide Frequency Range 78, 90
Wideband Antenna 226
WiMax and WLAN Bands 89
Wireless Personal Area 124, 127, 129
Wireless System 128, 129

X

X and KU bands 78
X-Ray Density 150, 151
X-ray Diffraction (XRD) 148, 171
X-Ray Diffractometer 149